Embedded Systems Design with Platform FPGAs

Embedded Systems Design with Platform FPGAs
Principles and Practices

Ron Sass
Andrew G. Schmidt

AMSTERDAM • BOSTON • HEIDELBERG • LONDON
NEW YORK • OXFORD • PARIS • SAN DIEGO
SAN FRANCISCO • SINGAPORE • SYDNEY • TOKYO

Morgan Kaufmann Publishers is an imprint of Elsevier

Acquiring Editor:	Todd Green
Development Editor:	Nate McFadden
Project Manager:	André Cuello
Designer:	Alisa Andreola

Morgan Kaufmann Publishers is an imprint of Elsevier.
30 Corporate Drive, Suite 400, Burlington, MA 01803, USA

This book is printed on acid-free paper. ∞

Library of Congress Cataloging-in-Publication Data

Sass, Ronald (Ronald R.)
  Embedded systems design with platform FPGAs: Principles and Practices / Ronald Sass.
    p. cm.
  Includes bibliographical references.
  ISBN 978-0-12-374333-6
  1. Embedded computer systems. 2. Field programmable gate arrays. I. Title.
  TK7895.E42S27 2010
  004.16–dc22

                                                                          2010010632

British Library Cataloguing-in-Publication Data

A catalogue record for this book is available from the British Library.

For information on all Morgan Kaufmann publications
visit our Web site at *www.mkp.com* or *www.elsevierdirect.com*

Typeset by: diacriTech, India

CONTENTS

PREFACE

Xilinx, Inc. introduced the Field-Programmable Gate Array (FPGA) in 1984 as an advanced programmable logic device. It is now part of a multi-billion dollar market and FPGAs have made their way into products as diverse as digital cameras, automobiles, and network switches that drive the Internet. FPGAs have even flown to Mars (Ratter, 2004).

Almost since its inception, people have recognized the potential of using these devices to build custom computing architectures, but to date the market is overwhelmingly "glue logic" and prototyping. Nonetheless, advances in process technology have yielded modern FPGAs with very large capacities and a wide range of features built into the chip. The confluence of these features — which include multiple processors, large amounts of memory, hundreds of multipliers, and high-speed I/O — have reached a critical mass: now more than ever, Platform FPGAs are poised to realize a more prominent role in computing systems.

This ability to deploy sophisticated computing systems on a single FPGA device is likely to make a significant impact on embedded computing systems. While small (indeed tiny) 8 and 16 bit computing systems are and will remain a very important segment of the embedded systems market, trends over the last several years suggest that the use of standard, off-the-shelf 32-bit processors for embedded systems is growing fast. These higher-end embedded systems come with high levels of integration, often incorporating a significant portion of the system on a (fixed and manufactured) chip. There are many benefits to this level of integration but one significant drawback is that much of the system architecture is predetermined and may not be optimal for the particular targeted application. The alternative, developing a system architecture on a custom System-on-a-Chip (SoC), is too costly except for high volume (millions of units) products. When the application fits the resources provided, all is well. But oftentimes the application does not fit and some of the integrated resources are wasted while additional discrete hardware is required to make up the deficiency. With the Platform FPGA, engineers gain all of the advantages of integration but retain the flexibility to engineer a balanced system architecture on a per-application basis.

Armed with multiple kinds of buses, various direct communication links, bridges, I/O components, and an assortment of other special-purpose Intellectual Property cores, Platform FPGA designers can readily customize their system architecture. For example, with hundreds of distributed block RAMs, a designer can configure one large addressable memory, disperse the RAM throughout the system as individual buffers, or design some combination of the two. Application-specific functionality (i.e., custom hardware cores) can be designed and incorporated. Although Application-Specific Integrated Circuits (ASICs) will always outperform the identical FPGA implementation, the FPGA solution avoids the expense, risk, and time-to-market issues associated with manufacturing of an ASIC. The FPGA is a virtual blank slate giving the engineer the ability to provision the resources to best fit the application after the manufacture, test, and verification of the physical device. This flexibility increases efficiency of the system solution and every discrete component saved reduces cost and increases reliability.

Of course, there is a price to be paid for this enormous degree of hardware flexibility. In addition to compilers, debuggers, and other conventional software tools used for processor-based embedded system development, the Platform FPGA designer has to be fluent in hardware design and synthesis as well as system integration tools. Decisions that previously were largely constrained by architecture choice — such as how to partition the application between hardware and software — now have a much larger solution space. Beyond understanding the characteristics of a particular processor's system bus, the designer has to weigh the strengths and weaknesses of multiple communication mechanisms. Balancing complex networks of on-chip components presents a novel challenge for those used to working with predetermined architectures.

In many ways, these challenges embody computer engineering. However, until the Platform FPGA emerged, practical issues (such as the cost of building custom silicon solutions) made a hands-on study prohibitively expensive. Students learned computer system architecture from textbooks and conceptual models. The practical aspects of engineering custom computer systems from the ground up was typically the purview of a few specialists. Knowledge and practical tools — such as how to create a Board Support Package for an embedded system — was typically learned on the job.

The aim of this textbook is to introduce the reader to system development on a Platform FPGA. The focus is on embedded systems but it also serves as a general guide to building custom computing systems. The text describes the fundamental technology in terms of hardware, software, and a set of principles to guide the development of Platform FPGA systems. The goal is to show how to systematically and creatively apply these principles to construction of application-specific embedded system architectures. There is a strong focus on using Free and Open Source software to increase productivity.

The organization of each chapter in the book includes two parts. The white pages describe concepts, principles, and general knowledge. The gray pages include a technical rendition of the main issues of the chapter and show the concepts applied in practice. This includes step-by-step details for a specific developer board and chain so that readers can carry out the same steps on their own. Rather than try to demonstrate the concepts on a broad set of tools and boards, the text uses a single set of tools (Xilinx Platform Studio, Linux, and GNU) throughout and uses single developer board (Xilinx ML-510) for the examples. The belief is that a single system, completely described, is more valuable to the reader than partial information about a range of systems.

How to Read This Book

This book was designed to make it easy for a number of different readers to quickly find the information they need. If you are an undergraduate with a software background who might not have had a course in electronics, the white pages in chapter two start with a basic transistor and describes how an FPGA — a solid state device — implements programmable hardware. If you are a practicing engineer with experience in embedded systems but are new to FPGAs, then reading just the blue pages will let you skip the theory and focus on the practical aspects of building Linux-based systems on an FPGA. If you are a student in a senior design course, the lectures probably focus on project management. You won't find that material here but you may find this text to be a handy practical guide to complete your project. On the other hand, if you have a very specific project involving cutting-edge technology, then you may not find that particular topic covered in sufficient depth here. However, if you are looking to get started, need a general overview of the concepts, and still want enough step-by-step details to implement a real, working system this may prove to be the perfect text for you!

Note to Instructors

There are several roles this text may play in a typical computer curriculum. In many departments, a single embedded systems course is a technical elective offered to Seniors. The core material is delivered within the context of a substantial, semester-long project. This text provides a suitable introduction to embedded systems and — because it provides all of the practical material students need to carry-out their projects — the instructor has the freedom to introduce select topics in class. For curricula with a focus area in embedded systems (offering multiple technical electives), this text may be used to teach a course on reconfigurable computing. Such a course would complement a comprehensive embedded systems course that gives a full treatment of the area including small systems and an expanded description of real-time issues. Finally, the IEEE-CS/ACM Joint Task Force on Computing Curricula (Ironman Draft) includes as a basic computer engineering component a "culminating project." This often manifests itself in current curricula as a Senior Design or Capstone course. As such, the course draws together subjects taught independently in the curriculum and helps bridge the students' academic and professional careers. FPGAs serve exceedingly well (and are often used) in such a course. Their flexibility allows a wide range of potential projects and those projects necessarily encompass both hardware and software components: the essence of computer engineering. For that reason, many instructors may find this text to be an excellent resource for the students in their culminating project.

Online Materials

Many of the commands, scripts, and URL links referenced in the text are included in the text. However, for the reader's convenience, the publisher maintains a Web site with all of these materials online including scripts too long to insert into the text. In addition to links pointing to the latest version of the open source software used in the text, the Web site also archives the exact (known to work) versions used in this text.

Reference

Ratter, D. (2004). FPGAs on Mars. *Xcell Journal,* **Q3**(50), 8–11.

ACKNOWLEDGMENTS

There have been many people that have contributed to this project and helped improve the book in numerous ways. We would like to acknowledge the current students in the Reconfigurable Computing Systems lab (University of North Carolina at Charlotte) for reading chapters and providing feedback – sometimes with very short notice. These include William V. Kritikos, Scott Buscemi, Bin Huang, Shanyuan Gao, Robin Jacob Pottathuparambil, Siddhartha Datta, Ashwin A. Mendon, Shweta Jain, Yamuna Rajasekhar, and Rahul R. Sharma. Other colleagues and previous students have also contributed; the authors would also like to thank Brian Greskamp (D.E. Shaw), Srinivas Beeravolu (Xilinx, Inc.), Parag Beeraka (AMD Inc.), and David Andrews (University of Arkansas) for their help. Many of the questions and examples have been adapted from our collaborations with these individuals. Several external reviewers provided valuable feedback. We would like to thank Roy Kravitz (Serveron, A Division of BPL Global), Duncan Buell (University of South Carolina), Cameron Patterson (Virginia Tech), and Jim Turley (Embedded Technology Journal). We would also like to thank the team at Morgan-Kaufmann; in particular, Nate McFadden and Andre Cuello, were instrumental in helping us complete the manuscript on time. Also, we would like to thank Xilinx — they encouraged us to start this project and then gave us the freedom to complete it as we saw fit. Finally, the authors would like to thank our families — Jennie, Joseph, and Hilary — for giving us the time to complete this project. You guys are the best!

1

INTRODUCTION

em·bed — *to make something an integral part of*

Merriam-Webster Online

From smart phones to medical equipment, from microwaves to antilock braking systems — modern embedded systems projects develop computing machines that have become an integral part of our society. To develop these products, computer engineers employ a wide range of tools and technology to assemble embedded systems from hardware and software components. One component — the Field-Programmable Gate Array (FPGA) — is becoming increasingly important. Informally, an FPGA can be thought of as a "blank slate" on which any digital circuit can be configured (Figure 1.1). Moreover, the desired functionality can be configured in the field — that is, after the device has been manufactured, installed in a product, or, in some cases, even after the product has been shipped to the consumer. This makes the FPGA device fundamentally different from other Integrated Circuit (IC) devices. In short, an FPGA provides programmable "hardware" to the embedded systems developer.

The role of FPGA devices has evolved over the years. Previously, the most common use of the technology was to replace a handful of individual small- and medium-scale IC devices, such as the ubiquitous 7400 series logic, with a single FPGA device. With well-known improvements and refinements in semiconductor technology, the number of transistors per IC chip has increased exponentially. This has been a boon for FPGA devices, which have increased dramatically in both programmable logic *capacity* and functional *capability*. By capacity, we are referring to the number of equivalent logic gates available; by capability we are referring to a variety of fixed, special-purpose blocks that have been introduced, both of which are discussed in more detail throughout this book.

As a result of this growth, modern FPGAs are able to support processors, buses, memory controllers, and network interfaces, as

1

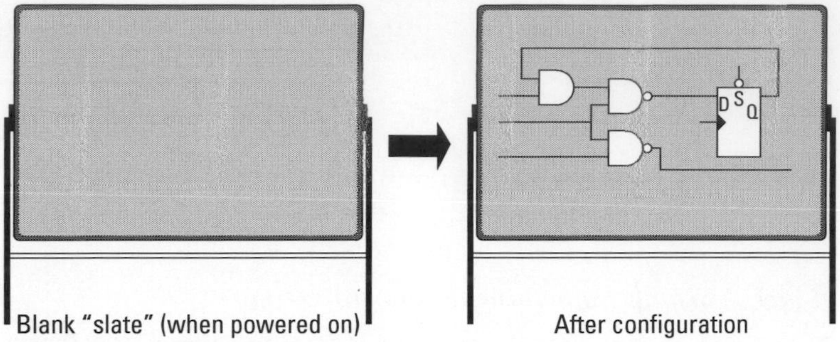

Figure 1.1. A 10,000-meter view of an FPGA; a blank slate that can be configured to implement digital circuits after the chip has been fabricated.

Blank "slate" (when powered on)

After configuration

well as a continually increasing number of common peripherals, all on a single device. With the addition of a modern operating system, such as Linux, these FPGAs begin to appear more like a desktop PC in terms of functionality and capability, albeit on a single IC chip. Furthermore, with Open Source operating systems comes a plethora of Open Source software that can be leveraged by FPGA based designs.

We use the term *Platform FPGA* to describe an FPGA device that includes sufficient resources and functionality to host an entire system on a single device. The distinction is somewhat arbitrary, as there is no physical difference between a large FPGA device and a Platform FPGA device. Rather it is a matter of perspective: how the developer intends to use the device. We would say that an ordinary FPGA is generally used as a peripheral and plays a supporting role in a computing system. In contrast, a Platform FPGA has a central role in the computing system.

Overall, FPGAs offer a great deal to the embedded systems designer. Beyond simply reducing the numbers of chips, the Platform FPGA offers an enormous degree of flexibility. As it turns out, this flexibility is extremely valuable when designing systems to meet the complex and demanding requirements of embedded computing systems that are becoming so universal today. Often it is this flexibility that makes FPGA technology more appealing than traditional microprocessor-based solutions, Structured ASIC solutions, or other System-on-a-Chip (SoC) solutions.

Along with its advantages, FPGA technology also comes with a new set of challenges. Previously, one of the most important questions for an embedded systems designer was the choice of the processor. That choice usually dictated much of the remaining architecture (or at least limited the range of design choices). With Platform FPGAs, in order to fully utilize their capabilities and cost savings, a much wider pool of system developers needs to be able to combine computer engineering, programming, system analysis, and technical skills.

The aim of this book is to draw this necessary information together into a single text so as to provide the reader with a solid and complete foundation for building embedded systems on Platform FPGAs. This includes the underlying engineering science needed to make system-level decisions, as well as the practical skills needed to deploy an assembled hardware and software system.

The organization of this book reflects that twofold mission: each chapter consists of a set of white pages followed by a set of gray pages. The white pages emphasize the more theoretical concepts and slow-changing science. The gray pages provide descriptions of state-of-the-art technology and emphasize practical solutions to common problems.

The specific learning objectives of this chapter are straightforward.

- We begin with an abstract view of a computing machine and define an embedded system by distinguishing it from a general-purpose computing system. Since we are using a Platform FPGA device to develop application-specific, custom computing machines, it is important to revisit the traditional concepts of hardware and software and, in particular, describe the two very different compute models used by each.
- Next, we consider some of the specific challenges that embedded systems designers face today. These include short product lifetimes leading to tight project development schedules, increased complexity, new requirements, and performance metrics that often define the degree of success. The complex interplay of these challenges presents one of the most important problems facing the next generation of embedded systems.
- The white pages of this chapter end with a section describing Platform FPGA characteristics and why these devices are well suited to meet these complex demands facing modern embedded systems designers today.

    In the gray pages, a practical example using a fictitious scenario is presented; we briefly describe how to set up the software tools needed throughout the book and show to how to create the obligatory "Hello, World!" on a Platform FPGA development board. Overall, the goals of this chapter are to establish a solid motivation for using Platform FPGAs to build embedded systems and to initiate the reader on how to use the development tools.

1.1. Embedded Systems

In the simplest sense, an embedded system is a specialized computing machine. A *computing machine* (or just *computer*) is

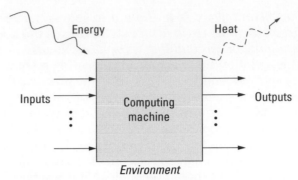

Figure 1.2. An abstract view of a computing system.

frequently modeled as a system that includes inputs, outputs, and a computation unit. The machine exists in some sort of *environment* that provides the energy to propel the machine. In addition to the manipulation of information, the computing machine produces heat that is returned to the environment. This organization is illustrated in Figure 1.2. (This very abstract figure is fleshed out in the next chapter.)

When the machine is being used, it is said to be executing and that period of time is called "run-time." When the machine is not executing, it is said to be "off-line." The inputs, which come from the environment, determine the outputs, which are conveyed from the machine to the environment. The inputs and outputs are often physical signals and as such are assigned meaning depending on the context of the problem, which is the point of building computing machines: to solve problems.

The personal computer (or desktop) is a well-known example of a computing machine but there are a multitude of others as well. The slide rule and the abacus are old-fashioned mechanical machines that were used to do arithmetic. These have since been replaced by the electronic calculator, which is an electronic computing machine.

1.1.1. Embedded and General-Purpose

Depending on how the computing machine is used, we classify it as either embedded or general-purpose. An *embedded computing system* (or simply an embedded system) is a computing machine that is generally a component of some larger product and its purpose is narrowly focused on supporting that product. In other words, it is a computing machine dedicated to a specific purpose. In terms of the abstract machine in Figure 1.2, the environment of an embedded computing system is the product it is part of. The end user of the product typically does not directly interact with the embedded system, or interacts with only a limited interface, such as a remote control. Even though the computer may be capable of more, the embedded system is typically restricted to perform a limited role within the enclosing product, such as controlling the behavior of the product.

There are numerous examples of embedded systems. They are ubiquitous and have permeated most aspects of modern life: from the assortment of consumer electronics (DVD players, MP3 players, game consoles) we interact with almost daily, to energy-efficient refrigerators and hotel key-card activated door handles. Even large, heavy earth-moving equipment typically has hundreds of sensors and actuators connected to one or more embedded systems.

In contrast, a *general-purpose computing system* is a product itself and, as such, the end user directly interacts with it. Another way to put it, the end user explicitly knows the product they are buying is a computer. General-purpose systems are characterized by having relatively few, standardized inputs and outputs. These include peripherals, such as keyboards, mice, network connections, video monitors, and printers. Embedded systems often have some of these standard peripherals as well, but also include much more specialized ones. For example, an embedded system might receive data from a wide range of special-purpose sensors (accelerometers, temperature probes, magnetometers, push button contact switches, and more). It may signal output in a variety of ways (lamps and LEDs, actuators, TTL electrical signals, LCD displays, and so on). These types of inputs and outputs are not typically found in general-purpose computers. Furthermore, how the inputs and outputs are encoded may be device specific in an embedded system. In contrast, general-purpose computers use standard encodings. For example, the movement of a mouse is typically transmitted using low-speed serial communication and many manufacturers' mice are interchangeable. Another example is each key on a keyboard has fixed standard ASCII encoding. While an embedded system may also use these standards, it is just as likely that it will receive its input encoded in the form of a pulse frequency. Both may look similar electrically but how meaning is attached to the signals is very different.

Ultimately, such a precise definition of an embedded system can be difficult to agree with depending on what an individual considers a "larger product." For example, is a handheld computer an embedded system? If so, what is its enclosing product? It is not used to control anything. Furthermore, the computing system is exposed to the user: the user can download general-purpose applications. So by the criterion just described, it is not an embedded system. A handheld computer shares many characteristics of embedded systems. (For example, like an embedded system, handhelds are very sensitive to size, weight, and power constraints.) For this reason some might be inclined to call it an embedded system. Now add a mobile phone application to the system and most would agree that this is an embedded system. With the addition of some specialized inputs and outputs, a general-purpose system can be viewed as an embedded system. So, the exact boundary between general-purpose and embedded is not black and white but rather a spectrum.

General-purpose systems strive to strike a balance, compromising performance in some areas in order to perform well over a broad range of tasks. More to the point, general-purpose computers are engineered to make the common case fast. Since embedded systems commonly have a more narrowly defined purpose, designers have the benefit of more precise domain information. This information can be used during the product development to improve performance. Often, this can lead to very substantial improvements.

For example, a general-purpose computer may be used to process 50,000 frames of video from a home camcorder and then later the same computer may need to handle spreadsheet calculations while serving World Wide Web content. For a single user, this computer has to handle these functions (and many others) reasonably well. As long as other issues, such as energy use, are reasonable, the user is happy. However, an embedded system designer might know that their system will never need to do spreadsheet calculations and the engineer can exploit this knowledge to better utilize the available resources.

Most computer courses target general-purpose computing systems. Operating systems, programming languages, and computer architecture are all taught with general-purpose computing machines in mind. That is appropriate because — embedded or not — many of the principles are the same. However, given the enormous presence of embedded computing systems in our day-to-day lives, it is worthwhile to accentuate the unique characteristics of embedded systems and focus a program of study on those.

1.1.2. Hardware, Software, and FPGAs

One of the differences between general-purpose computing and embedded computing is that embedded systems often require special-purpose hardware and software. It is worthwhile to define these terms before we proceed. Furthermore, Platform FPGAs begin to blur the distinction between hardware and software. Before we get into the details of FPGA technology, though, let's revisit the definitions of hardware and software.

Hardware refers to the physical implementation of a computing machine. This is usually a collection of electronic circuits and perhaps some mechanical components. Simply put, it is made of matter and is tangible. For example, consider a computing machine implemented as a combinational circuit of logic gates (AND, OR, NOT) and assembled on a breadboard. The hardware is visible and the design physically occupies space. Another characteristic of hardware is that all of the components are active concurrently. When inputs are varied, changes in the circuit propagate through the machine in a predictable but not necessarily synchronized manner.

Software, however, is information and as such does not manifest itself in the physical world. *Software* is a specification that describes the behavior of the machine. It is generally written in a programming language (such as C, MATLAB, or Java) and this representation of desired machine behavior is called a *program*. While it is possible to print a program on paper, it is a representation of the program that exists in the physical world. The software is information expressed on the printout and is inherently intangible. (Note that a person who is composing software is called a *programmer* and that the act is called *programming*.)

These traditional definitions of hardware and software have served us well over the years, but with the advent of FPGAs and other programmable logic devices in general, these definitions and the distinction of what is hardware and what is software become less clear.

1.1.3. Execution Models

One of the most challenging aspects of embedded systems design is the fact that one has both hardware and software components in the design. These have fundamentally different execution models, which are highlighted.

Several programming paradigms exist for High-Level Languages (functional, object-oriented, etc.). However, at the machine level, all commercial processors use an imperative, fetch-execute model with an addressable memory. In software, this sequential execution model serializes the operations, eliminating any explicit parallelism. As a result, operations are completed in order, alleviating the software programmer from the arduous task of synchronizing the execution of the program. In contrast, hardware is naturally parallel and computer engineers must explicitly include synchronization in their design. This concept of sequential, implicitly ordered operations versus the unrestricted compute model is especially important in FPGA designs and is a topic that we will continue to address throughout this book.

A subtle point is that it is commonplace to refer to some components of our Platform FPGA system as "hardware" when, in fact, these components are written like software. That is, the component is a specification of how the device is to be configured. The real, physical hardware is the FPGA device itself. Thus, in place of the conventional definition of hardware given earlier, we will distinguish hardware and software components of our system based on their model of execution. Earlier we said that software is characterized by a sequential execution model and that the hardware execution model is distinctly nonsequential. In other words, software will refer to specifications intended to be executed by a processor, whereas hardware will refer to specifications intended to be executed on the fabric of an FPGA and, on occasion, in a broader sense, the physical components of the computing system. To make this important distinction more concrete, consider the following computation:

$$R0+R1+R2 \rightarrow Acc$$

Assume that all four names are registers. The computation implemented in the sequential model is illustrated in Figure 1.3. It shows the traditional fetch-execute machine. The dashed box highlights on the main mechanism. The ALU circuit is time-multiplexed to perform each addition sequentially. A substantial part of the circuit is used to decode and control the time steps. This is commonly referred to as the von Neumann stored-program compute model. We will continue to refer to it as the sequential execution model. In contrast, a nonsequential execution model can take a wide variety of forms, but is clearly distinguished from sequential execution by the absence of a general-purpose controller that sequences operations and the explicit time-multiplexing of named, fixed circuits. A nonsequential execution implementation of this computation is illustrated in Figure 1.4. This is generally known as a *data-flow model* of execution, as it is the direct implementation of a data-flow graph. (We use the broader sense of the term throughout this book; the data-flow architectures studied in the 1980s and 1990s incorporate more detail than this.)

In terms of how these two computing models work, consider their operation over time. In Figure 1.5, time advances from top to bottom. Each "slot" delineated by the dotted lines indicates what happens during one clock cycle. It is important to note that the speed of the two computing models cannot be compared by counting the number of cycles — the clock frequencies for each

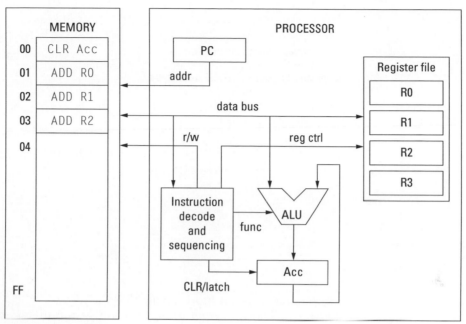

Figure 1.3. A sequential model.

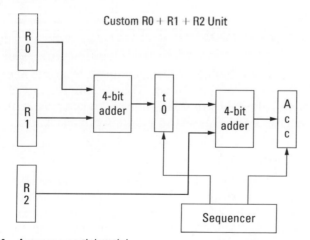

Figure 1.4. A nonsequential model.

are different. A typical FPGA clock period might be 5× to 10× longer than a processor implemented in the same technology. Also, modern processors typically operate on multiple instructions simultaneously, which has become crucial to their performance.

With this understanding, we can say that software runs on a processor where a *processor* is hardware that implements the

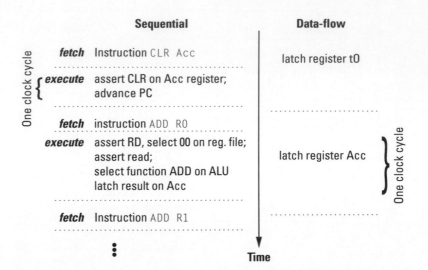

Figure 1.5. A cycle-by-cycle operation of sequential and data-flow models.

sequential execution model. Note that the terms Central Processing Unit (CPU) and microprocessor are common synonyms for processor. Hardware will be the specification that we use to configure the FPGA and does not use the sequential execution model.

1.2. Design Challenges

Now that we have a better understanding of what an embedded system is, it is worthwhile to understand a little bit about how embedded systems projects work. This section discusses the life cycle of a typical project, describes typical measures of embedded system success, and closes with a summary of costs.

1.2.1. Design Life Cycle

There are many books on how to manage an engineering project. Our goal in this section is simply to highlight a couple of terms and describe the design life cycle of a project to provide context for remaining sections. This life cycle, illustrated in Figure 1.6, is sometimes referred to as the "waterfall model." It is meant to suggest that going upstream (i.e., returning to a previous stage because of a bug or mistake in the design) is considerably more challenging than going downstream.

Whether initiated by a marketing department or by the end user, the first stage is called *requirements*. As the name suggests, the step is to figure out what the product is required to do. Often this will start as a very broad, abstract statement, and it is the developer's job to establish a concrete list of user measurable

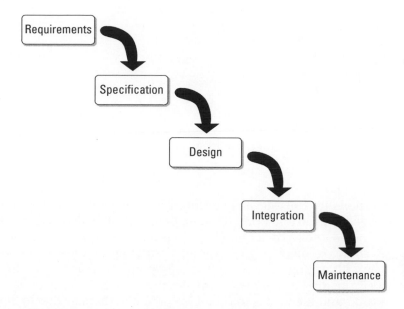

Figure 1.6. A "waterfall model" description of a project life cycle.

capabilities that, if met, the end user would agree the project was a success.

The next stage in is called *specifications*. In contrast to requirements, specifications are usually written for the developer, not the end user. Created from the requirements, they are used to judge whether a particular piece of code or hardware is correct. Often a specification comes from a particular requirement, but some specifications may result as the combination of several requirements or as the result of preplanning the design of the product. It is reasonable to develop a set of metrics at this stage that define the correct behavior (in terms of implementation).

The *design* stage follows next. Usually a developer is contemplating a working design throughout all of the previous stages; this is where the design is formalized. Formats, division of functionality into modules, and algorithm selection are all part of this stage. There are numerous methodologies for how to design software and some include developing a prototype, whereas others call for reiterating over specifications and requirements. Others espouse coding a solution as fast as possible under the adage that "get your first design finished as soon as possible because you always throw out the first one." For FPGA designs, a developer is typically designing the physical components, electrical and mechanical, at this stage as well.

The fourth stage — called *integration* (or in some methodologies, "testing") — begins when the physical components become available. All of the components are combined, and the

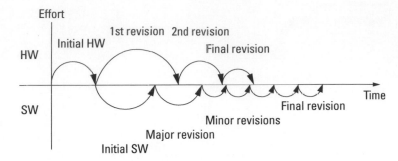

Figure 1.7. Rate of revisions in developing hardware and software.

functionality tests developed during the specification stage are applied. For custom Printed Circuit Boards (PCBs), there is usually an incremental process of (1) testing the electrical connectivity, (2) applying low-voltage power and verifying the digital components, (3) testing the high-voltage components (if any), and then finally loading the software. The process is called "bringing up a board," and the idea is to minimize potential damage due to any mistakes in the design or fabrication process. By testing incrementally, it also helps locate the mistakes that inevitably will exists.

After integration, the product is generally ready to be delivered (if it is a one-off custom design) or manufactured in mass. The final stage, *maintenance*, begins once the product has left integration. As the name suggests, it deals with fixing problems that were discovered after the product begins to ship. For FPGA-based designs, there is much that can be done in this stage because, after all, the device is programmable in the field! More to come on that later.

An alternative view of the design and integration stages is shown in Figure 1.7. The point of this illustration is to show the hardware (top) and software (bottom) revisions of the designs. Often, before software development can begin, some initial design effort is needed to produce the first hardware (first silicon). This can also be the time waiting for a development board to be fabricated. Once the hardware arrives, the software development can swing into full gear. Each arc represents another revision, where the large area under the arc indicates a major revision and smaller arcs represent minor revisions. A project may only have so much time to revise hardware, so the relatively few major revisions are shown. Software may be able to make many small revisions, especially toward the end of the design process.

1.2.2. Measures of Success

For the last several decades in the general-purpose computer world, the top design consideration has been speed (rate of execution). Computing machines were compared based on how fast they completed a task. In the embedded systems world, success is a combination of *requirements* and *performance metrics*. The requirements are capabilities that the system must meet to be functionally correct. Metrics are measured on a continuum. Also, an embedded systems design is, by definition, constrained by its environment (i.e., the larger product it is part of). This necessarily means that success is a multifaceted measure.

Clearly, the requirements and specifications guide the design of an embedded system. The requirements will define capabilities that must be met. Often there exist additional criteria (which may or may not be explicitly provided) that help judge the success of the product. We call these extra criteria performance metrics. Metrics are generally measured on a scale, such as "faster is better" or "lighter is better." For example, a six-week battery life might be a requirement but a product may also be judged by its battery life metric. A product that typically runs for eight weeks between charges is generally considered better than one that simply meets its six-week requirement.

Of course, a developer does not want to blindly "over engineer" a product because there are real costs associated with many decisions. In order to make sensible decisions given a fixed amount of resources, a developer needs quantifiable metrics as a guide. There are a number of quantifiable metrics that embedded systems developers may use; three common metrics are considered here: speed, energy, and packaging.

Speed

Throughout most computer-related undergraduate curricula, the idea of speed is theoretical. That is to say an algorithm is evaluated in terms of the number of time steps it takes to complete. Usually this is written in terms of the size of its input, n. This is a very effective tool for comparing the relative merits of algorithms, but for embedded systems, it is not as useful. Embedded systems, for example, typically run until the power is turned off — so picking a fixed n as an input is often impossible! Instead, it is more useful to measure the speed of the individual components.

If the computing machine has a dedicated purpose, then the speed may be characterized as a *minimum speed* requirement. In this case, the speed can be thought of as a "cliff" function: if you are heading toward a cliff, being able to stop in time is sufficient but not stopping in time has dramatic consequences. For many dedicated systems, there are limited benefits in exceeding the target speed, but falling short of the target does not gradually degrade the performance of the system.

For scientific applications, the speed — that is, the rate of computation — is usually the metric that trumps all others. Heat and packaging are important; however, as long as the machine can be physically built and the heat generated by the machine removed is at a reasonable cost, then the key factor that distinguishes one system from another is how fast it completes a given task. Likewise, for many business applications, reducing the amount of time that an end user spends waiting for a result provides a competitive edge over another machine. This is sometimes characterized as "how fast the time piece goes away" in reference to a mouse pointer icon that indicates the machine is working. In these and other general-purpose cases, performance increases as the speed of the machine increases. For such systems, an *unbounded speed* requirement means that faster is better.

Unlike many general-purpose systems, embedded systems incorporate both speed requirements. When the computing machine is used for controlling the product and the speed is not directly observable by the end user, then there is usually just a minimum speed requirement. The machine just has to be "fast enough" in order for the product to function correctly. If the end user *can* observe the speed, then an unbounded speed requirement may be in effect. For example, suppose the task is to skip forward to the next song in the playlist. An unbounded speed requirement is reasonable here because if the target is two seconds, then exceeding this results in a better end-user experience. Faster response times can translate into a measurable advantage. If the embedded system is designed to increase the energy efficiency of a household appliance, then exceeding the minimum speed actually degrades the overall performance (because a faster system presumably consumes more energy).

Speed requirements are usually specified for each use case. This often results in multiple tasks that, to some extent, share resources in the system. Operating systems can be used to manage those shared resources by creating a virtual environment for each task and then time-multiplexing the tasks. Ordinary operating systems schedule these tasks in real time with an assumption that the tasks have unbounded speed requirements. However, embedded system applications often need a mechanism to relate tasks' progress in their individual virtual times to the events in real time. An operating system that provides these mechanisms is called a real-time operating system.

Scheduling in real-time operating systems is challenging because, for a variety of reasons, the amount of time a particular task is going to take is not always known. Hence, before we leave the speed component of performance, it is worthwhile to note that *predictability* is valuable. Predictability can improve the scheduler of a real-time operating system, which allows us to more carefully specify minimum speed requirements.

Energy and Power

All machines — whether they are mechanical or solid state — are propelled by energy and it is hard to imagine any human activity in our 21st-century society that does not consume energy. Twenty years ago, small (slow) stationary computing machines (desktops) did not consume much energy. Hence, power — the rate at which energy is consumed — was a relatively insignificant issue. However, with rising energy prices and ever-faster microprocessors, power has become a first-class design metric. Moreover, all of the energy that goes toward computation is eventually converted into heat. For low-power systems, the ambient environment may be sufficient to dissipate the heat. As power consumption increases, active thermal measures need to be taken, which further increases the total power used by the system. In some very large high-end computing installations today, 2 MWatts are required to power the computing machine and 1 MWatt is needed to run the chillers! For embedded systems, there are two energy-related components to this issue. First is the total energy used to complete some task or application. The second is power.

Total Energy

Many computing systems are application specific; that is, there is a well-defined lifetime for the machine or there is a duty cycle. Examples of the former might include disposable greeting cards that play a musical tune when the card is opened or a wireless sensor that is built into a bridge to take structural integrity measurements. Examples of the latter include most rechargeable portable devices — mobile phones, personal digital assistants, and so on.

In both cases, the computing system must include some energy storage unit — a battery. The functional specification will probably be expressed in terms of time (i.e., a cell phone must operate for two days between charges). Battery technology is improving; however, it has not kept up with increased power demands of general-purpose microprocessors. Further complicating this picture is the fact that batteries tend to be an expensive and physically large component of the system. Consequently, for many projects the range of battery capacities is relatively restricted and it is up to the engineer to design the computational unit to fit a total energy budget.

For some applications, the total energy over the lifetime of the product is all that matters, since operating system cost is proportional to energy. If the product has a very short lifetime (imagine a greeting card with a small battery-powered audible message) or if the product is wired (like your desktop computer), then the analysis ends there. Ultimately, most embedded systems don't fall into either of these situations and so managing the total energy consumed is an important performance metric.

Power and Heat

While total capacity of a battery constrains the lifetime (or charge cycle) of a computing system, there is another dimension to the problem. Specifically, most energy storage units are limited in terms of how quickly they can release their stored energy. In other terms, energy may be available but the battery cannot meet the power demands at a particular moment in time.

To make these ideas more concrete, a quick review of electrical properties is in order. Starting with some power source, such as a battery, there is a voltage difference between its two terminals. This difference, measured in volts (v), is the potential energy that will be dissipated in an electrical circuit connected to the battery. A circuit provides a path for electricity to flow from one terminal to the other, and the amount of electrical charge that flows is current, measured in amps (i). Depending on how the path from one terminal to another is constructed, the circuit will limit the current. This is the resistance r of the circuit. These three variables are related by Ohm's law, which is $v = ir$. Once the circuit is connected and current flows, the circuit begins to convert the potential energy of the battery into heat, which is the rate of energy consumption, which we already know is power. Power (measured in Watts, p) is a product of the change in voltage and current, that is, $p = vi$. (For computing devices, this only is only part of the picture — without taking switching into account this only measures the static power.)

The rate at which energy is consumed by the circuit is proportional to the rate at which heat is released into the environment. Passive thermal management techniques are limited in how fast they can remove heat. If heat is not removed, then the temperature rises, which has a series of problems, including device failure and discomfort for the end user, who may have the device on their person. To keep the temperature low, we have to resort to removing the heat with active thermal management (fans and more exotic techniques). These, of course, require power themselves, complicate the design, and raise the cost.

To bring this discussion to the point: power is dictated by the circuit technology we use and by how we organize our machine.

While the total energy used is one factor, the rate of energy consumption is much more significant because it affects so many of our performance metrics, such as size and mass. Poor power budgeting can raise the cost significantly.

Size and Packaging

Often size (and mass) is a requirement: the product has to fit within some space or cannot weigh more than a certain amount, but once met, they can also be metrics. For example, thinner mobile phones and lighter earpieces are better. For embedded systems design, size matters. If a designed circuit board is too large to fit in the packaging, it may result in a redesign of the circuit or a reorder of the packaging. Both are likely to be expensive mistakes. Therefore, it is important to consider what packaging or size constraints exist prior to circuit layout or fabrication. Clearly, these size and packaging issues are easy to quantify, so we will not belabor it here.

1.2.3. Costs

The previous two subsections are, strictly speaking, performance related. Cost is usually held in opposition to performance. An implicit goal of nearly every engineering project is maximizing performance while minimizing cost. Furthermore, cost constrains all of the metrics discussed previously, thus cost fundamentally limits the range of potential solutions. The cost of developing a product and selling it includes development and Nonrecurring Engineering (NRE) costs and then manufacturing and distributing costs.

While the business aspects of bringing an embedded systems product to market are better left to people schooled in business, it is important for the developer to at least be aware of the business aspects. In particular, there are two points we want to emphasize. The first is related to delivery dates, market demand, and how projects are financed. The second point is how technology — especially Platform FPGAs — can be used to reduce development/ NRE costs and the risk of escalating development costs.

Costs Incurred

Personnel

Perhaps the most obvious cost to an engineer is the development cost. These costs can be divided into two categories: direct and indirect. Direct costs include things such as salaries and hourly wages for employees developing the product, consumable materials and supplies, and education costs. Special equipment needed for development is also a direct cost.

Indirect costs are sometimes called overhead and are basically "the cost of doing business." More precisely, overhead pays for infrastructure. This cost is determined as a percentage of the direct costs. Assuming the project is part of some bigger organization, the project will be charged indirect costs based on what the institution provides. For example, a research project housed in an educational institution might be charged anywhere from 40 to 100% of the indirect costs. Typically, large industrial organizations charge upward of 100 to 200% overhead. (That's correct — if the direct costs are $50,000, total budget might be $150,000!) The justification for indirect cost depends on the type of institution as well. Organizations charge indirect costs to cover building capital costs and staff salaries.

It may not be immediately obvious why indirect cost is charged as a percentage. Why not simply pay for all services directly? The answer lies in the fact that most organizations have multiple ongoing projects and it is inefficient to dedicate a resource (such as an administrative assistant) to a single project. Moreover, the demands on that resource are likely to vary over time. The idea is that an active project would require more services, so it generates more overhead for the organization. By treating expenditures (direct costs) as a reasonable approximation of project activity, different projects can be charged based on their approximate demands on the infrastructure.

The general cost formula is:

$$TOTAL = DIRECT + RATE \times (DIRECT - EQUIPMENT)$$

Because the cost of large equipment purchases would generally skew the indirect costs, it is usually not included. For example, a project with $100,000 direct costs (of which $25,000 is a special-purpose milling machine) would be charged overhead on $75,000.

Nonrecurring Engineering Costs

After the product has been developed, there may be a prototype or just a construction plan. The next step is manufacture, distribution, and marketing. While this is primarily a business concern, it is important for an engineer to understand how products are manufactured, as the design decisions can have a significant impact on these costs. Generally, these costs are divided into two components. The first, NRE costs, are upfront charges that are required to build the first unit. The second are material and labor charges, which are basically the per-unit cost for some lot size.

Nonrecurring engineering costs are usually some sort of tooling charge or one-time software development costs, for example, the production of ASICs (Application-Specific Integrated

Circuits). This process was originally designed to make inexpensive product-specific integrated circuits widely available. The idea is to keep many parts of the process identical, but leave some layers customizable by the designer. The customizable parts are used to generate a mask that can be used to make silicon wafers quickly and easily. Generating a mask is a steep NRE cost that is proportional to the IC technology used. While individual IC devices might cost less than $1 per unit to manufacture, the initial mask cost might be over $250,000!

Finance and Consumer Demand

Most consumer electronic devices have an imaginary "consumer demand curve." This curve represents the total number of potential sales of a product over a period of time. Unlike older technology (such as hammers), consumer electronics change rapidly and newer devices replace older devices (such as game consoles). Whereas a consumer will use a hammer until it breaks, many consumers will stop using a functional mobile phone and replace it with a newer phone. Depending on when your product arrives relative to the customer demand curve and how quickly it is adopted dictate the total income derived from the product. We adopt Vahid and Givargis' model (Vahid & Givargis, 2002) and extend it to explain how development delays hurt the potential income but also bear significant costs that may not be immediately obvious.

The on-time model is illustrated in Figure 1.8. Time is on the x axis and the expected delivery date is at the origin. Everything to the left is development, manufacture, and delivery. To the right of the origin is after an on-time delivery. The y axis represents an account balance — below the x axis, the project costs have exceeded its income, above the x axis indicates that the project is profitable. In an ideal world, the consumer demand curve begins at the origin, will peak, and then asymptotically approach the x axis again. (It is easiest to think of "demand" as the number of potential units sold per time period. If there are no discounts and all of the costs are carefully accounted for, the number of units can be translated in potential profit.)

With an on-time delivery, we can see that a certain amount of monthly development costs will accrue between the start of the project (point A) and the origin. Once the embedded systems product has been been completed, a factory is contracted to produce a lot. This lot is then distributed to retailers at point B. If the product has no competition (and we contracted for a big enough lot), then the income will follow the consumer demand curve and, at some point, our account balance comes out of the

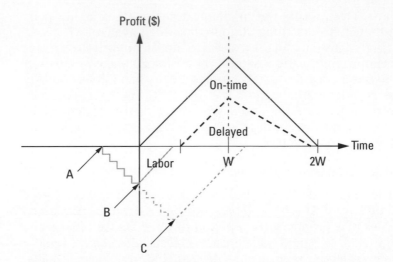

Figure 1.8.   The importance of time-to-market in terms of break even and profitability.

"red" and the project becomes profitable. What we want to learn from this illustration is: what is the impact of delays during development? Delays during development that change the delivery date doubly hurt profitability. Suppose the actual delivery date is at point C. We continue to incur monthly development costs, and the consumer demand between points B and C is lost income that will never be recovered. (Furthermore, the financing charge continues on the original development cost estimate plus the additional costs.) Thus, missing a deadline for a consumer electronic device is not just delayed profits. In fact, it means that a diminished potential income will be available to recover a larger amount of debt.

While this section could continue into a variety of different product/project management and finance discussions, it is not the intention of this book to dwell deeply into these topics. Likewise, the brevity of the discussion is not to diminish the significance of the role they play in embedded systems design. In short, embedded systems design is a multifaceted endeavor and it is in the best interest for the reader to become familiar with as many aspects to the design process beyond the engineering solution. Meanwhile, this book will continue with a more engineering, science, and technology focus of embedded systems design.

1.3. Platform FPGAs

As highlighted earlier, embedded systems designers have increasingly turned to the use of FPGAs in their designs. While the remainder of this book addresses the specifics of "how" to use FPGAs, we first must explain "what" are and "why" use FPGAs. Before FPGAs there were programmable logic devices (PLD). Early PLDs were designed to implement arbitrary digital (combinational) logic circuits. They were implemented as ICs with an array of gates (inverters, AND, and OR) and wires. Various mechanisms were used to make connections between the wires and the gates. The key feature of these devices was their ability to be configured after the manufacture of the device. A single PLD package is able to replace several packages of small-scale integrated circuits (such as the 7400-series parts). Hardware engineers began using the verb *program* to describe the process of setting the configuration. Hardware engineers would typically figure out the settings, and only the last step of transferring the configuration was the act of programming the device.

The FPGA is an advancement on these earlier programmable logic devices and is generally distinguished by two features. First, the gates are no longer physically implemented, but rather the logic is implemented with function generators and discrete memories. Second, most FPGAs use static RAM cells to hold the configuration information as opposed to more permanent ways of earlier devices (such as antifuses). This allows the FPGA to be configured and reconfigured after the device has been installed in a product. That is, an FPGA can be reconfigured in the field.

Field-Programmable Gate Arrays permit an arbitrary digital circuit to be realized after the physical silicon chip has been manufactured, tested, and installed. The digital circuit that is programmed (or configured) into an FPGA is called a *design*. Design entry is a generic term for creating a machine-readable representation of the desired digital circuit. We will return to this in the next chapter. A *Hardware Description Language* (an HDL) is a programming language used to describe the behavior of hardware and is the most common form of design entry today. Originally, HDLs were only intended to simply describe hardware for documentation purposes. Later, it was recognized that the HDL could be used to simulate hardware for testing the design before implementing it. Before long, engineers began to refer to the intangible HDL *design* as hardware because everyone understood that eventually it would be physically implemented. The definition of hardware was further co-opted when HDLs began to be used as a synthesis language. At this point, the common practice of referring to the specification as hardware becomes misleading: in an FPGA those gates described are never physically realized. The FPGA device is the hardware, and programming it does not manifest any physical change. (As shown in Chapter 2, the FPGA device is simply constructed to mimic the desired digital circuits.) Thus, even though the design is, strictly speaking, intangible like software and the hardware is the physical FPGA device, the current convention is to refer to the design as "hardware." This broadens the definition of hardware given earlier to include not only the tangible physical components of the computing system, but also the intangible specifications that use a nonsequential execution model.

We are now ready to give a more refined definition of a Platform FPGA. In 2002, Xilinx introduced the Virtex-II family of FPGA devices. The Virtex-II Pro members of this family incorporated one or more PowerPC processors as diffused intellectual property (IP) cores, to be explained shortly. Coupled with other diffused IP, such as RAM and high-speed transceivers, this meant that it is possible to deploy a complete, fully functional computing system on a single FPGA package. Moreover, with the large amount of configurable resources on the die, standard and nonstandard peripherals can be readily incorporated. Because it is configurable, the resources can be used to efficiently implement application-specific architectures. Naturally, this makes an excellent foundation (or platform) upon which to build an embedded system. We use Platform FPGA to designate an FPGA device that is deployed as the central component of a computing system. Indeed we will see that the presence of a diffused processor core is not always strictly required: with large FPGAs, the processor can be implemented in the reconfigurable resources of the FPGA.

It is useful to define some of these "hardware" components more carefully now. First, we begin with an Intellectual Property core. An IP core (or simply a *core*) refers to a hardware specification that, depending on how it is expressed, can be used to either physically manufacture an integrated circuit or configure the resources of an FPGA. For example, a pipelined multiplier could be packaged as a core. A *diffused core* in an FPGA means that the hardware is physically part of the silicon device and its functionality is realized in transistors when the device is manufactured. Because it is, by the traditional definition, truly hardware, it is more often called a *hard core*. In fact, the term hard core has come to replace diffused IP so much so that we will go forward with this text using hard core in place of diffused IP core.

In contrast, a *soft core* is implemented in the reconfigurable resources of an FPGA. Finally, some communities refer to cores as either *blocks* or *modules*. A hard block is a core that has been implemented in CMOS transistors, and a soft block is a core that has been implemented in the function generators and memories of an FPGA device. The design automation community often refers to modules, and modules are always destined to be hard blocks on an IC chip. Details regarding synthesis, mapping, and routing are covered in detail in Chapter 2 where the inner workings of an FPGA are discussed in detail.

In summary, the Platform FPGA is simply an FPGA that an engineer uses to deploy a custom computing system. Instead of its role as a support device, it becomes the central compute device. In our case, we will be using it to develop custom computing machines for embedded systems designs. The rest of this book is spent covering details regarding Platform FPGAs and embedded systems design; this section is meant to introduce and motivate their use as an alternative to a microprocessor-based system.

Chapter in Review

The overall aim of this chapter was to set the stage for building embedded systems with Platform FPGAs. To accomplish this, we addressed three key questions.

- *What is an embedded system?*
- *Why are embedded systems different?*
- *How do Platform FPGAs help?*

We answered the first question by revisiting the concept of a computing machine and then distinguishing an embedded computing machine from a general-purpose computing machine. We answered the next question by showing the complex constraints that embedded systems impose on the developers. Finally, we gave a quick overview of the characteristics of a Platform FPGA and described how this technology can be useful to embedded systems developers.

In the gray pages that follow, we present an embedded systems scenario and contrast two approaches. We also describe the setup and installation of a popular set of commercial tools and then do a simple demonstration.

Practical Expansion

Every chapter of this book is divided into two sections: white pages describe the *science* of the chapter's topic and gray pages focus on specific *technology*. The white pages contain information that has been true for twenty years and will probably be true for the next decade. The gray pages (these pages) contain more details about state-of-the-art technology and examples introduced by the white pages. The reader might ask, "why not put it at the end of the book?" In short, the gray pages are specifically relevant to the material covered in each chapter, which covers practical exercises and examples that, if placed at the end of the book, the reader may miss. While the white pages refer to the science that is considered to be "timeless," the gray pages contain specific details on software and hardware that do age with time and, as is common with any technology, can become dated quickly. We have selected materials for these gray pages that are toward the beginning of their shelf life, but it may be evident that these technologies are less relevant as the book ages. However, the concepts of embedded systems development on an FPGA are far less vendor and hardware specific. Being too abstract would leave the reader without any solid, tangible examples, just as covering every FPGA vendor and family would result in far too much information for a reader to digest in a timely manner. Therefore, we have chosen the Xilinx ML-510 development (Xilinx, Inc., 2009c) board for its available resources that an embedded systems designer may use in their design. A photo of the ML-510 is shown in Figure 1.9. The reader is by no means required to purchase this board; there are less expensive development boards with the same FPGA available. As a result, we will do our best to present the material with an emphasis on FPGA generality and highlight what is ML-510 specific. Overall, the material we wish to cover is more specific to FPGA design and should be applicable over a range of FPGAs.

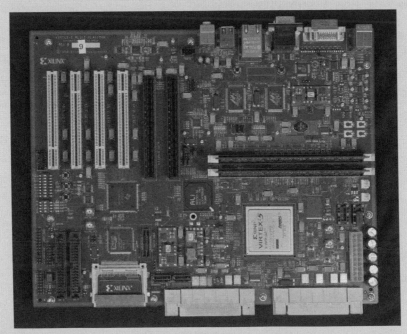

Figure 1.9. A photo of the Xilinx ML-510 development board.

Embedded Systems developers wear many hats. As Section 1.2 detailed, much more than just designing operational hardware or functionally correct software goes into a successful project. So far in the white pages we have discussed some of the benefits of Platform FPGAs, but the reader may still be asking "why use Platform FPGAs over commodity off-the-shelf components?" So let us start these gray pages with an instructive end-to-end practical example to motivate the use of Platform FPGAs. Then, we will discuss the technical details necessary to begin embedded systems design with Platform FPGAs.

1.A. Spectrometer Example

First, the general scenario and key requirements are defined. Next, two designs are described — one based strictly on commodity components and the other uses Platform FPGAs. This section concludes with a discussion of strengths and weaknesses of each approach.

1.A.1. Scenario

Suppose a hospital is preparing to conduct a large study involving a large number of experiments and thousands of volunteer subjects over a three-year period. This means every subject will make routine visits to the hospital to have various specimens collected depending on the subject. In order for the study to be conclusive, subjects have a specific time frame in which they have to report for various experiments.

Based on the large amount of data being collected and the risk of logistical mistakes corrupting the massive study, the principal investigators have budgeted money for an automated information system. Over the three years it is probable that new staff members will be hired and others will leave. As a result, the system needs to be simple enough to train new employees quickly. Also, the subject's experience needs to be pleasant and not take up much additional time. Once at the hospital it is important that subjects are not frustrated by long lines. (Every subject that fails to complete the study represents a loss of investment by the study; if enough subjects leave the study, the results of the whole study are at risk.)

Based on this, a systems analyst has designed a system where each volunteer carries an RFID tag with them while at the hospital. Many of the experiments will be custom-built so that when a subject's specimen is taken and analyzed, the subject and data are automatically associated. The results are transmitted over the campus network back to a central server. The study bears the cost of such an elaborate system because they are very interested in features that will allow them to track their subjects and eliminate paperwork transcription errors.

For our Platform FPGA case study, we will focus on one experiment within this larger context. The project requires that a specimen be taken and analyzed with a spectrograph. The embedded system needs to read the subject's RFID tag and process the spectrograph's output to determine the presence of various particular chemical elements. Together with the date and time, a database record is constructed and transmitted to the server. A minimal user interface is needed for the professional collecting the specimen and for station diagnostics. This is the minimum functionality required, but as it is anticipated that additional projects for this study are under consideration, it is worthwhile to spend some time designing the system for reuse.

1.A.2. Two Solutions

Commodity Components Solution Many of the components needed for this station are available as commodity off-the-shelf components, one easy solution is to buy a commodity desktop machine (monitor, main board, disk, keyboard/mouse, network card, off-the-shelf OS). Spectrographs come in a wide variety of packages. The one we need comes in two flavors — one is bare-bones hardware with an electrical interface and the other is hardware plus a built-in computer with preinstalled software. The hardware itself is not very expensive, but for many stand-alone lab users, the all-in-one solution is well worth the price. Finally, a USB-based RFID wand, with ergonomic handle and software that makes recording an ID "as easy as cutting and pasting," can be purchased.

From this information, we could buy the RFID wand, all-in-one spectrograph, and the desktop computer with a network interface. For this approach, what remains is simply a large software integration problem. The spectrograph has a graphical user interface, but processed data can be exported to an ASCII file. This file can be transmitted over a serial port to the desktop computer.

A graphical user interface on the desktop computer could have a window with database fields for the subject's ID (which can be pasted in with RFID wand), date, time, and a button that initiates a transfer of the data file exported by the spectrograph. Our software application can then combine data and transmit it over the network. A short user's manual can be written for the staff person, explaining how to operate the spectrograph, how to export data, how to launch the desktop application, and how to use the RFID wand to fill in the the subject ID's field. We'll call this the off-the-shelf solution.

Platform FPGA Solution   Alternatively, we might consider using a Platform FPGA approach. For a cost-effective solution it will still be necessary to leverage off-the-shelf components; however, in this case, we will use simpler subsystems with a higher degree of integration. At the heart of the system we have a Platform FPGA that can be purchased as part of an evaluation or development board. (A custom-printed circuit board is possible, but the initial layout of the board may not be cost-effective for a few boards. We'll explore the relationship between the number of units and the cost in the next chapter.) Interfaced to the FPGA is the bare-bones spectrometer device. Likewise, small RFID-printed circuit boards (about the size of a quarter) can also be purchased. The interface of the spectrometer is fairly complex, requiring that a sequence of digitally encoded amplitudes be correlated to known frequencies, and other electrical signals are used to control the behavior of the device. The RFID has a simple RS-232 serial interface. A case, power supply, and an inexpensive two-line LCD display round out our hardware needs.

Note that we do not need a monitor, a video adapter, or even a system disk. The programmable logic in the FPGA can be used to build a custom hardware solution that will correlate frequencies coming from the spectrometers. Likewise, three standard UART cores are easily instantiated in the FPGA's programmable logic (one for the RFID, one for the LCD display, and one that we'll reserve for development testing). A simple keyboard interface and a standard Ethernet network interface can be realized in programmable logic as well. Because modern FPGAs have transceivers integrated, neither an external MAC or a PHY chip are necessary — the whole network interface is on-chip. FPGA development boards typically use flash to store the FPGA configuration and most allow the flash to store the OS and application as well. With this hardware, the main tasks of the software are to (i) recognize the presence of a subject by their RFID, (ii) interact with the staff and operate the spectrometer, and (iii) communicate with the central server over the network.

1.A.3. Discussion

Both solutions are reasonable approaches and both meet the functional requirements of the problem, but which one is better? To answer this question, we will consider a number of factors, including quantitative measures (such as cost) and qualitative measures (such as how well does it serve the customer's goals). Indeed, we'll see that this complex evaluation is typical for embedded systems.

We begin with cost. Cost is the price of all of the hardware and intellectual property plus the development cost. In accounting the total cost, we can discount solutions that provide components that realistically can be reused in the

Commodity Costs		Platform FPGA Costs	
Desktop PC	$400	FPGA, Flash, PCB	$500
Spectrometer Unit	$2500	Bare Bones Spectrometer	$500
RFID Wand	$200	RFID Device	$100
TOTAL	$3100	TOTAL	$1100
(a)		(b)	

Figure 1.10. (a) All-commodity parts solution and (b) custom FPGA solution.

future. No adjustment will be made for the learning experience (i.e., the benefit of learning a new technology that will improve a future project). The hardware costs for each solution are shown in Figure 1.10. Even though commodity components are much less expensive than custom components, the all-commodity solution uses more components. Furthermore, packaging and application software associated with each subsystem contribute to the hardware cost of solution. Of course, hardware is only one part of the total cost. The software/development costs for the solutions are heavily in favor of the commodity solution. Rather than estimating an exact cost, we might simply compare ratios. Chances are that for most software/applications programmers, the time it takes to learn and debug hardware development will eat through the $2400. However, experienced computer engineers might break even if they can integrate the system in a week.

There are two other factors that may be included in the cost. First is the risk of failure. Of the two solutions, the all-commodity approach probably has a lower risk of failure, as each subsystem is already an independent entity. For example, the all-in-one spectrograph and data it exports are warrantied. The only risk is in making sure it is calibrated properly. In contrast, the hardware solution has more unknowns. Correlating frequencies in hardware is fairly straightforward in an academic setting. However, measurements from physical devices can often have "messy" data with invalid samples and skewed constants. These are normally corrected with software, but, if unanticipated, this can lead to longer development times.

Now consider scalability. With the cost of the all-commodity solution over 2.5× greater, as the number of data collection stations grows, the Platform FPGA solution becomes the clear winner. Of course, a careful reader may point out that these numbers are suspect and that a case could be made to argue that a commodity solution would be better. Our goal with this example is to show how these two technologies can both be used to implement a viable solution. A reader familiar with the commodity parts can begin to connect the dots between how a similar solution is constructed from Platform FPGAs. We can then extend the solution to highlight some of the unique features the Platform FPGA can support, such as reduced chip count and flexible scalability.

Of course this is only one specific scenario and it is impossible to make an argument that Platform FPGAs will always be the practical solution. What is important is understanding that Platform FPGAs do provide much of the same functionality as all-commodity solutions, but often at a reduced total cost. Therefore, it is important to consider the custom solution when designing systems.

1.B. Introducing the Platform FPGA Tool Chain

With the practical example complete, it is time to shift our focus away from the theory and science of Platform FPGA development and instead begin to understand the necessary technology behind developing embedded computing systems with Platform FPGAs. This section exposes the reader to a number of graphical user interfaces, as well as their corresponding command line tools to generate complete Platform FPGA designs. The end goal of this first practical example is to build a basic "hello world" system and run it on the FPGA. It may not be technically challenging, but there is something gratifying about running a design on the hardware within the first chapter.

An important aside, this book focuses on the use of Linux for the development environment as well as the run-time environment on the Platform FPGA. Our goal is not to force the reader to switch from their current development environment, but to describe what the authors and many research institutions and companies have been or are beginning to use. (The Xilinx tools are supported under Microsoft Windows, and we highlight any applications, tools, or commands that are specific to Linux development.)

So we begin by using a graphical user interface that makes FPGA development exceedingly easy. Here we will look at Xilinx Platform Studio, (Xilinx, Inc. 2009d) (or XPS for short). Other FPGA vendors have similar tools with slightly different features and operating modes. XPS provides a multitude of wizards, tools, and premade design templates that greatly simplify the initial design process. Our focus for this technical section is to begin to familiarize the reader with the tool chain through a simple Platform FPGA design.

1.B.1. Getting Started with Xilinx Platform Studio

The first step is to acquire and install the Xilinx software package: Integrated Software Environment (ISE), (Xilinx, Inc., 2009b). The ISE install should consist of at least the ISE Design Suite, Embedded Development Kit (EDK), (Xilinx, Inc. 2009a), Software Development Kit (SDK), and ChipScope Pro. We will be focusing our attention initially on the EDK and SDK tools (which are part of the XPS tool chain). Many digital logic courses use ISE; therefore, we will spend less time discussing ISE until later chapters. ChipScope is a very useful hardware development tool that we will use in future chapters. If you are installing the tools yourself on a Linux system, the directory structure shown in Figure 1.11 is a convenient way to keep the multiple versions organized. (The directory /opt is often used to install packages that have a custom organization; i.e., they don't follow the standard GNU package format and filesystem hierarchy standard discussed in the next chapter.)

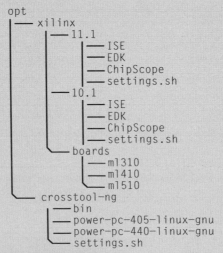

Figure 1.11. A directory structure for installing Xilinx ISE.

As can be seen, under the directory `/opt/xilinx` there is a subdirectory for each version of the tools. At the time of this writing, 11.1 is currently the latest release. (There have been minor updates available at the time of this writing, which would make this release technically 11.4, but major release version number 11 is the important identifier.) In that directory, there are two subdirectories: `ISE` and `EDK`. This is where the Integrated Software Environment and Embedded Development Kit were installed (respectively). (Other tools, such as ChipScope and PlanAhead, can also be installed here.)

Now that the install locations have been identified, the second step is to update your user environment variables. This depends on which of two families of command line shells you use. For GNU/Linux systems, most distributions default to BASH, which is part of the Bourne-shell family of command line shells (others include the original Bourne shell `sh` as well as Korn shell `ksh` and others). The second family of shells includes C shell `csh` and `tcsh`. If you are working on a Solaris Unix workstation, C shell is common.

There are several ways to determine which shell you are using, perhaps the easiest way is to type:

```
echo $SHELL
```

If the result is `/bin/csh` or `/bin/tcsh`, then you set your environment variables by typing:

```
source /opt/xilinx/11.1/ise/settings.csh
source/opt/xilinx/11.1/edk/settings.csh
```

Note that this *only* changes the current shell and will be lost when you logout.

For the rest of the book, we will use the syntax for the Bourne-shell family of command line shells and, specifically, we will assume that the reader is using BASH. (If you use C shell, you'll need to occasionally adjust the examples.) Assuming that you are using BASH, then the commands are:

```
source /opt/xilinx/11.1/edk/settings.sh
source/opt/xilinx/11.1/ise/settings.sh
```

1.B.2. Using Xilinx Platform Studio

Now we are ready to launch the graphical user interface for XPS. Type `xps` at the command line shown in Figure 1.12 (or navigate to the Xilinx Platform Studio icon in the Window's Start Menu). A brief Xilinx splash screen is displayed before the application starts and an initial dialog pops up (shown in Figure 1.13).

Base System Builder Wizard   We will begin by creating a new project and using the Base System Builder (BSB) wizard to create a basic template design for the Xilinx ML-510 development board. If you have a different board that is supported by the BSB wizard, select it and then following the rest of these directions.

The BSB wizard's welcome screen (Figure 1.14) allows you to either create a new design or load an existing BSB configuration file. We will start with a new design.

```
% xps
Xilinx Platform Studio
Xilinx EDK 11.1 Build EDK_LS2.6
Copyright (c) 1995-2009 Xilinx, Inc.  All rights reserved.

Launching XPS GUI...
```
Figure 1.12. Launching XPS from the command line.

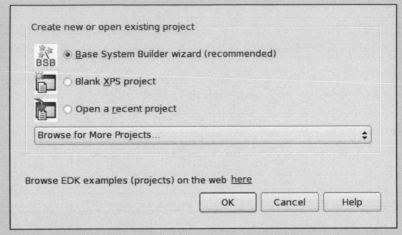

Figure 1.13. Initial XPS dialog.

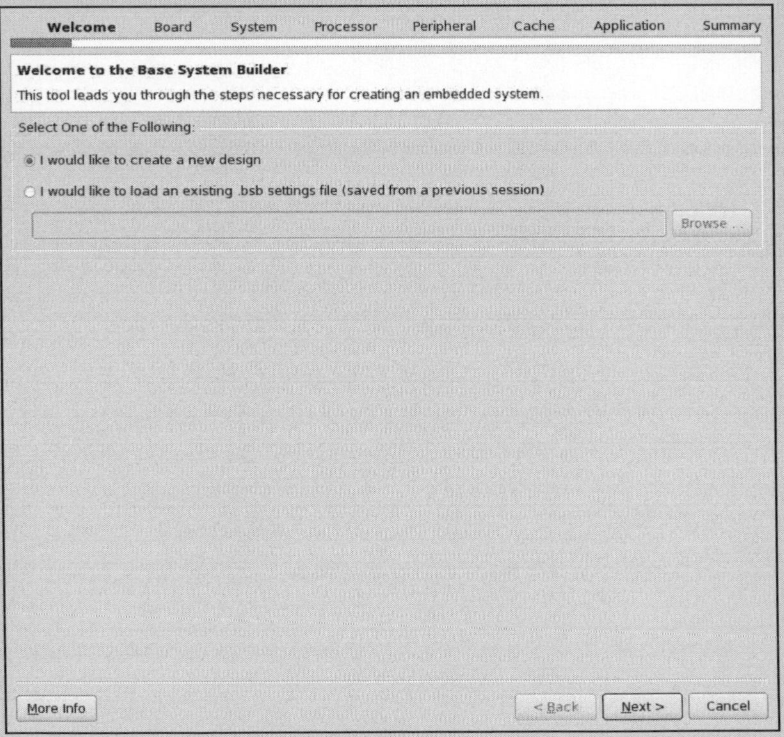

Figure 1.14. The base system builder wizard's welcome screen.

Figure 1.15. The base system builder wizard's board selection screen.

The BSB wizard's board selection screen (Figure 1.15) allows you to select which development board you will be designing for. This book uses the Xilinx Virtex 5 ML-510 Evaluation Platform.

The BSB wizard's system configuration screen (Figure 1.16) allows you to select a single-processor or dual-processor system. (Note: this option may not exist for all development boards.) For this first design we will select a single-processor system.

The BSB wizard's processor configuration screen (Figure 1.17) allows you to set the processor type and operating frequency for the processor and bus. Most of the designs presented in these gray pages focus on using the embedded PowerPC common in the Virtex series parts. The specific operating frequencies are less important for this initial design.

The BSB wizard's peripheral configuration screen (Figure 1.18) allows you to add or remove peripherals to the design. These are the basic peripherals that the particular development board supports. In this design, add only the *RS232_Uart* (xps_uartlite) and the *xps_bram_ctlr*. The UART will be used to connect the printed output from the development board to the control computer's terminal. The block RAM (BRAM) controller is on-chip memory used to store the application that will be written shortly.

The BSB wizard's cache configuration screen (Figure 1.19) allows you to add/configure the PowerPC's embedded cache. At this time we will not add cache.

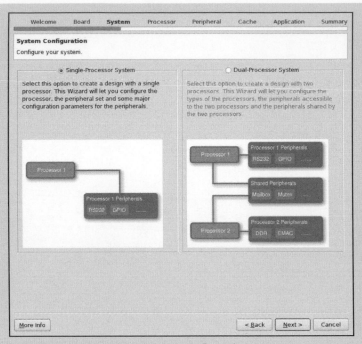

Figure 1.16. The base system builder wizard's system configuration screen.

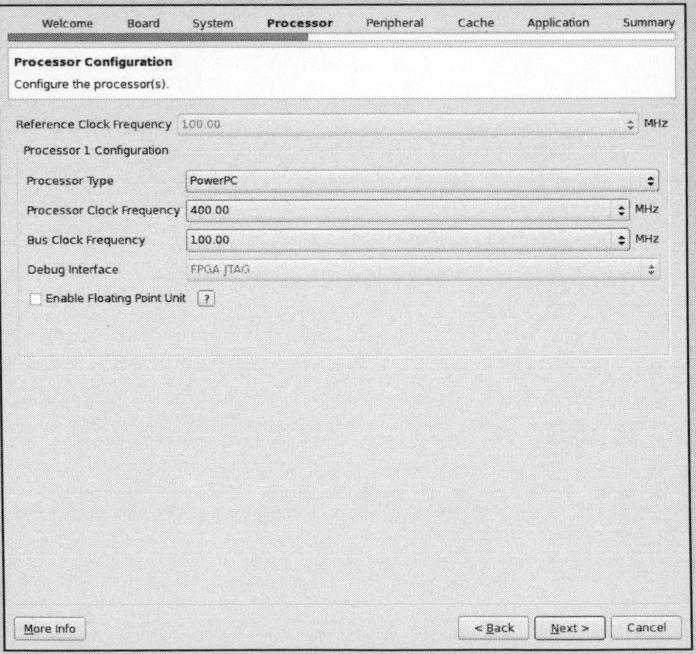

Figure 1.17. The base system builder wizard's processor configuration screen.

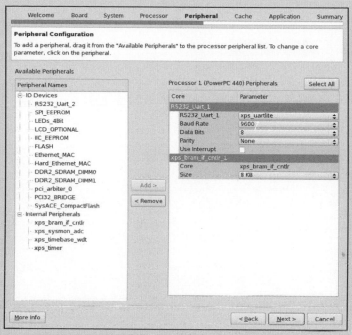

Figure 1.18. The base system builder wizard's peripheral configuration screen.

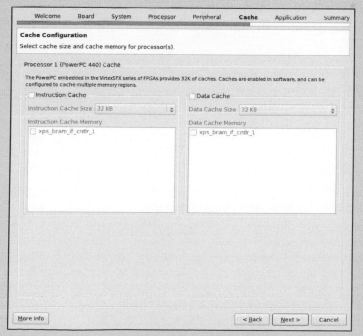

Figure 1.19. The base system builder wizard's cache configuration screen.

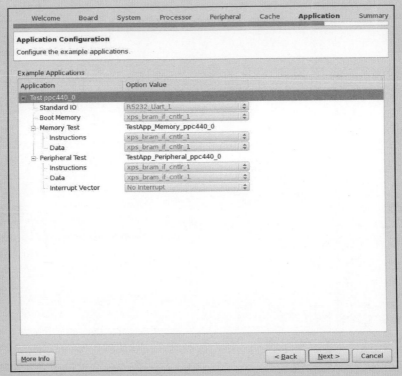

Figure 1.20. The base system builder wizard's application configuration screen.

The BSB wizard's application configuration screen (Figure 1.20) allows you to set the memory location for the Memory Test application and Peripheral Test application that are standard with the Xilinx BSB wizard. We will be designing our own application shortly so it is unnecessary to modify these configurations.

The BSB wizard's summary screen (Figure 1.21) presents a summary of the design that has been created with the wizard. It provides the designer with an opportunity to double-check that all configurations have been set correctly.

XPS Overview The BSB wizard simplifies the design process into a sequence of button clicks, but now XPS can be used to further customize the design. In this example we will hold off on the discussion of these customizations and instead focus on the XPS application. Future chapters will explore more customized designs. Details of the software and hardware flows appear in the next two chapters (respectively). It is worth noting that XPS does not do any of the actual work — it provides a graphical front end to command line tools that do the actual assembly and synthesis. Throughout the book we will cover these command line tools to provide the reader with a more thorough understanding of their role in FPGA development.

By default, the XPS GUI has four main areas, which are highlighted in Figure 1.22. The menu bar and buttons span across the top of the application. The menu system has every function that XPS can invoke, and the buttons are associated with specific menu items. Thus the buttons can be seen as shortcuts to navigating the menus. In the middle and on the left of the application screen is a tabbed project information area. This is where basic information about the current project is stored, such as options, location of various configuration files, and the IP Catalog. The IP Catalog provides a list (organized

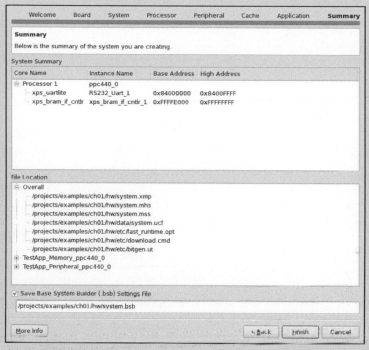

Figure 1.21. The base system builder wizard's summary screen.

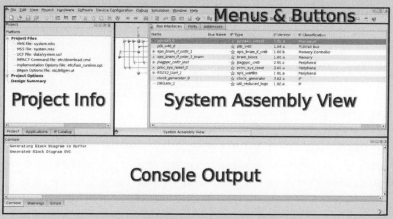

Figure 1.22. XPS menus, buttons, project information, and console areas.

in a tree) of IP cores available to the designer. By right-clicking on a leaf in the tree, one can add an instance of the IP core to their project. At the bottom of the application is the console output window. This window shows a filtered list of all output generated by the command line applications invoked by XPS. The command line tools generate a large amount of information while they process a design. One handy feature of this window is that by clicking on the "Errors" tab, only error messages are displayed. After a long run, this is a quick way to see what, if any, error messages appeared in the copious output.

The fourth area is the system assembly window. This window allows the designer to connect and configure IP cores. For example, by connecting a DDR RAM controller to the Processor Local Bus, the designer is, in effect, putting memory on the bus. Double-clicking on an instance in the design will bring up an IP-specific configuration tool. For those familiar with VHDL already, these configuration tools allow the designer to assign values to all of the generics and parameters in the IP core.

Once all of the configuration changes have been made and the system is ready to be run through the toleration to produce a configuration file (bitstream) for the FPGA, we can go ahead and synthesize the design. This is accomplished through the `Hardware` menu. Select `Generate Bi'tstream` to begin the synthesis process. This may take between 10 to 30 minutes depending on the speed of the computer running XPS. Once the bitstream has been generated, it is now time to switch from hardware development to software development. The hardware design must be exported to the software development kit (SDK). From the `Project` menu select `Export Hardware Design to SDK.`

Software Design Kit Overview The Xilinx Software Design Kit (SDK) is used to develop applications for the hardware specified within XPS. Xilinx has stated that beginning with the release of the 12.1 tools the SDK will replace the application's tab within the XPS. Meaning, Xilinx will no longer support software development in the hardware environment development kit. Abstracting the software development from the hardware development is important because, as mentioned earlier in this chapter, the distinction between hardware and software becomes less clear with FPGA development. For developers more familiar with the Eclipse C/C++ Development Toolkit (CDT), the SDK should be a welcome improvement for project management over the previous application process. This section aims to describe the SDK in sufficient detail to generate a stand-alone software platform and a simple "hello world" application. In future chapters, more details will be given regarding the specifics of the application development and among the stand-alone, xilkernel, and linux kernel platforms. The SDK can be launched from the command line with the `xps_sdk` command.

The SDK will prompt you to select and import a workspace from a previous design. Browse to the workspace created when you exported the design from XPS. (Note: this step can be skipped if the reader is continuing directly from the previous section.) With the SDK open, a few steps need to be taken to create a software platform and an application. We have chosen a simple stand-alone platform as we do not require any complex operating system support. All that is necessary for our application is to print "hello world!"

The first step is to create a stand-alone software platform. A software platform consists of the necessary libraries and drivers for the application to link to during compile time. The stand-alone software platform is the bottom layer of the software stack and provides a single-threaded environment with support for input and output streams. We will take advantage of the output (via `print`) support for the "hello world" example. This step can be accomplished in a variety of ways, but we will use the `File` menu option and select new followed by `software platform...` A new software platform project window opens, shown in Figure 1.23. Name the project "standalone_project" and set the platform type to be "standalone," as shown in Figure 1.24.

The second step is to create an application. The application will use the stand-alone software platform. To create a new application, use the `File` menu option and select new followed by `Managed Make C Application Project`, as seen in Figure 1.25. Under this project, the Makefile will be updated by the SDK. This is one of the default options and requires the least amount of effort to maintain. Set the project name to "hello_world" and under `Sample Applications` select `Hello World`, shown in Figure 1.26. This will create a simple C hello world program.

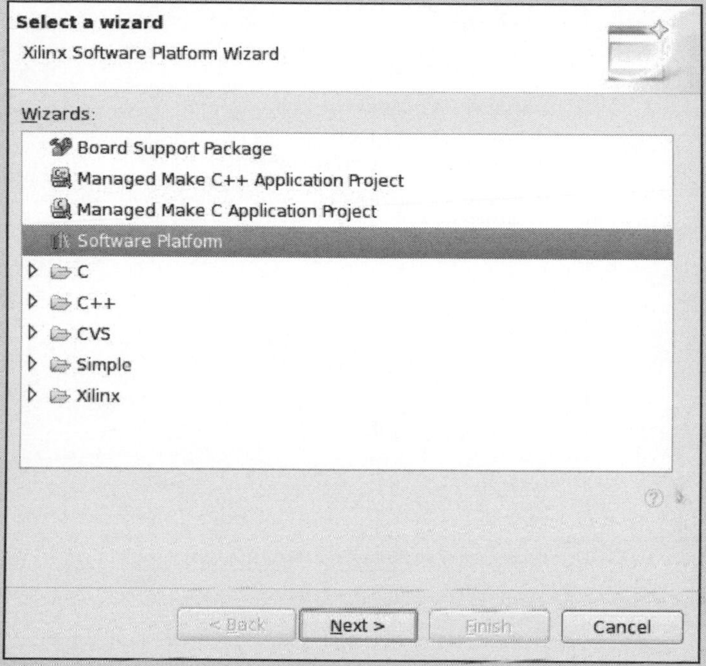

Select a wizard

Xilinx Software Platform Wizard

Wizards:

- 🐿 Board Support Package
- 📇 Managed Make C++ Application Project
- 📇 Managed Make C Application Project
- ▓ Software Platform
- ▷ 📂 C
- ▷ 📂 C++
- ▷ 📂 CVS
- ▷ 📂 Simple
- ▷ 📂 Xilinx

< Back   Next >   Finish   Cancel

Figure 1.23. A new software platform project window.

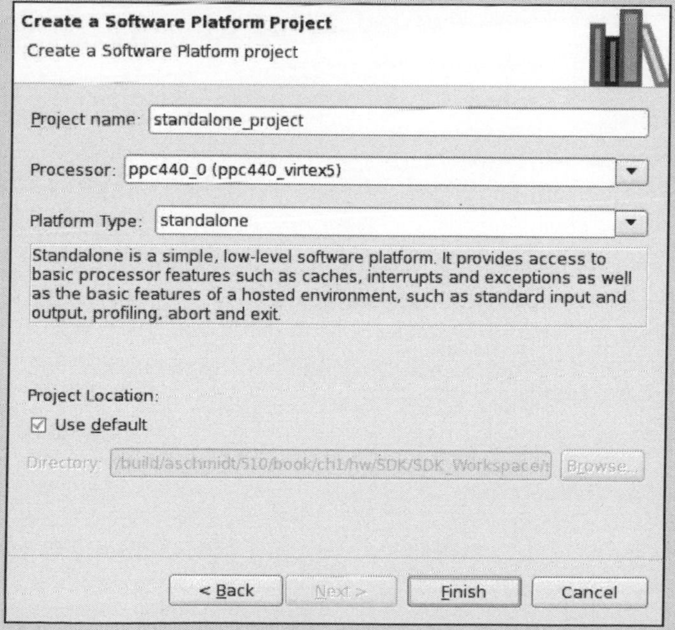

Create a Software Platform Project

Create a Software Platform project

Project name: standalone_project

Processor: ppc440_0 (ppc440_virtex5)

Platform Type: standalone

Standalone is a simple, low-level software platform. It provides access to basic processor features such as caches, interrupts and exceptions as well as the basic features of a hosted environment, such as standard input and output, profiling, abort and exit.

Project Location:

☑ Use default

Directory: /build/aschmidt/510/book/ch1/hw/SDK/SDK_Workspace/s   Browse

< Back   Next >   Finish   Cancel

Figure 1.24. A software platform configuration window.

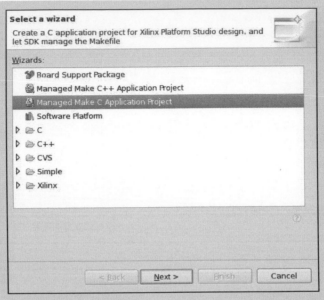

Select a wizard

Create a C application project for Xilinx Platform Studio design, and let SDK manage the Makefile

Wizards:

- 🐚 Board Support Package
- 🖺 Managed Make C++ Application Project
- 🖺 Managed Make C Application Project
- 📖 Software Platform
- ▷ 📂 C
- ▷ 📂 C++
- ▷ 📂 CVS
- ▷ 📂 Simple
- ▷ 📂 Xilinx

[< Back] [**Next >**] [Finish] [Cancel]

Figure 1.25. A new software application window.

Managed Make C Project

Create a new Managed Make C project.

Project Name: hello_world

Software Platform: standalone_project ▼

Project Location

☑ Use Default Location for Project (recommended)

Location: /build/aschmidt/510/book/ch1/hw/SDK/SDK_Worksp. [Browse]

Sample Applications

- 📂 Empty Application
- 📂 Dhrystone
- 📂 Hello World
- 📂 Memory Tests
- 📂 Peripheral Tests
- 📂 Xilkernel POSIX Threads Demo
- 📂 lwIP Echo Server

Description

C Program to print Hello World

[< Back] [Next >] [Finish] [Cancel]

Figure 1.26. A new software configuration window.

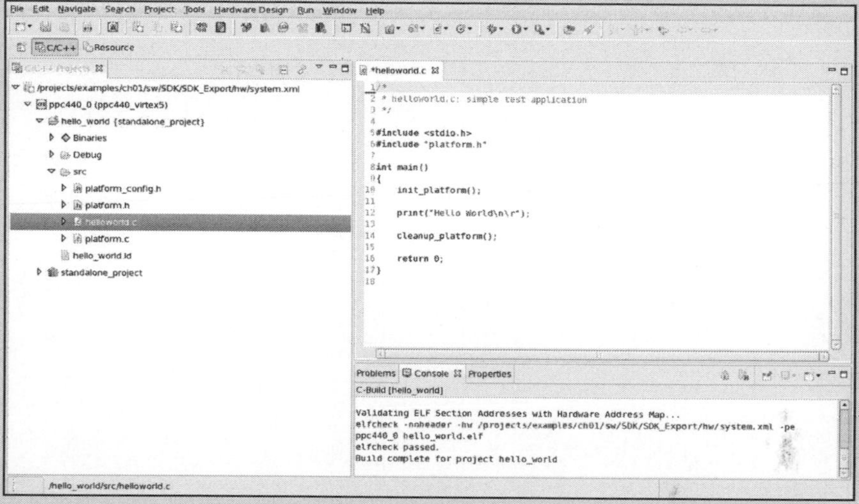

Figure 1.27. SDK GUI.

File	Location
genace.tcl	$XILINX_EDK/xmd/data/
download.bit	xps_project_directory/implementation/
hello_world.elf	sdk_workspace/hello_world/Debug/

Table 1.1 Files needed to generate ACE file.

To view the source code for the hello world application, expand the hello world arrow followed by the src arrow (as shown in Figure 1.27). Changes made to the source code will automatically be compiled when the file is saved. Alternatively, use the Project menu and select Build All to force a recompile.

Once the application has been created and compiles without errors it is time to download hardware and software to the FPGA. This can be done in a variety of ways. The simplest is through the JTAG cable that may come with the development board (or can be purchased separately). The ML-510 development board includes a CompactFlash card to program the FPGA. We will create a Xilinx Advanced Configuration Environment (ACE) file. The ACE file is used on power-up to program the FPGA (both the hardware bitstream and application). These details are discussed in future chapters; for now we will focus on getting the design to run on the FPGA and "hello world" output.

To generate the ACE file you will need three files, located in the following directories, shown in Table 1.1.

Using the Xilinx Microprocessor Debugger (XMD), we will run the genace.tcl script to build the ACE file from the download.bit and hello_world.elf files.

```
% xmd -tcl genace.tcl -jprog -hw download.bit \
    -elf hello_world.elf -board ml510 -ace hello.ace
```

Next, to copy the `hello.ace` file to the development board, pull the CompactFlash card out of the development board (when the board is off!) and put it in any standard CompactFlash card reader connected to a computer. The directory structure is:

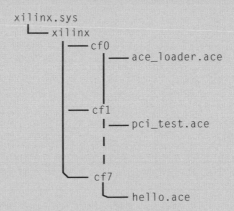

Each `cfn` directory holds a configuration, and the development board comes with a Compact Flash with slots 6 and 7 free for user designs. Simply remove any ACE file in the cf7 directory and copy the `hello.ace` into that directory on the CompactFlash. Eject/unmount the CompactFlash card, and return the card to the development board.

Finally, connect the development board to a computer over an RS-232 serial line. On Linux machines, many people use either `minicom` or `kermit` as a terminal emulator. Be sure to set the proper serial line and parameters (9600 baud, 8 bits, no parity, 1 stop bit). Start up the terminal emulator, turn on the development board, and, after the power-on self-test, type '7' (to boot the configuration in slot 7). The results are shown in Figure 1.28.

With the "Hello World" application running on the board, you have successfully completed your first Platform FPGA project! We will rely on some of this basic knowledge going forward when we begin to assemble more complex hardware cores and base systems. At this point, we recommend spending some time exploring these tools and becoming familiar with many of their options and functions as possible.

```
ML510 ACE-loader
----------------
 Enter Desired System ACE CF Configuration <0-7>.
   0: ACE-loader.
   1: Configuration 1.
   2: Configuration 2.
   3: Configuration 3.
   4: Configuration 4.
   5: Configuration 5.
   6: Configuration 6.
   7: Configuration 7.
Select <0-7>: Rebooting to System ACE Configuration Address 7...
Hello World!
```
Figure 1.28. Output displayed on terminal for the HelloWorld example.

Exercises

P1.1. Name three consumer electronics products that have embedded systems. Include a justification.

P1.2. Name three consumer electronics products that do not contain embedded computer systems. (You may need to consider very old products or products that are not consumer electronics.)

P1.3. In addition to the explicit inputs and outputs of a computing system, what is implicitly consumed and produced by (electronic) computing machines?

P1.4. How is the specification phase of a project different from the design phase?

P1.5. What is the difference between design and integration in the waterfall model?

P1.6. Is "access to a wireless network" a capability or a performance metric?

P1.7. Suppose two devices both have wireless access, but one is faster? Is this a new capability?

P1.8. Device A uses a network standard that is 100 Mb/s and device B uses a newer, revised standard that is 100 or 1000 Mb/s. Is this a capability or a functional improvement?

P1.9. Consider two solutions. One uses a 4-bit processor with a clock frequency of 1000 MHz; the other is a 64-bit processor with a clock rate of 100 MHz. Both are executing an infinite loop of 64-bit ADD operations. Contrast the two solutions in terms of:

- latency
- results/second
- energy
- development cost

P1.10. What is the advantage of using a standard — such as ASCII, binary, and BCD — versus an application-specific format?

P1.11. What is the disadvantage of using an application-specific format versus using a standard such as an IEEE 754 floating-point format?

P1.12. Suppose the dynamic range of an instrument flying on a satellite is -192 to $+191$ in discretized (integer) steps. How many bits are needed to represent a sample from the instrument?

P1.13. Assuming the same instrument just given: If a 32-bit processor (with a 100-MHz 32-bit system bus) was used to read data, one sample at a time, what is the peak theoretical bandwidth of the information moved from the instrument to the processor?

P1.14. What is the bandwidth if multiple samples were buffered on the instrument and multiple samples could be read each time?

P1.15. What is the bandwidth if the bus was 64 bits?

P1.16. Suppose the NRE on the fabrication of an ASIC is $150,000 and the unit cost is $0.15. The unit cost of an FPGA is $15. All other costs going into the product are equal. If the anticipated demand is expected to be X units the first year and will decay by 50% each subsequent year ($X/2$ units the second, $X/4$ units the third, and so on), how many units have to be sold the first year for the ASIC to be more profitable than the FPGA?

P1.17. Explain the difference between hardware and software in terms of a traditional processor-based system.

P1.18. How does the "hardware" of a Platform FPGA-based system differ from the hardware of a traditional processor-based system?

P1.19. What is the difference between a soft IP core and a hard IP core in a Platform FPGA system?

References

Berger, A. (2002). *Embedded systems design: An introduction to processes, tools, and techniques.* San Francisco, CA, USA: CMP Books.

Catsoulis, J. (2003). *Designing embedded hardware.* Sebastopol, CA, USA: O'Reilly & Associates, Inc.

Hollabaugh, C. (2003). *Embedded Linux: Hardware, software, and interfacing.* Boston, MA, USA: Addison/Wesley Publishing.

Vahid, F., & Givargis, T. (2002). *Embedded systems design: A unified hardware/software introduction.* New York, NY, USA: John Wiley & Sons, Inc.

Wolf, W. (2001). *Computers as components: Principles of embedded computing system design.* San Francisco, CA, USA: Morgan Kaufmann Publisher.

Xilinx, Inc. (2009a December). *EDK concepts, tools, and techniques: A hands-on guide to effective embedded system design.*

Xilinx, Inc. (2009b June). *ISE in-depth tutorial (UG695) v11.2.*

Xilinx, Inc. (2009c June). *ML510 reference design user guide (UG355) v1.2.*

Xilinx, Inc. (2009d June). *Platform specification format reference manual (UG642) v11.2.*

Yaghmour, K. (2003). *Building embedded Linux systems.* Sebastopol, CA, USA: O'Reilly & Associated, Inc.

THE TARGET

"What is the use of a book," thought Alice, *"without pictures or conversations?"*

Lewis Carroll
Alice's Adventures in Wonderland (1865)

The intention of this book is to show how to build embedded systems where the FPGA is the central computing device. The goal is not to teach how to design and fabricate an FPGA or show how to develop computer-aided design (CAD) tools for FPGA devices. Nonetheless, to meet the complex performance metrics described in the previous chapter, a good system designer has to understand how the programmable logic devices are manufactured, their basic electrical characteristics, and their general architectures. Unfortunately, the 10,000-meter view of FPGA devices of the first chapter is insufficient. Certain underlying details of FPGA technology are needed to understand what the software tools are doing when they transform a hardware description into a configuration. Although high-level synthesis tools have made huge strides over the years, basic knowledge of the target device almost always leads to more efficient designs.

The learning objectives of this chapter include:

- a short review of CMOS transistors and their central role as the building blocks of digital circuits.
- a brief discussion of programmable logic devices that have paved the way for modern FPGAs.
- the specifics of the FPGA, their low-level components, and those that are used to distinguish Platform FPGAs. This is a more detailed description than found in Chapter 1.
- a brief overview of hardware description languages; a useful tool to aid in complex digital circuit design.
- a description of the process of transforming a hardware description of the desired circuit into a stream of bits suitable to configure the FPGA device.

The examples used throughout this chapter were purposely chosen for their simplicity and brevity to help the reader quickly establish (or refresh) a foundation of knowledge to build upon in the remaining chapters.

2.1. CMOS Transistor

Because digital circuits, such as FPGAs, rely on CMOS transistors, being aware of their construct and, more specifically, their functionality can help aid the designer in making better overall design decisions. For example, power consumption is emphasized within this section to encourage readers to think about how their designs will behave based on these CMOS principles. The long-term result can be a more power-efficient design, a welcome addition to any embedded systems designs.

The details we do include are related to what an FPGA developer needs to know to make design decisions. First we cover the CMOS transistor. The intention is not to try to cover everything related to electronic circuits or even transistors. In fact, we rely on the reader already having taken at least one course in electronic circuits. We try to cover transistors with enough detail to act as a refresher or to help those less familiar with the material. As already stated, we are not setting out to design FPGAs, we aim to use them, and as such we should be familiar with their constructs. We then cover the progression from transistors to the programmable logic device, the precursor in many ways to the FPGA. Finally, we dwell into great detail regarding FPGAs.

Transistors have several distinct modes that make them useful in a variety of applications. However, here we simplify them to just one role: a voltage-controlled switch. The most common transistor in digital circuits is in the form of a Metal Oxide Semiconductor Field Effect Transistor (MOSFET). The MOSFET is created from two-dimensional regions in a slab of silicon substrate. The regions are constructed to have either an excess of positive or an excess of negative charge. A layer of silicon dioxide is placed over the substrate to insulate it from a metal layer placed above the silicon dioxide. An n-channel MOSFET transistor (or NMOS for short) is illustrated in Figure 2.1 with (a) a cut-away view of the silicon substrate and (b) the NMOS transistor's schematic symbol. There are three electrical terminals on this transistor: source, drain, and gate.

As silicon is a semiconductor, the gate can be used to control whether current flows from the drain to the source. For an NMOS transistor, if the voltage difference between the gate and the source is close to 0, then current is able to flow from the drain to the source. But if the voltage difference is significant, say

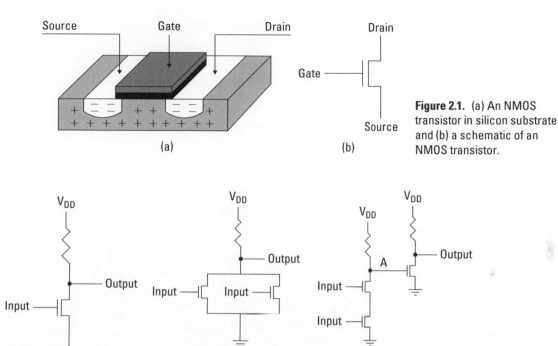

Figure 2.1. (a) An NMOS transistor in silicon substrate and (b) a schematic of an NMOS transistor.

Figure 2.2. An NMOS (a) NAND gate, (b) NOR gate, and (c) AND gate.

5 volts, then the transistor is in "cut-off" mode. In cut-off mode it behaves like an open switch and current cannot flow from the drain to the source. Using schematic symbols, this circuit is shown in Figure 2.3. Note that the input is a voltage on the gate and the output is a voltage on the conductor connected to the drain.

With a low voltage producing current and a high voltage producing no current, this can be considered a form of an inverter (a Boolean NOT gate), and without much work the reader will recognize the NAND, NOR, and AND gates illustrated in Figure 2.2. At first glance, we now have what is necessary to build digital circuits from these basic gates. However, there are a number of problems with these circuits. The most important is power consumption. When the voltage drops over an electric component, potential energy is converted into heat. So whenever the output of Figure 2.3 is 0, current is flowing and the resistor is generating heat. (To a small degree, the transistor is, too.) It turns out that this use of power is unacceptable in practice.

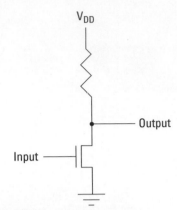

Figure 2.3. Schematic symbol of an NMOS inverter.

Figure 2.4. (a) A PMOS transistor in silicon substrate and (b) a schematic of a PMOS transistor.

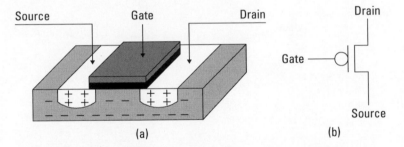

To address the power problem we consider the opposite of the n-channel transistor, the p-channel (or PMOS) transistor. PMOS rearranges the areas on the substrate so that regions that had an excess of negative charge now have an excess of positive charge and vice versa, shown in Figure 2.4. However, the PMOS inverter draws the most power when its output is high (because current is flowing and there is a voltage drop over the resistor). The PMOS unfortunately presents the same problem as the NMOS, only at the opposite time.

We can get the best of both worlds if we combine a PMOS transistor with an NMOS transistor such that in the steady state, no appreciable current is flowing. The Complementary MOS (CMOS) transistor combines a simple CMOS NAND circuit and is illustrated in Figure 2.5. It also leads to an important rule of thumb: the greatest amount of energy in a design is consumed when a CMOS gate is changing state. In steady state, no (appreciable) current is flowing. However, when the inputs start to change and the MOSFET transistors have not completely switched on or off, current flows and power is dissipated.

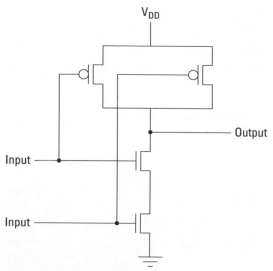

Figure 2.5. A simple CMOS NAND circuit.

2.2. Programmable Logic Devices

The transistors described in the last section are fixed at manufacture. The p- and n-type areas on the substrate, the silicon dioxide insulator, and metal layer(s) are created from masks. Once the chip is fabricated, those positions are fixed and, by extension, the transistor functionality is fixed, which leads to the question: how can devices be built that can be configured after manufacture/fabrication? To answer this, we need to build up a structure that is capable of implementing arbitrary combinational circuits, a storage cell (memory), and a mechanism to connect these resources.

The first programmable logic devices were organized around the sum-of-products formulation of a Boolean function. These devices, called Programmable Logic Arrays (PLAs), had an array of multi-input AND gates and an array of multi-input OR gates. The organization is shown in Figure 2.6. (Typically, the inputs x, y, and z would also be inverted, and, in this case, six-input AND gates would be used.) To configure this device, all of the unwanted inputs to the AND gates are broken and all of the unwanted inputs to the OR gates are broken. However, for manufacturing reasons, this approach fell out of favor. The PLA can be simplified by hard-wiring some AND gate outputs to specific OR gates. This constrains some inputs to use dedicated gates, but the PLA manufacturer compensates for this by increasing the number of gates. The PLA device is programmed by breaking unwanted connections at the inputs to the AND and OR gates.

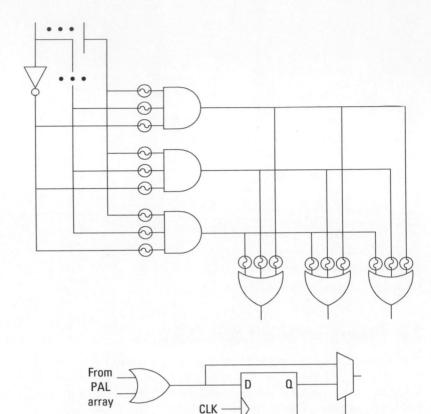

Figure 2.6. PLA device organization.

Figure 2.7. Adding storage to a PLA logic cell.

To these PLA structures, several programmable logic companies have introduced memory elements, such as D-type flip-flops. In this case they added a MUX and a D-Type Flip-Flop to the output of the OR gates. Hence, the output of the function can either be stored on a rising clock edge or the signal can bypass the flip-flop as illustrated in Figure 2.7. Connected to the outputs of the OR gates, this structure allows a designer to create state machines on programmable logic devices.

The final intermediate step in the progression from PLAs to FPGAs was the introduction of Complex Programmable Logic Devices (CPLDs). These devices start with a PLA block (with a storage element). These blocks are then replicated multiple times on a single Integrated Chip. A configurable routing network allows the different PLA blocks to be connected to one another and the

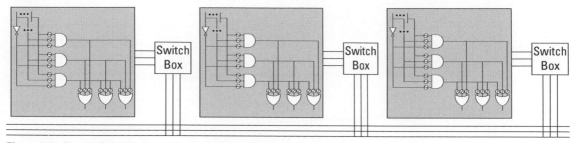

Figure 2.8. Simple CPLD device organization example.

outside pins. (The routing network is similar to those found on an FPGA, which is described next.)

2.3. Field-Programmable Gate Array

A modern FPGA consists of a 2-D array of programmable logic blocks, fixed-function blocks, and routing resources implemented in the CMOS technology. Along the perimeter of the FPGA are special logic blocks connected to external package I/O pins. Logic blocks consist of multiple logic cells, while logic cells contain function generators and storage elements. These general terms are discussed in more detail throughout this chapter. In Section 2.A specific details regarding the Xilinx Virtex 5 FPGA device constructs are given.

2.3.1. Function Generators

The programmable logic devices described thus far use actual gates implemented with CMOS transistors directly in the silicon to generate the desired functionality. FPGA devices are fundamentally different in that they use *function generators* to implement Boolean logic functionality rather than physical gates.

For example, suppose we want to implement the Boolean function:

$$f(x, y, z) = xy + z'$$

Using a 3-input function generator, we first create the eight-row Boolean truth table for this function (as shown in Figure 2.9a). For each input the truth table represents what the output of the Boolean function will be. If each of the function's output bits were stored into individual static memory (such as SRAM) cells and connected as inputs to an 8×1 multiplexer (MUX), the three inputs (x,y,z) would be the select lines for the MUX. The result is commonly what is known as a look-up table (LUT).

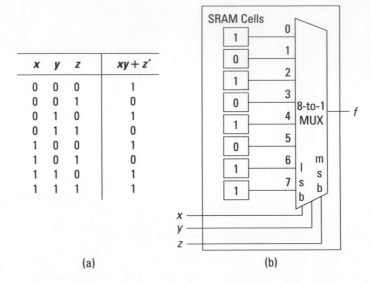

x	y	z	xy + z′
0	0	0	1
0	0	1	0
0	1	0	1
0	1	1	0
1	0	0	1
1	0	1	0
1	1	0	1
1	1	1	1

(a)   (b)

Figure 2.9. Three-input function generator in (a) truth table and (b) look-up table form.

We can draw the basic 3-input LUT circuit (Figure 2.9b) taking care to label the inputs 0 through 7 and identify the most significant select bit (MSB) and least significant select bit (LSB). The last step is to assign x, y, and z to the select bits and fill in the SRAM cells with the corresponding entries in the truth table. As drawn, when x, y, and z are 0,0,0 then the bottom SRAM cell is selected, so '1' is stored there. When x, y, and z are 0, 0, 1 the second from the bottom SRAM cell is selected, so '0' is stored there. This continues through 1, 1, 1 where '1' is stored at the top SRAM cell.

This example illustrates, at a basic level, the process of mapping Boolean logic to FPGAs. Soon, we will discuss in more detail how to map more meaningful logic to the FPGA fabric. One additional important lesson to learn from this example is that unlike a digital circuit implemented within logic gates, the propagation delay from a LUT is fixed. This means, regardless of the complexity of the Boolean circuit, if it fits within a single LUT, the propagation delay remains the same. This is also true for circuits spanning multiple LUTs, but instead the delay depends on the number of LUTs and additional circuitry necessary to implement the larger function. More on that topic later.

To generalize the aforementioned example, the basic n-input function generator consists of a 2^n-to-1 multiplexer (MUX) and 2^n SRAM (static random access memory) cells. By convention, a 3-LUT is a 3-input function generator. The 3-input structure is shown for demonstration purposes, although 4-LUTs and 6-LUTs are more common in today's components.

To implement a function with more inputs than would fit in a single LUT, multiple LUTs are used. The function can be decomposed into subfunctions of a subset of the inputs, each subfunction is assigned to one LUT, and all of the LUTs are combined (via routing resources) to form the whole function. There are some dedicated routing resources to connect neighboring LUTs with minimal delay to support low propagation delays.

An important observation is that SRAM cells are volatile; if power is removed the value is lost. As a result, we need to learn how to set the SRAM cell's value. This process, called *configuring* (or programming), could be handled by creating an address decoder and sequentially writing the desired values into each cell. However, the number of SRAM cells in a modern FPGA is enormous and random access is rarely required. Instead, the configuration data are streamed in bit by bit. The SRAM cells are chained together such that, in program mode, the data out line of one SRAM cell is connected to the data in line with another SRAM cell. If there are n cells, then the configuration is shifted into place after n cycles. Some FPGA devices also support wider, byte-by-byte transfers as well to support parallel transfers for faster programming.

2.3.2. Storage Elements

While function generators provide the fundamental building block for combinational circuits, additional components within the FPGA provide a wealth of functionality. As in a PLA block, D-type flip-flops are incorporated in the FPGA. The flip-flops can be used in a variety of ways, the simplest being data storage. Typically, the output of the function generator is connected to the flip-flop's input. Also, the flip-flop can be configured as a latch, operating on the clock's positive or negative level. When designing with FPGAs, it is suggested to configure the storage elements to be D flip-flops instead of latches. A latch being level-sensitive to the clock (or enable) increases the difficulty to route clock signals within a specific timing requirement. For designs with tight timing constraints, such as operating custom circuits at a high operating frequency that span large portions of the FPGA, D flip-flops are more likely to meet the timing constraints.

2.3.3. Logic Cells

By combining a look-up table and a D flip-flop the result is commonly what is referred to as a ***logic cell***. Logic cells are really the low-level building block upon which FPGA designs are built. Both combinational and sequential logic can be built from within a logic cell or a collection of logic cells. Many FPGA vendors will compare the capacity of an FPGA based on the number of logic cells (along with other resources as well). In fact, when comparing designs, it is no longer relevant to describe an FPGA circuit in terms of "number of equivalent gates (or transistors)." This is because a single LUT can represent very modest equations, which would only require a few transistors to implement, or a very complex circuit such as a RAM, which would require many hundreds of transistors. While we have not described the mapping of larger circuits for logic cells (a process that is now a software compilation problem), we are able to identify, based on the number of logic cells, how big or small a design is and whether it will fit within a given FPGA chip.

2.3.4. Logic Blocks

While logic cells could be considered the basic building blocks for FPGA designs, in actuality it is more common to group several logic cells into a block and add special-purpose circuitry, such as an adder/subtractor carry chain, into what is known as a *logic block*. This allows a group of logic cells that are geographically close to have quick communication paths, reducing propagation delays and improving design implementations. For example, the Xilinx Virtex 5 families put four logic cells in a *slice*. Two slices and carry-logic form a *Configurable Logic Block* or CLB. Abstractly speaking, logic blocks are what someone would see if they were to "look into" an FPGA. The exact number of logic cells and other circuitry found within a logic block is vendor specific; however, we are now able to realize even larger digital circuits within the FPGA fabric.

Logic blocks are connected by a routing network to provide support for more complex circuits in the FPGA fabric. The routing network consists of switch boxes. A *switch box* is used to route between the inputs/outputs of a logic block to the general on-chip routing network. The switch box is also responsible for passing signals from wire segment to wire segment. The wire segments can be short (span a couple of logic blocks) or long (run the length of the chip). Because circuits often span multiple logic blocks, the carry chain allows direct connectivity between neighboring logic blocks, bypassing the routing networking for potentially faster implemented circuits.

Because competing vendors and devices often have different routing networks, different special-purpose circuitry, and different size function generators, it is difficult to come up with an exact relationship for comparison. The comparison is further complicated because it also depends on the actual circuit being implemented. In Section 2.A the Xilinx Virtex 5 FPGA architecture is described in more detail to help the reader take advantage of its resources when designing for embedded systems.

2.3.5. Input/Output Blocks

We augment our logic block array with Input/Output Blocks (IOBs) that are on the perimeter of the chip. These IOBs connect the logic block array and routing resources to the external pins on the device. Each IOB can be used to implement various single-end signaling standards, such as LVCMOS (2.5 V), LVTTL (3.3 V), and PCI (3.3 V). IOBs can also support double data rate signaling used by commodity static and dynamic random access memory. The IOBs can be paired with adjacent IOBs for differential signaling, such as LVDS.

Taken all together, the high-level structure of a simple FPGA is seen in Figure 2.10. That is, an *FPGA* is a programmable logic device that consists of a two-dimensional array of logic blocks connected by a programmable routing network.

2.3.6. Special-Purpose Function Blocks

So far we have focused on internals of the FPGA's configurable (or programmable) logic. A large portion of the FPGA consists of logic blocks and routing logic to connect the programmable logic. However, as semiconductor technology advanced and more

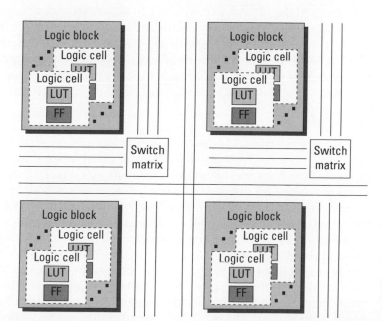

Figure 2.10. A high-level block diagram of a simple FPGA consisting of logic blocks, logic cells, storage elements, function generators, and routing logic.

transistors became available, FPGA vendors recognized that they could embed more than just configurable logic to each device.

Platform FPGAs combine programmable logic with additional resources that are embedded into the fabric of the FPGA. A good question to ask at this point is "why embed specific resources into the FPGA fabric?" To answer this we must consider FPGA designs compared to ASIC designs. An equivalent ASIC design is commonly considered to use fewer resources and consume less power than an FPGA implementation. However, ASIC designs are not only often found to be prohibitively expensive, the resources are fixed at fabrication. Therefore, FPGA vendors have found a compromise with including some ASIC components among the configurable logic. The exact ASIC components are vendor specific, but we will briefly cover in this section a few of the more commonly seen components. The components range in complexity from embedded memory and multipliers to embedded processors. The following paragraphs introduce some of these components with more detail to be found in Section 2.A and throughout the rest of the book.

The block diagram of a Platform FPGA, seen in Figure 2.11, shows the arrangement of these special-purpose function blocks placed throughout the FPGA. Which function blocks are included and their specific placement are determined by the physical device; this illustration is meant to help the reader understand a general construct of modern FPGAs. The logic blocks still occupy a majority of the FPGA fabric in order to support a variety of complex digital designs; however, the move to support special-purpose blocks provides the designer with an ASIC implementation of the block as well as removes the need to create a custom design for the block. For example, a processor block could occupy a significant portion of the FPGA if implemented in the logic resources.

Block RAM

Many designs require the use of some amount of on-chip memory. Using logic cells it is possible to build variable-sized memory elements; however, as the amount of memory needed increases, these resources are quickly consumed. The solution is to provide a fixed amount of on-chip memory embedded into the FPGA fabric called **Block RAM** (BRAM). The amount of memory depends on the device; for example, the Xilinx Virtex 5 XC5VFX130T (on the ML-510 developer board) contains 298 36 Kb BRAMs, for a total storage capacity of 10,728 Kb. Local on-chip storage such as RAMs and ROMs or buffers can be constructed from BRAMs. BRAMs can be combined together to form larger (both in terms of data width and depth) BRAMs. BRAMs are also dual-ported, allowing

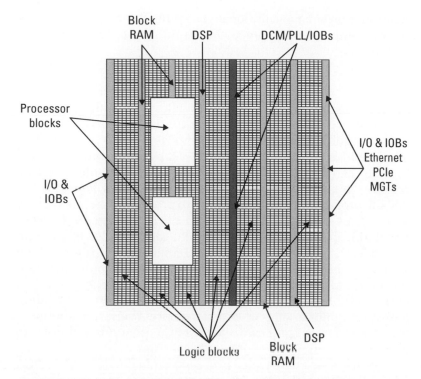

Figure 2.11. A high level view of a Platform FPGA.

for independent reads and writes from each port, including independent clocks. This is especially useful as a simple clock-crossing device, allowing one component to produce (write) data at a different frequency as another component consuming (reading) data.

Digital Signal Processing Blocks

To allow more complex designs, which may consist of either digital signal processing or just some assortment of multiplication, addition, and subtraction, *Digital Signal Processing Blocks* (DSP) have been added to many FPGA devices. As with the Block RAM, it is possible to implement these components within the configurable logic, yet it is more efficient in terms of performance, and power consumption to embed multiple of these components within the FPGA fabric. At a high level, the DSP blocks of a multiplier, accumulator, adder, and bitwise logical operations (such as AND, OR, NOT, and NAND). It is possible to combine DSP blocks to perform larger operations, such as single and double precision floating point addition, subtraction, multiplication, division, and square root. The number of DSP blocks is device dependent;

however, they are typically located near the BRAMs, which is useful when implementing processing requiring input and/or output buffers.

Processors

Arguably one of the more significant additions to the FPGA fabric is a processor embedded within the FPGA fabric. For many designs requiring a processor, often choosing an FPGA device with an embedded processor (such as the FX series part for the Xilinx Virtex 5) can simplify the design process greatly while reducing resource usage and power consumption. The IBM PowerPC 405 and 440 processors are examples of two processors included in the Xilinx Virtex 4 and 5 FX FPGAs, respectively. These are conventional RISC processors that implement the PowerPC instruction set. Both the PowerPC 405 and the 440 come with some embedded system extensions but do not implement floating-point function units in hardware. Each come with level 1 instruction and data cache and a memory management unit (MMU) with a translation look-aside buffer to support virtual memory. A variety of interfaces exist to connect the processors to the FPGA programmable logic to allow interaction with custom hardware cores.

Not all FPGAs come with a processor embedded into the FPGA fabric. For these devices the processor must be implemented within the FPGA fabric as a soft processor core. This text is not affected adversely by the decision to include or remove processors from within the FPGA fabric. Instead this book is aimed at providing the reader with the necessary skill set to use the available resources as efficiently as possible to solve the given problem. It would be unwise to ignore a discussion of hard processor cores merely on speculation.

Digital Clock Manager

Most systems have a single external clock that produces a fixed clock frequency. However, there are a number of reasons a designer might want to have different logic cores operating at different frequencies. A Digital Clock Manager (DCM) allows for different clock periods to be generated from a single reference clock. You could use logic to divide an existing clock signal down to generate a lower frequency clock — say 100 to 25 MHz. While possible, the correct design practice is to use a DCM to generate the slower clock. The advantage of using a DCM is that the generated clocks will have lower jitter and a specified phase relationship. Section 2.A gives more specific details on how to use DCMs in a Xilinx design.

Multi-Gigabit Transceivers

Over the last 20 years, digital I/O standards have varied between serial and parallel interfaces. Serial interfaces time-multiplex the bits of a data word over fewer conductors while parallel interfaces signal all the bits simultaneously. While the parallel approach has the apparent advantage of being faster (data are being transmitted in parallel), this is not always the case in the presence of noise. Many recent interfaces — including various switched Ethernet standards, Universal Serial Bus (USB Implementers Forum (USB-IF), 2010a,b), SerialATA (Grimsrud, Knut and Smith, Hubbert, 2003), FireWire (1394 Trade Association, 2010), and InfiniBand (Futral, 2001) — are now using low-voltage differential pairs of conductors. These standards use serial transmission and change the way data are signaled to make the communication less sensitive to electromagnetic noise.

High-Speed Serial Transceivers are devices that serialize and deserialize parallel data over a serial channel. On the serial side, they are capable of baud rates from 100 Mb/s to 11.0 Gb/s, which means that they can be configured to support a number of different standards, including Fiber Channel, 10G Fiber Channel, Gigabit Ethernet, and Infiniband. As with the other aforementioned FPGA blocks, the transceivers can be configured to work together. For example, two transceivers can be used to effectively double the bandwidth. This is called channel bonding (multilane and trunking are common synonyms).

This initial overview does not encompass all of the available resources on every FPGA device. It does provide a comprehensive overview of the significant components often used in FPGA designs. More details are forthcoming for these components and their applicability and implementation within embedded systems designs. For the interested reader, a more detailed explanation of how to construct these basic FPGA elements from CMOS transistors is presented in Chow *et al.* (1991) and Compton & Hauck (2002).

2.4. Hardware Description Languages

Now that we know what is inside of an FPGA, the next step is to learn how to "configure" them. We use a *hardware description language* (HDL) as a high-level language to describe the circuit to be implemented on the FPGA. Most courses in digital logic nowadays introduce HDLs in the context of designing digital circuits. Unfortunately, some very popular HDLs were never intended to be used as they are today. The origins of hardware description languages were rooted in the need to document the behavior of hardware. Over time, it was recognized that the descriptions could be used to simulate hardware circuits on a general-purpose processor. This process of translating an HDL source into a form suitable for a general-purpose processor to mimic the hardware described is called *simulation*. Simulation has proven to be an extremely useful tool for developing hardware and verifying the functionality before physically manufacturing the hardware. It was only later that people began to *synthesize* hardware, automatically generating the logic configuration for the specified device from the hardware description language.

Unfortunately, while simulation provided a rich set of constructs to help the designer test and analyze the design, many of these constructs extend beyond what is physically implementable within hardware (on the FPGA) or synthesize inefficiently into the FPGA resources. As a result, only a subset of hardware description languages can be used to synthesize designs to hardware. The objective of this section is to present two of the more popular hardware description languages, VHDL and Verilog. It is arguable to which is better suited for FPGA design; both are presented here for completeness, but it is left to the readers to decide which (if any) is best suited for their needs. We will focus on VHDL throughout the remainder of this book as it becomes redundant to give two examples, one in VHDL and one in Verilog, for every concept. That being said, we will also focus within this section on synthesizable HDL and how it maps to the previous section's components.

2.4.1. VHDL

VHDL, which stands for VHSIC[1] Hardware Description Language, describes digital circuits. In simulation, VHDL source files are analyzed and a description of the behavior is expressed in the form of a netlist. A *netlist* is a computer representation of a collection of logic units and how they are to be connected. The logic units are typically AND/OR/NOT gates or some set of primitives that makes sense for the target (4-LUTs, for example). The behavior of the circuit is exercised by providing a sequence of inputs. The inputs, called *test vectors*, can be created manually or by writing a program/script that generates them. The component that is generating test vectors and driving the device under test is typically called a *test bench*.

Synthesizable VHDL

In VHDL, there are two major styles or forms of writing hardware descriptions. Both styles are valid VHDL codes; however, they

---

[1] Very high-speed integrated circuit.

model hardware differently. This impacts synthesis, simulation, and, in some cases, designer productivity. These forms are:

Structural/data flow circuits are described in terms of logic units (either AND/OR/NOT gates or larger functional units such as multipliers) and signals. Data flow is a type of structural description that has syntactic support to make it easier to express Boolean logic.

Behavioral circuits are described in an imperative (procedural) language to describe how the outputs are related to the inputs as a process.

A third style exists as a mix between both structural and behavioral styles. For programmers familiar with sequential processors, the behavioral form of VHDL seems natural. In this style, the process being described is evaluated by "executing the program" in the process block. For this reason, often complex hardware can be expressed succinctly and quickly — increasing productivity. It also has the benefit that simulations of certain hardware designs are much faster because the process block can be executed directly. However, as the design becomes more complex, it is possible to write behavioral descriptions that cannot be synthesized. Converting behavioral designs into netlists of logic units is called *High-Level Synthesis*, referring to the fact that behavioral VHDL is more abstract (or higher) than structural style.

In contrast, because the structural/data-flow style describes logic units with known implementations, these VHDL codes almost always synthesize. Also, assembling large systems (such as Platform FPGAs) requires structural style at the top level as it is combining large function units (processors, peripherals, etc.). Also, it is worth noting that some structural codes do not synthesize well. An example of this is using a large RAM in a hardware design. A RAM is not difficult to describe structurally because of its simple, repetitive design. However, in simulation, this tends to produce a large data structure, which needs to be traversed every time a signal changes. In contrast, a behavioral model of RAM simulates quickly because it matches the processor's architecture well.

VHDL Syntax

All components of a design expressed in VHDL consist of an entity declaration followed by one or more declared architectures. The entity declaration describes the component's interface — its inputs, outputs, and name. By analogy, it is a function prototype in C or a class header file in C++.

The architecture block, in contrast, provides an implementation of the interface. VHDL allows for more than one implementation. Why? One reason is the simulation/synthesis issue just mentioned. One architecture might be used for simulation and another for synthesis. A second reason is different architectures might be needed for different targets (ASIC versus FPGA).

Starting with an empty source file, the program usually begins by including some standard library definitions. VHDL uses Ada's syntax for this by first stating a library's name and then which modules from that library to use via the `library` and `use` keywords. Next comes the entity declaration followed by one or more architecture blocks. The simplest way to learn this syntax is to study some examples. Consider the 3-input, 2-output full adder circuit in Listing 2.1.

The entity block says that the name of this component is `fadder` and that it has five ports. A port can be an input (in), output (out), or both (inout) to the component. In this example there are three inputs and two outputs. The signals can be of any primitive type defined by the VHDL language or a complex type. In our example, we use a complex type that was defined by the library "ieee" called `std_logic`. We use this to model a single Boolean variable with the binary value of either '0' or '1'.

The architecture block defines how the entity is to be implemented. It has a name as well (`imp_df`) and, in this example, only uses the data-flow style to describe how the outputs are related to its inputs. The following are some initial notes regarding VHDL. First, VHDL is not sensitive to case. The strings `cin` and `CIN` are the same variable; `begin` and `bEgIn` are both legal ways of writing the begin keyword. Second, the input and output ports can be

```
library ieee;
use ieee.std_logic_1164.all;

entity fadder is port (
   a, b : in std_logic;
   cin  : in std_logic;
   sum  : out std_logic;
   cout : out std_logic);
end fadder;

architecture beh of fadder is
begin
   sum  <= a xor b xor cin;
   cout <= (a and b) or (b and cin) or (a and cin);
end beh;
```

Listing 2.1. A 1-bit full adder in VHDL.

described on the same line like a, b have been defined or they can be placed on separate lines like cin. Inputs and outputs cannot be specified on the same line. Also, ports of different types (even if they are the same direction) cannot be specified on the same line. Third, the name of the architecture (imp_df) is user specific and does not have to follow any specific convention. Each subsequent architecture for the same entity must have unique architecture names. Finally, signal assignment is denoted by the <= operator. This is often a cause for syntax errors for those more familiar with C/C++ or Java syntax.

So how does the full adder map to hardware? With two outputs both depending on three inputs, two function generators are needed. Depending on the FPGA device, this may be reduced to a single LUT (if using 6-LUTs) or two LUTs. No D flip-flops are needed as there is no clock source, so the circuit represents a purely combinational circuit.

Now that we have built one component, we can use it to design a slightly more sophisticated component. In Listing 2.2 a 2-bit full adder is implemented. In this design, additional logic has been added for academic purposes; this may not be the ideal implementation of such a component. We are using it to highlight a number of VHDL-specific constructs.

```vhdl
library ieee;
use ieee.std_logic_1164.all;

entity fadder2bit is port (
    a, b : in  std_logic_vector(1 downto 0);
    cin  : in  std_logic;
    clk  : in  std_logic;
    rst  : in  std_logic;
    sum  : out std_logic_vector(1 downto 0);
    cout : out std_logic );
end fadder2bit;

architecture beh of fadder2bit is
  -- 2-bit Register Declaration for A, B, Sum
  signal a_reg, b_reg, sum_reg : std_logic_vector(1 downto 0);
  -- 1-bit Register Declaration for Cin, Cout
  signal cin_reg, cout_reg     : std_logic;
  -- 1-bit Internal signal to connect carry-out to carry-in
  signal carry_tmp             : std_logic;
  -- 1-bit Full Adder Component Declaration
  component fadder is
  port (
    a, b : in  std_logic;
    cin  : in  std_logic;
    sum  : out std_logic;
    cout : out std_logic);
  end component fadder;
```

```vhdl
begin
  -- VHDL Process to Register Inputs A, B and Cin
  register_proc : process (clk) is
  begin
    if ((clk'event) and (clk = '1')) then
      -- Synchronous Reset
      if (rst = '1') then
        -- Initialize Input Registers
        a_reg    <= "00";
        b_reg    <= "00";
        cin_reg  <= '0';
        -- Initialize Outputs
        sum      <= "00";
        cout     <= '0';
      else
        -- Register Inputs
        a_reg    <= a;
        b_reg    <= b;
        cin_reg <= cin;
        -- Register Outputs
        sum      <= sum_reg;
        cout     <= cout_reg;
      end if;
    end if;
  end process register_proc;
  -- Instantiate Bit 1 of Full Adder
  fa1 : fadder port map (
    a    => a_reg(1),
    b    => b_reg(1),
    cin  => carry_tmp,
    sum  => sum_reg(1),
    cout => cout_reg );
  -- Instantiate Bit 0 of Full Adder
  fa0 : fadder port map (
    a    => a_reg(0),
    b    => b_reg(0),
    cin  => cin_reg,
    sum  => sum_reg(0),
    cout => carry_tmp );
end beh;
```

Listing 2.2. A 2-bit full adder in VHDL.

In this example we have added clock and reset inputs and changed a, b and sum from std_logic to std_logic_vector(1 downto 0), a 2-bit signal. Next, we add internal signals to register the inputs and outputs. This is not a requirement, but it is recommended to buffer these signals to meet tighter timing constraints as signals are passed from a top-level component down to the low-level 1-bit full adder components. This is followed by the fadder component declaration, which indicates which components will be used within this design. Within the beh architecture, a VHDL process named `register_proc` is added. This process is used to register the inputs and outputs on the rising edge of the clock signal. This code infers all of the D flip-flops within the logic cells. The reset signal `rst` is a synchronous reset, which is recommended over asynchronous resets. This simplifies the reset logic needed to route the asynchronous signal across the chip. Finally, two 1-bit full adder component instances are included. Note that the internal signal `carry_tmp` connects the carry-out from bit 0's full adder to the carry-in of bit 1's full adder.

The resulting 2-bit full adder circuit consumes more resources than the single bit full adder. By adding registers to the input and outputs, we have extended the full adder's resource count by eight D flip-flops. Two flip-flops are needed for each input `a`, `b` and output `sum` and one flip-flop is needed for the input `cin` and output `cout`. Also, a third LUT is included for the carry-in/out logic.

The previous example was used to show the use of a VHDL process statement to infer D flip-flops and how to instantiate other VHDL components. It is not the intention of this book to teach the reader all of the details of VHDL, as there are far better books available for this purpose. Instead we are "priming the pump," giving the reader some VHDL now to help get the process started. Throughout the rest of the book, additional examples will be provided that are suited for specific embedded systems designs or recommendations on how to write VHDL to correctly infer the desired components that lead to more efficient designs. We conclude this VHDL subsection with one final example, a finite state machine (FSM). A finite state machine can be used for a variety of purposes, such as sequence generators and detectors, and is well suited for embedded systems design.

In this example we use an FSM to perform 8-bit addition. This is similar to what was done earlier except that we will add the constraint that the two operands are not guaranteed to be aligned. That is, the operands may or may not arrive at the same time and we must keep track of when both have arrived before adding the results together. This may be due to single access to data memory or data arriving from two different sources. Either way, the FSM will wait for one of the operands to arrive, register it, and then wait for the second operand to arrive, register it, and then add them together. The primary goal of this example is to explain how to use finite state machines in VHDL; there may be more effective ways to implement this example, but we will leave those more elegant solutions to the reader.

The FSM example in Listing 2.3 contains several new features beyond the first two examples. The most significant is how to use VHDL to describe a finite state machine. There are different

```vhdl
-- Traditional Library and Packages used in a hardware core
library ieee;
use ieee.std_logic_1164.all;
use ieee.std_logic_arith.all;
use ieee.std_logic_unsigned.all;

entity adderFSM is
  generic (
    C_DWIDTH : in  natural := 8);
  port (
  a, b        : in  std_logic_vector(C_DWIDTH-1 downto 0);
  a_we, b_we : in  std_logic;
  clk, rst   : in  std_logic;
  result     : out std_logic_vector(7 downto 0);
  valid      : out std_logic);
end adderFSM;

architecture beh of adderFSM is
  -- 8-bit Register/Next Declaration for A and B
  signal a_reg, b_reg   : std_logic_vector(7 downto 0);
  signal a_next, b_next : std_logic_vector(7 downto 0);
  -- Finite State Machine Type Declaration
  type FSM_TYPE is (wait_a_b, wait_a, wait_b, add_a_b);
  -- Internal signals to represent the Current State (fsm_cs)
  -- and the Next State (fsm_ns) of the state machine
  signal fsm_cs : FSM_TYPE := wait_a_b;
  signal fsm_ns : FSM_TYPE := wait_a_b;

begin
  -- Finite State Machine Process to Register Signals
  fsm_register_proc : process (clk) is
  begin
    -- Rising Edge Clock to infer Flip Flops
    if ((clk'event) and (clk = '1')) then
      -- Synchronous Reset / Return to Initial State
      if (rst = '1') then
        a_reg      <= (others => '0');
        b_reg      <= (others => '0');
        fsm_cs     <= wait_a_b;
      else
        a_reg      <= a_next;
        b_reg      <= b_next;
        fsm_cs     <= fsm_ns;
      end if;
    end if;
  end process fsm_register_proc;

  -- Finite State Machine Process for Combination Logic
  fsm_logic_proc : process (fsm_cs, a, b, a_we, b_we, a_reg, b_reg) is
  begin
    -- Infer Flip-Flop rather than Latches
    a_next <= a_reg;
    b_next <= b_reg;
    fsm_ns <= fsm_cs;
```

```vhdl
        case (fsm_cs) is
          -- State 0: Wait for either A or B Input
          --    denoted by a_we = '1' and/or b_we = '1'
          when wait_a_b =>
               -- Clear Initial Context of Registers/Outputs
               a_next     <= (others => '0');
               b_next     <= (others => '0');
               result     <= (others => '0');
               valid      <= '0';
               if ((a_we and b_we) = '1') then
                 a_next     <= a;
                 b_next     <= b;
                 fsm_ns     <= add_a_b;
               elsif (a_we = '1') then
                 a_next     <= a;
                 fsm_ns     <= wait_b;
               elsif (b_we = '1') then
                 b_next     <= b;
                 fsm_ns     <= wait_a;
               end if;

          -- State 1: Wait for A Input
          --    Being in this state means B already arrived
          when wait_a =>
               result     <= (others => '0');
               valid      <= '0';
               if (a_we = '1') then
                 a_next     <= a;
                 fsm_ns     <= add_a_b;
               end if;

          -- State 2: Wait for B Input
          --    Being in this state means A already arrive
          when wait_b =>
               result     <= (others => '0');
               valid      <= '0';
               if (b_we = '1') then
                 b_next     <= b;
                 fsm_ns     <= add_a_b;
               end if;

          -- State 3: Perform Add Operation and return to wait_a_b
          when add_a_b =>
               result     <= a_reg + b_reg;
               valid      <= '1';
               fsm_ns     <= wait_a_b;
        end case;
    end process fsm_logic_proc;
end beh;
```

Listing 2.3. A finite state machine example in VHDL.

ways to go about describing the FSM, but we have chosen what is considered a two-process finite state machine. There are also one-process and three-process FSMs, but the two-process FSM is arguably the most common. For those familiar with the traditional two-process finite state machine, this particular implementation may seem a little different. We present a variation on the popular two-process FSM because we find it more efficient and effective for the programmer. Specifically, we eliminate the redundant description of states in separate processes as it is a common place for design errors. Before we get too far ahead of ourselves, let's take a step back for those less familiar with finite state machines in VHDL.

To begin, a special VHDL signal type must be declared to differentiate the states in the finite state machine. This is done by adding the line:

```
type FSM_TYPE is (wait_a_b, wait_a, wait_b, add_a_b);
```

This line creates a new type (similar to a `typedef` or `enumerate` in C), which contains four states: (`wait_a_b`, `wait_a`, `wait_b`). Then, two signals are added of the type `FSM_TYPE` and initialized to the first state `wait_a_b` called `fsm_cs` and `fsm_ns` for the finite state machine's current state and next state indicators.

In addition to the state signals, we include additional registers for intermediate inputs. As was done with the second example (2-bit adder), we will register specific signals to help alleviate potential timing constraints, yielding higher operating frequencies (clock rates) for the component. These two signals are `a_reg` and `b_reg`, which have accompanying signals `a_next` and `b_next`. These signals behave just like the current/next state registers and will be used to hold the operands *a* and *b* inputs when they are valid, denoted by `a_we` and `b_we` being asserted.

To help illustrate how the `fsm_cs` and `fsm_ns` signals are used, consider Figure 2.12. A sequential circuit is built from both combinational logic and state registers. This is covered in an introductory digital logic course; we expand on it a little to include registering internal signals (`a_reg`, `b_reg`). Each clock cycle the state register is updated based on the combination logic next state output. In VHDL, this is done with one, two, or three processes. We are using a two-process FSM to help differentiate the state register and combinational logic. In the `fsm_register_proc` process, the state signal is registered.

In the `fsm_logic_proc` process, the combinational logic is described, including state transitions and each state's output. Here, a VHDL case statement is used for the finite state machine.

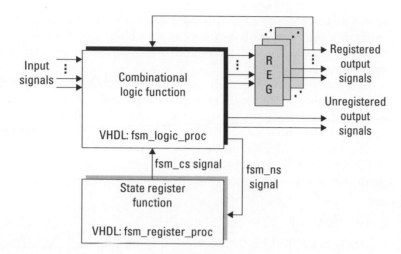

Figure 2.12. The VHDL process with the corresponding sequential digital circuit for finite state machines.

The initial state is wait_a_b. The state transition is based on if the input a and/or b arrive. There are three possible conditions: both a and b arrive in unison, a arrives before b, or b arrives before a. If a and b arrive at the same clock cycle the inputs are registered and the FSM transitions to the add_a_b state. If a arrives before b, then a is registered and the FSM transitions to the wait_b state. Likewise, if b arrives before a, b is registered and the FSM transitions to the wait_a state. If neither a or b arrives, then the FSM stays in this initial state.

In both the wait_a and the wait_b states, we already know that one of the two operands has already arrived and is registered so we are just waiting for the opposite input. Once it arrives, we register it and transition to the add_a_b state, where the two operands are added and the output is set. The valid signal is used to notify whatever component is dependent on the result output that is now ready.

There are some important notes to address with this example. First is the *sensitivity list*; that is, the list of signals that a process waits on for changes that cause the process to be re-evaluated. In simulation, if a signal (say a_we) is omitted, then even if that signal changes, it will not necessarily cause the process to be re-evaluated. This can cause the system to behave differently from what is expected or desired. In hardware the process is *always running*; hardware does not get invoked like a C function call. As a result, the sensitivity list will be automatically generated by the synthesis tools. Care must be taken by the designer when simulating the design, and sensitivity list omissions are often an example of a design error that will cause simulation and synthesis results to not match. We say all of this because while we find simulation extremely useful as both a design and a debug tool, the designer must use simulation tools intelligently with the mindset of writing synthesizable VHDL.

Clearly this was only intended to be an introduction to VHDL for the reader. The examples chosen were picked to give the reader a sampling of code to begin to play with when designing new components. Often the hardest part of any language is getting past the first hump of practical examples. For further VHDL reading we recommend Ashenden, Peter J. (2008). If you instead are interested in seeing what else is commonly used, we present Verilog next.

2.4.2. Verilog

Another common hardware description language is Verilog. Verilog has many similarities to VHDL, as both were originally intended to describe hardware circuit designs. Verilog is considered to be less verbose than VHDL, often making it easier to use, especially for designers more familiar with an imperative coding style such as C++ or Java. As with VHDL, Verilog became more than just a textual representation of a circuit. Designers used Verilog to simulate circuits that eventually led to a subset of the language supporting hardware synthesis. This section is written to provide the reader with a sufficient set of examples to begin to learn Verilog. The examples are based on those presented in the VHDL section, but with Verilog syntax. As was noted previously, this book focuses primarily on VHDL, but the reader should take care to familiarize themselves with both VHDL and Verilog, as they are both used within industry and academia. As with VHDL, there is a significant amount of resources already available for Verilog and should the reader find Verilog to be the more comfortable HDL, we encourage the reader to use Verilog. While this book may be more VHDL centric in its examples, the underlying ideas and concepts are applicable to both HDLs.

Synthesizable Verilog

In Verilog, there are three major styles or forms of writing hardware descriptions. Both styles are valid VHDL codes; however, they model hardware differently. This impacts synthesis, simulation, and, in some cases, designer productivity. These forms are:

Gate-Level Modeling circuits are described in terms of logic units such as AND/OR/NOT gates.

Structural circuits are described in terms of modules.

```
module fadder(a, b, cin, sum, cout);
   input a, b, cin;
   output sum, cout;

   assign sum = a ^ b ^ cin;
   assign cout = (a & b) | (a & cin) | (b & cin);
endmodule
```

Listing 2.4. A 1-bit full adder in Verilog.

Behavioral circuits are described in an imperative (procedural) language to describe how the outputs are related to the inputs.

Verilog Syntax

All components of a design expressed in Verilog consist of modules. Starting with an empty source file, the program usually begins with the keyword `module` followed by the module's name and ports. The ports are then defined in terms of input and output along with individual types, such as single bits or bit vectors. Unlike VHDL, there are no explicit libraries to include at the beginning of each file. For demonstration purposes, consider the same 1-bit full adder example presented in the VHDL section written in Verilog, shown in Listing 2.4.

The module is named `fadder`, which has five ports. A port can be an input, output, or both (inout) to the module. In this example, there are three inputs and two outputs. The data type can be a wire, register, integer, or constant. We will give specific examples shortly. Each of the inputs and outputs in this example is a single bit representing binary '1' or '0'. One important note regarding syntax is that Verilog is case sensitive and all keywords are lowercase; that is, `wire` is a Verilog data type, whereas `Wire` is a unique variable name.

Unlike VHDL, there is no formal distinction between the module interface and the design's logic. In the full adder example, the keyword `assign` is used with combination circuit assignment. The line

```
assign sum = a ^ b ^ cin
```

infers a three input `xor` gate with the output connected to the sum output. The symbols &, |, ∧, ∼ represent bitwise `and`, `or`, `xor`, `not`, respectively.

So how does the Verilog full adder map to hardware? Not surprisingly, it maps to the same resources as VHDL implementation. With such a simple example it is easy for the synthesis tool to infer the same logic, correctly.

Extending the Verilog 1-bit full adder to a 2-bit full adder as was done with VHDL will provide additional Verilog constructs and the conciseness of Verilog becomes more apparent. Listing 2.5 shows the 2-bit full adder example. The `reg` keyword is used for register declarations, and the `wire` keyword is used for internal combinational interconnects. The equivalent of a VHDL `process` block is the `always` block. In this example, inputs and outputs are registered to the input clock's rising edge, inferring D flip-flops. Two instances of the 1-bit fadder component follow, connecting the inputs, outputs, and carry logic accordingly. The components

```verilog
module fadder2bit(a, b, cin, clk, rst, sum, cout);
  input [1:0] a, b;
  input cin, clk, rst;
  output [1:0] sum;
  output cout;
  // 2-bit Register declaration for A, B, Sum
  reg [1:0] a_reg, b_reg, sum_reg;
  // 1-bit Register declaration for cin_reg, cout_reg
  reg cin_reg, cout_reg;
  // 1-bit internal wires for carry-out
  wire carry_tmp, cout_wire;
  // 2-bit internal wire for sum
  wire [1:0] sum_wire;

  // Always Block to Register Inputs A, B, Cin
  always @(posedge clk)
  begin
    if (rst == 1'b1) begin
      a_reg    = 2'b00;
      b_reg    = 2'b00;
      cin_reg  = 1'b0;
      sum_reg  = 2'b00;
      cout_reg = 1'b0;
    end else begin
      a_reg    = a;
      b_reg    = b;
      cin_reg  = cin;
      sum_reg  = sum_wire;
      cout_reg = cout_wire;
    end
  end

    // Instantiate Bit 1 of the 2-bit Adder
    fadder fa1 (a_reg[1], b_reg[1], carry_tmp, sum_wire[1], cout_wire);
    fadder fa0 (a_reg[0], b_reg[0], cin_reg, sum_wire[0], carry_tmp);

    // Assign Output Signals
    assign sum = sum_reg;
    assign cout = cout_reg;
endmodule
```

Listing 2.5. A 2-bit full adder in Verilog.

for bit 1 and bit 0 are named `fa1` and `fa0`, respectively. Finally, the module's outputs (`sum` and `cout`) are assigned the registered output values. The resulting circuit utilizes the same number of resources as the VHDL version.

Finally, we present the finite state machine example in Verilog in Listing 2.6. Here is where the biggest difference between

```verilog
module adderFSM (a, b, a_we, b_we, clk, rst, result, valid);
  // Input Ports
  input [7:0] a, b;
  input a_we, b_we, clk, rst;
  // Output Ports
  output [7:0] result;
  output valid;
  // Internal Registers
  reg [7:0] a_reg, b_reg;
  reg [7:0] a_next, b_next;
  reg [1:0] fsm_cs;
  reg [1:0] fsm_ns;
  // Output Registers
  reg [7:0] result;
  reg valid;
  // FSM State Encoding
  parameter [1:0] wait_a_b=2'b00,wait_a=2'b01,wait_b=2'b10,add_a_b=2'b11;

  // Initialize current state to wait_a_b state
  initial begin
    fsm_cs = 2'b00;
  end

  // Finite State Machine Register Always Block
  always @(posedge clk)
  begin
    // Synchronous Reset
    if (rst == 1'b1) begin
      a_reg  = 8'b00000000;
      b_reg  = 8'b00000000;
      fsm_cs = wait_a_b;
    end else begin
      // Register current signal to next signals
      a_reg  = a_next;
      b_reg  = b_next;
      fsm_cs = fsm_ns;
    end
  end

  // Finite State Machine Combinational Logic Always Block
  always @(fsm_cs or a or b or a_we or b_we or a_reg or b_reg)
  begin
    a_next = a_reg;
    b_next = b_reg;
    fsm_ns = fsm_cs;
    // Finite State Machine
```

```verilog
case (fsm_cs)
  // State 0: Wait for Inputs a and/or b
  //   register input and either go to do addition or
  //   transition to wait for other operand
  wait_a_b : begin
    // Initialize registers and set outputs to 0
    a_next = 8'b00000000;
    b_next = 8'b00000000;
    result = 8'b00000000;
    valid  = 1'b0;
    if ((a_we & b_we) == 1'b1) begin
      a_next = a;
      b_next = b;
      fsm_ns = add_a_b;
    end else if (a_we == 1'b1) begin
      a_next = a;
      fsm_ns = wait_b;
    end else if (b_we == 1'b1) begin
      b_next = b;
      fsm_ns = wait_a;
    end
  end

  // State 1: Wait for A Input
  wait_a : begin
    result = 8'b00000000;
    valid  = 1'b0;
    if (a_we == 1'b1) begin
      a_next = a;
      fsm_ns = add_a_b;
    end
  end

  // State 2: Wait for B Input
  wait_b : begin
    result = 8'b00000000;
    valid  = 1'b0;
    if (b_we == 1'b1) begin
      b_next = b;
      fsm_ns = add_a_b;
    end
  end

  // State 3: Perform Addition of a_reg + b_reg
  add_a_b : begin
    result = a_reg + b_reg;
    valid  = 1'b1;
    fsm_ns = wait_a_b;
  end
  endcase
  end
endmodule
```

Listing 2.6. A finite state machine example in Verilog.

Verilog and VHDL can lie. It is possible to code FSMs similar to the VHDL implementation or completely different. We have chosen to closely match the two implementations, but we recommend those with an interest in Verilog to purchase one of the many Verilog HDL books available to learn the other possible implementations.

The most noticeable difference is the syntax; in fact, with this implementation you could set both files side by side and see an almost one-to-one mapping between the two languages. We chose to continue to use a two-process (always block) finite state machine and a case statement for its implementation. One difference is rather than using a *type* to define the FSM as was done in VHDL, we use a *parameter* and encode each state.

The reader may notice that the Verilog section is shorter than VHDL. This is not by chance. The authors favor VHDL, we make no attempt to hide this fact. We do not fault Verilog; in actuality we find Verilog a delightful language and offer the fact that we learned VHDL first (and Ada) as our only definite reason for our tendency toward VHDL. This section tried to provide some examples of Verilog that a reader can use with additional material to further pursue Verilog as a hardware description language. For those interested in a additional Verilog reading, we recommend Ashenden, Peter J. (2007).

Regardless of the language choice, we strongly suggest the reader take note of what is considered *synthesizable* HDL. When creating components, take the time to synthesize each component individually and understand how the component is being mapped to hardware. This will give you a better understanding of the resources needed by both the component and the final design.

2.4.3. Other High-Level HDLs

While the majority of hardware engineers use either VHDL or Verilog, there are other HDLs available. Because we are focusing on VHDL, we will not go into specific details about these additional languages at this time. In an effort to increase productivity, SystemC (Open SystemC Initiative (OSCI), 2010), HandelC (Agility, 2010), and Impulse (Impulse Accelerated Technologies, 2010) are a few of the attempts at building hardware and software systems together by providing the designer a higher level language to work in. Some are extensions to the C/C++ library, while others use C-like syntax to provide a more programmer-friendly environment. Some do provide designers with an alternative to the verification of a complete hardware/software system, which can be carried out before further hardware/software partitioning. Moreover, when translating existing algorithms into hardware-implementable architecture, preserving the source programs in C or C++ becomes possible with the support of an intelligent compiler, which may only involve a minor effort of code modification. Overall, it is debatable whether these languages are mature enough to support the desired goal of providing the design with the ability to model large systems, which incorporate both hardware components and software applications.

2.5. From HDL to Configuration Bitstream

So far in this chapter we have covered the internals of the FPGA, along with hardware description languages that provide a mechanism to describe complex digital circuit designs. As mentioned earlier, the process of converting an HDL into logic gates is called *synthesis* and the result of this process is a *netlist* file. This section describes how to convert a digital circuit to a netlist. Then, it describes how to configure an FPGA device from the netlist. A more detailed example of this process with vendor-specific tools is provided in Section 2.A.

Let's start with an example. Consider the Boolean function:

$$f(w, x, y, z) = w + x + yz$$

The schematic is shown in Figure 2.13. This circuit is easily expressed in a netlist. Each of the gates (two OR gates and one AND gate) becomes a "cell." Each primitive cell will include the type of gate, a name for this cell, and a set of ports. Ports will have names and direction (input or output). Complex cells can be formed by adding contents, which are instances of other cells and nets. Nets describe how the cells are "wired up"; that is, which output ports are connected to which input ports. Since complex cells can include other complex cells, a netlist forms a hierarchy with a single *top-level cell* encompassing all other cells. Ports on the top-level cells are then associated with external pins on the FPGA device.

There are many netlist formats and most are binary (i.e., not human readable) but some, such as EDIF, use a simple syntax and ASCII text to describe a cell hierarchy. EDIF (pronounced *E-diff*) stands for Electronic Design Interchange Format and is a vendor-neutral standard. Different tools use different file extensions. Some use .edif but others have adopted various three-letter suffixes to be compatible with DOS file name restrictions such as .edf and

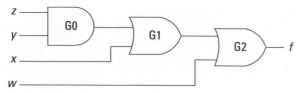

Figure 2.13. Schematic of a simple Boolean function.

```
(edif gates
  (edifVersion 2 0 0)
  (external UNISIMS
    (cell LUT4
      (cellType GENERIC)
      (interface
        (port I0
          (direction INPUT)
        )
        (port I1
          (direction INPUT)
        )
        (port I2
          (direction INPUT)
        )
        (port I3
          (direction INPUT)
        )
        (port O
          (direction OUTPUT)
        )
      )
    )
  )
)
```

Figure 2.14. Portions of a Boolean function in EDIF.

.edn. To give the reader a flavor for its format, a stripped down sample is shown in Figure 2.14. One last comment, the primitive cells used in a netlist are not universal. They may be basic logic gates (AND, OR, etc.) or they can be vendor-specific components (like a TBUF) that a vendor's back-end tool flow knows about.

The process we want to explain is how to go from a computer-readable expression of a digital circuit to a sequence of bits suitable to configure an FPGA. Using the example described earlier,

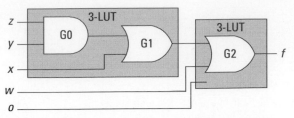

Figure 2.15. Example of a four-input function mapped to two 3-LUTs.

$f(w, x, y, z)$, the first thing to recognize is that we have four inputs — so f cannot be implemented in a single 3-LUT function generator. With only three select lines, an 8-to-1 MUX can only handle three variables. The solution is to use *two* 3-LUTs and and the routing resources to connect them. For example, f can be decomposed into two functions: $f(w, x, y, z) = f_2(w, f_1(x, y, z))$, where $f_1(x, y, z) = x + yz$ and $f_2(w, t) = w + t$. Now f_1 and f_2 can be used to configure two 3-LUTs. The output of f_1 is routed to the second input of f_2. This decomposition is illustrated graphically in Figure 2.15.

Note that G1 could be part of either 3-LUT — the result is the same (two function generators are required). Also note that if a 3-LUT only has two inputs, then the third select line is connected to ground and half of the SRAM entries are *don't cares*.

The process of grouping gates and assigning functionality to FPGA primitives is called **mapping** or MAP for short. The result of mapping a netlist is another netlist. The difference is that the cells of the input are generic Boolean gates; the cells of the output are FPGA primitives (3-LUTs, Flip-Flops, TBUFs, etc.).

The next step in the back-end tool flow is to assign the FPGA primitives in the netlist to specific blocks on a particular FPGA device. This process is called **placement**. In our simple example, f uses two 3-LUTs. During placement, the tools will decide which two 3-LUTs to use and add that information to the netlist.

Because every 3-LUT is the same as every other 3-LUT, the real challenge here is to place the cells relative to one another such that routing the output of one 3-LUT to the next one has the lowest propagation delay. Recall from Section 2.3 an output from the logic block goes to a connect box. From the connect box, a pass transistor decides where the signal propagates, which may be an adjacent logic block or to a switch box. A *pass transistor* is able to either propagate or block a signal by connecting the gate of the transistor to an SRAM cell, allowing the transistor to control the signals in a general-purpose circuit. If the signal goes to a switch box, another pass transistor determines which wire segment to use and so on to the next switch box and eventually to a connect box and a logic block. Every transistor and, to some extent, the length of the wire segments used contribute to the propagation delay. The longest propagation delay in the design will determine the maximum clock frequency. The process of picking wire segments and setting passing transistors is called *routing*.

If the project requires something like a 100-MHz clock frequency, but the design has a propagation delay that limits the clock to say 50 MHz, we say that the design does not meet timing. For very large netlists that use a large fraction of the resources, it is possible (and fairly common) that bad placement will use all of the routing resources in one area of the chip, making it impossible to connect all of the nets in the netlist. When this happens, we say the design failed to route.

In either case — not meeting timing or failing routing — the solution is to try a different placement to see if that produces less congestion or shorter routes. In fact, because placement and routing are so tightly coupled, they are considered a single step and one place-and-route tool (PAR). Early place-and-route algorithms did things like (1) randomly place all of the cells and then (2) route each signal sequentially. If a signal cannot be routed, throw away all the routes and start with a different signal. If it appears routing is going to always fail, throw out everything and randomly place all of the cells again. The process continues until all of the signals are routed or some threshold for giving up is met! Today, the tools are more sophisticated but the specific algorithms are trade secrets that vendors keep to themselves. The drawing in Figure 2.16 illustrates the concepts of our running example after PAR. The SRAM cells are organized in a 2-D array (additional logic not shown).

The final step is to convert this 2-D collection of SRAM cell settings into a linear stream of bits. One can imagine all of the cells arrayed into two dimensions. One column might be associated with a column of 3-LUTs on the chip. All of the SRAM cells that make up each 3-LUT would be stacked on top of each other. The next column might have all of the SRAM cells used to configure the connect and switch boxes. The next would be another column of 3-LUTs. All of these SRAM cells are connected vertically as shown in Figure 2.17 to form a giant shift register. To set the SRAM cells, the first column of data is shifted into place, followed by the next column and so on. So the last step, called *bitgen*, takes a placed and routed netlist and sets the SRAM cells of this large 2-D array accordingly. Along with a header, each of the columns is written out sequentially to a binary file called a *bitstream*.

Again, this is not exactly how any vendor configures an FPGA but this illustrates enough details that a system designer needs to know.

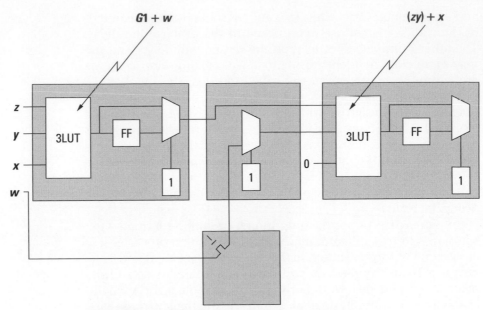

Figure 2.16. Illustration of $f(w, x, y, z)$ placed and routed.

Chapter in Review

This chapter covered a wide range of material, some of which may be considered background reading or act as a refresher for the reader. Our intention in this chapter was to emphasize material that we believe the reader should be aware of and, if not, provide enough detail to quickly catch the reader up to speed for the rest of the book.

At the heart of the FPGA is the CMOS transistor. We are keenly interested in this technology not because we are physically specifying and implementing transistors in our designs, but because all of the reconfigurable technology rests on top of it. Therefore, being aware of its characteristics, such as power consumption, will benefit the embedded systems designer. In addition to transistors, a brief overview of programmable logic devices was given with the intention to explain how FPGAs evolved into what they are today and to possibly provide some insight into the direction FPGAs may go in the future.

The cornerstone of this chapter was the discussion on FPGAs and Platform FPGAs. In Chapter 1 we covered some details about FPGAs, but here we dwelled into the specific constructs of the device. Just as a painter must be familiar with the available colors, so should the designer be familiar with the low-level components of the device.

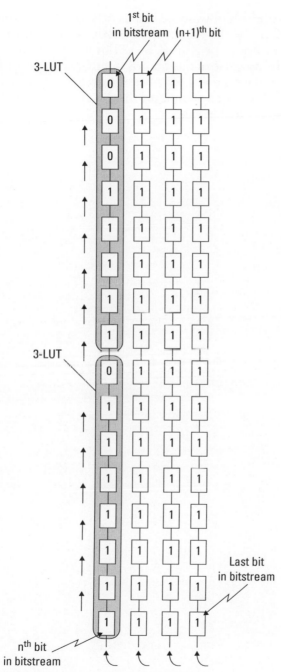

Figure 2.17. Illustration of $f(w, x, y, z)$ placed and routed.

This led into hardware description languages where we discussed how to describe hardware in both VHDL and Verilog. This was again meant to be a quick review or introduction to the language, as the purpose of this book is not to teach the reader everything about HDLs. Finally, we covered how to generate a bitstream from the HDL so we can configure the FPGA and run our hardware design on the actual device.

Practical Expansion: The Target

This chapter's white pages covered a wide range of material that is essential for designing systems with FPGAs. Throughout the gray pages we provide more technical details with respect to Xilinx FPGAs, Xilinx tool chain, and VHDL. For some readers, the white pages are sufficient to illustrate the point; however, we feel that practical examples with existing FPGA devices are more often what solidifies the readers, understanding of the material.

Toward this end we use a single device from a family of FPGAs, the Xilinx Virtex 5 (XC5VFX130T-FF1768). The goal is to learn the details of this device, build up a vocabulary of terms, explore several low-level ways of configuring an FPGA, and cover the design process from concept to implementation to running system. From there, it is relatively easy to learn a different device by comparing their features and functionality.

There are two more advantages to learning how to configure the low-level internal details of an FPGA. First, when the need arises, it is possible for an engineer to forgo the high-level tools and build a device-specific, efficient hardware module. (By analogy, most embedded systems are programmed in high-level languages but there are rare occasions that call for subroutine to be written in assembly.) The second reason — and this is very important with state-of-the-art high-level tools — is that oftentimes an engineer needs to understand how a small change in a high-level language may affect the eventual implementation in an FPGA. That is, with today's tools it is easy to break a working design by making a simple change to a high-level construct that ends up requiring twice as many FPGA resources.

These gray pages begin with a more detailed view of the Xilinx Virtex 5 FX130 FPGA. This material is available from the Xilinx Web site; however, aggregating the material is useful for someone new to the area or unfamiliar with how all of the technical documents and details are related. Next, a more detailed overview of the Xilinx Integrated Software Environment (ISE) tool chain is presented to help clarify each step in the FPGA configuration process. Finally, through the use of a few Xilinx tools, a custom hardware core and components will be developed and downloaded to the board.

2.A. Xilinx Virtex 5

To start, the Xilinx Virtex 5 FPGA series consists of several related but different devices. The groups are denoted LX, SX, TX, and FX. What do they all mean? In short, they refer to different mixes of configurable logic blocks and hard cores on the device. The LX (and any variants such as LXT) refers to devices that have a large number of configurable logic blocks relative to other hard cores on the device. The resources on the FPGA are more heavily allocated to configurable logic than say a processor embedded into the FPGA fabric. The SX refers to signal processing series parts, allocating more resources toward digital signal processing applications. The TX/HX refers to parts that have additional high-speed serial transceivers for high bandwidth interconnection capacity. The FX refers to FPGAs that have one or more PowerPC's embedded into the FPGA fabric.

It is possible to build Platform FPGAs from devices that do not include a hard processor. In that case, one or more soft processor cores can be used within the configurable logic. We focus on the FX series parts with two PowerPC 440 hard processor cores. The reason for this is because a majority of our embedded systems designs rely on at least a single processor, using a hard processor is a better use of FPGA resources as it consumes less power, has a higher operating frequency, and requires less physical space than an equivalent FPGA implementation.

2.A.1. Look-Up Table

Xilinx refers to the function generators within the FPGA fabric as *look-up tables* (LUT). The Virtex 5 FPGAs are built from 6-input LUTs. The white pages refer to 3-input LUTs for demonstration purposes (an 8-row truth table/Boolean circuit is much easier to represent than a 64 row truth table in a book); however, today's devices include larger 4-LUTs and 6-LUTs.

In practice, using hardware description languages (HDL) to describe the digital circuit and then using synthesis tools to map the textual description, an equivalent look-up function is more common than the designer defining the LUTs logic itself. What is important as far as the designer is concerned is how to represent a circuit efficiently to utilize the available resources. The 6-LUT on the Virtex 5 can be used either as a single 6-LUT or as two 5-LUTs as long as both 5-LUTs share the same inputs. A designer can take advantage of this when building digital circuits by not including the unnecessary inputs, which the synthesis tools may infer to a larger LUT. Figure 2.18 represents the block diagram of a 6-LUT. In the event that all six inputs are used for the LUT, the bottom output O5 is not used.

2.A.2. Slice

In the white pages we presented the term *logic cell* to represent a look-up table and a storage element (D flip-flop). The Virtex 5 combines four of these logic cells to create a *slice*, as is shown in a simple block diagram in Figure 2.19. With four 6-LUTs and D flip-flops contained within close proximity, it is possible to use these components to design more complex circuits. In addition to Boolean logic, a slice can be used for arithmetic and RAMs/ROMs. Some slices are connected in such a way that they can be used for data storage as distributed RAMs or shift registers. This is accomplished by combining multiple LUTs in the slice. The distributed RAM can be configured as single, dual, and, in some cases, quad ports providing independent read-and-write access to the RAM. The depth of the RAMs varies based on the number of ports, but can range from 32 to 256 1-bit elements. The distributed RAMs' data width can be increased beyond 1-bit; however, there will be a trade-off between the width and depth and the resource usage. For example, a 64 × 8 (64 8-bit elements) RAM is implemented in nine LUTs (1-LUT per 64 × 1 RAM and an additional LUT for logic). However, a 64 × 32 extends beyond an efficient use of the configurable resources and is moved into what will be discussed shortly, Block RAM (BRAM).

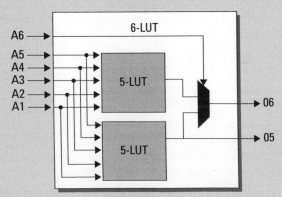

Figure 2.18. The Virtex 5 6-input LUT actually is built from two 5-input LUTs with the sixth input controlling a mux between the two LUTs.

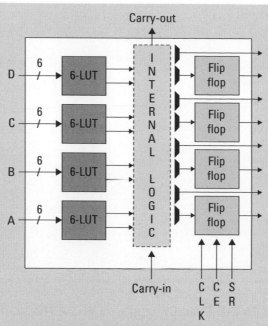

Figure 2.19. A block diagram of a Virtex 5 logic slice with four 6-LUTs and four flip-flop/latch storage elements.

In addition to logic and memory, slices can be used as shift registers. A shift register is capable of delaying an input x number of clock cycles. Using a single LUT, data can be delayed up to 32 clock cycles. Cascading all four LUTs in one slice, the delay can increase to 128 clock cycles. This is useful for small buffers that would traditionally be implemented within a more valuable resource, such as a Block RAM.

In all of these possible uses, a D flip-flop can be added to provide a synchronous read operation. With the additional D flip-flop, a read will be subject to an additional latency of one clock cycle. This may or may not impact a design, but for designs with high timing constraints, adding the synchronous operation can relax the constraint.

2.A.3. Configurable Logic Block

In the Virtex 5, a configurable logic block (CLB) contains two slices and the carry logic to connect neighbor slices. A CLB is considered the highest level of abstraction for the FPGA's configurable fabric. The two slices within the CLB are not directly connected, instead they sit in what is considered separate slice columns. Each CLB is connected to a switch matrix, providing configurable designs to span many CLBs. The switch matrix consists of long and short wires to provide more direct, point-to-point connections between CLBs in close proximity. Figure 2.20 depicts the basic structure and interconnection of CLBs, slices, and switch matrices.

2.A.4. Block RAM

Block RAM is dedicated random access memory grouped together in 36 Kbit blocks on Virtex 5 FGPAs. This is in addition to the distributed RAM mentioned earlier. BRAM provides on-chip memory, which can be used as large look-up tables, local storage, or data buffers (FIFOs). Each BRAM has two independent access ports (A and B) that only share data in the memory cells. Clocks are relative to each port. Care must be taken when using BRAMs in dual-port mode where the two

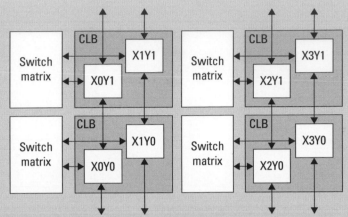

Figure 2.20. Virtex 5 CLB block diagram of the interconnection between each slice in the CLBs and their connection to the switch matrix.

ports write to the same address at the same time instance. This will not cause any ill effects to the physical BRAM, but effectively a race condition occurs between the two ports and it is nondeterministic which port's data will be actually stored in the memory cells. It is not difficult to design around this problem, but being aware of the problem before writing any designs makes the problem all the easier.

BRAMs are configurable in the width and depth supported. Each 36-Kb BRAM can be configured as 32K × 1 (32,768 by 1 data bit), 16K × 2, 8K × 4, 4K × 9, 2K × 18, or 1K × 36 RAMs. The depth and width are related to produce a maximum of 36-Kb. A BRAM can be combined with an adjacent BRAM to provide deeper or wider BRAMs.

Likewise, a BRAM can be split as two independent 18-Kb RAMs. Each 18-Kb RAM can be configured as a 16K × 1, 8K × 2, 4K × 4, 2K × 9, or 1K × 18. This provides the Virtex 5 FPGAs with an ability to utilize their resources more efficiently.

One common use of BRAMs in FPGA designs is for FIFOs. FIFOs, or simply data queues, are primitives designers can take advantage of, rather than building their own out of BRAM logic, reducing design and debugging time. Recently, FPGAs have started to include FIFOs as separate components within the FPGA fabric. The Virtex 5 and 6 are two such devices, although the physical limitations on the functionality may rule out their use in a design. We cover FIFO primitives in more detail in Section 6.A. For now, we focus on the use of BRAMs for storage and large buffers within Platform FPGA designs.

2.A.5. DSP Slices

A common use for FPGAs is with digital signal processing (DSP). The need to perform operations in parallel and to customize the operations based on the application has increased in the community of FPGA-based DSP designers. To improve the performance, many FPGAs come with special DSP blocks. In the Virtex 5, the DSP slices are known as DSP48E (48-bit DSP element) slices.

The DSP slices include a 25 × 18 two's complement multiplier, 48-bit accumulator (for multiply accumulate operations), an adder/subtractor for pipelined operations, and bitwise logical operations. Embedding this functionality into the slice provides a significant savings in FPGA resources, as implementing the equivalent resources in LUTs is quite expensive.

For applications needing filters, such as a comb and finite impulse response, to transforms, such as fast and discrete Fourier, to CORDIC (coordinate rotational digital computer) algorithm, DSP slices are used when available. The Virtex 5

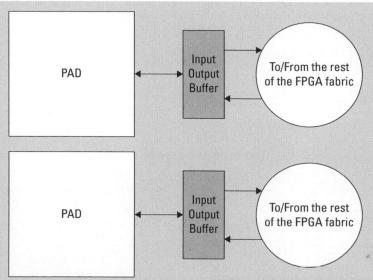

Figure 2.21. Xilinx Select Input/Output buffer (IOB) tile for connections between FPGA fabric and the pad on the FPGA device.

FX130T on the ML-510 contains 320 DSP slices. Compared to 20,480 regular slices, this seems like a disproportionate amount, yet not all designs require the use of DSP slices.

For designs requiring a higher percentage of DSP slices, the Xilinx SX series FPGAs include more DSP resources. There are tools, one of which we will introduce shortly, that help the designer quickly implement customized DSP components. They are also useful for resource and performance approximation.

2.A.6. Select I/O

Interfacing off the FPGA is another important issue when designing for embedded systems. In most cases, there will be a need to interface with some physical device(s). Depending on the number of I/O pins required, some devices are better suited than others, but they are all built around Input/Output Blocks.

From Figure 2.21 each I/O tile spans two pads (which connect to physical pins). Each Pad connects to a single IOB, which connects to the input and output logic. Xilinx uses the term Select I/O to refer to configurable inputs and outputs, which support a variety of standard interfaces (LVCMOS, SSTL, LVDS, etc.). Select I/O also take advantage of Digitally Controlled Impedance (DCI) to eliminate adding resistors close to the device pins, which are needed to avoid signal degradation. DCI can adjust the input or output impedance to match the driving or receiving trace impedance. Some advantages include a reduction in the number of parts, which simplifies the PCB routing effort. It also provides a way to correct for variations in manufacturing, temperatures, and voltages.

2.A.7. High-Speed Serial Transceivers

Along with Select I/O, some FPGAs include additional high-speed transceivers (transmit/receive) to connect devices serially at low latency and high bandwidth. The transceivers are coupled within the FPGA fabric to provide direct access to and from soft cores. Line rates are programmable from 100 Mb/s to 6.5 Gb/s. In the Virtex 5 series FPGAs, two types of transceivers exist, RocketIO GTX and GTP. GTX transceivers are capable of a higher bandwidth, whereas GTP transceivers are lower bandwidth and require less power. The number of transceivers varies from part to part. For example, the Virtex 5 FX130T includes 20 GTX transceivers. As with DSP slices and SX series FPGAs, there are applications that require a higher percentage of transceivers to configurable logic. The TX (also known as TXT or HX) series FPGAs include these additional transceivers. Both GTX and GTP transceivers are bidirectional, providing independent transmit and receive at the same time.

The transceivers can support parallel data through use of the serialize/deseralize (SERDES) logic, which connects to the input/output blocks. To support higher bandwidth needs, it is possible to bond the channels together. For example, two transceivers operating at 5.0 Gb/s bonded together can offer 10.0 Gb/s bandwidth at the cost of the additional transceiver resource. In high bandwidth-sensitive applications, the extra resource is a necessary expense.

Xilinx includes a customizable GTX/GTP wizard to expedite adding the transceiver logic to a design or specific component. In short, it is possible to specify the data width (parallel data), frequency, and channel bonding needed by the design. Design considerations are needed for systems with tight timing or resource constraints; however, much of the headache typically associated with high-speed integration can be eliminated.

2.A.8. Clocks

In FPGA designs it is common to operate different cores at different frequencies. In traditional design, any clocks needed would have to be generated off-chip and connected as input to the system. With FPGA designs it is possible to generate a wide range of clock rates from a single (or a few) clock source(s). While it is easier to design systems with only a few different clock rates, having the flexibility to incorporate clock rates after board fabrication is compelling to designers. However, there are limitations on the number of clocks that can be generated and routed to various parts of the FPGA. A number of clock regions exist on the FPGA (varying from 8 to 24 based on the Virtex 5 FPGAs), which support up to 10 clocks domains. The FPGA is split in half and on each half a clock region spans 20 CLB. The Virtex 5 FX130T has 20 clock regions. Figure 2.22 is a simple example of the clock regions on the FPGA.

To help design, use, and manage these clocks, Xilinx uses digital clock managers (DCM). Generally speaking, a DCM takes an input clock and can generate a customizable output clock. By specifying the multiply and divider values, the

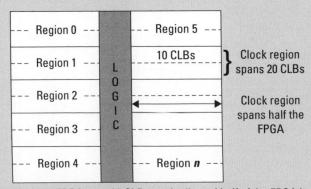

Figure 2.22. Clock regions on an FPGA span 20 CLBs vertically and half of the FPGA horizontally; the number of clock regions varies by chip.

frequency-synthesis output clock `clkfx` can generate a custom clock. Given an input clock `clkin` the equation:

$$clkfx = M/D * clkin$$

is used to generate the output clock. However, the DCM provides more than just generating different clock rates. A DCM is also capable of phase shifting the input clock by 90, 128, and 270 degrees. The DCM also provides a 2× the input clock rate and can phase shift this input clock by 180 degrees. It is easy to see without going into more detail that DCMs are a useful tool to generate the necessary clock(s) in FPGA designs. In short, meeting timing when designing with FPGAs can be a difficult task unless design considerations for timing and clocks are included from the beginning of the design process.

2.A.9. PowerPC 440

The Virtex 5 FX series FPGAs include one or two PowerPC 440 (Xilinx, Inc., 2009b) processors embedded within the FPGA fabric. The PowerPC 440 is a RISC processor with an operating frequency of up to 550 MHz. It contains a seven-stage pipeline with out-of-order execution capabilities. Included are data and instruction 32 KB, 64-way set associative level 1 caches.

The PowerPC connects to a system bus known as the Processor Local Bus (PLB), which is part of IBM's CoreConnectIBM (2009) suite of soft cores. The PLB supports 128-bit data transfers at user-define frequencies (in most cases, 200 MHz is the upper limit, but more discussion on the PLB will come in Section 3.A). A example of the processor's block diagram is shown in Figure 2.23.

In addition to the PLB, the PowerPC 440 includes a crossbar switch. The switch is used to connect the processor to a memory controller for a high bandwidth, low latency connection to off-chip memory. The switch also supports up to four direct memory access (DMA) ports, enabling more efficient transfers to and from memory to soft compute cores implemented within the FPGA fabric. We go into more detail regarding DMA in Chapter 4.

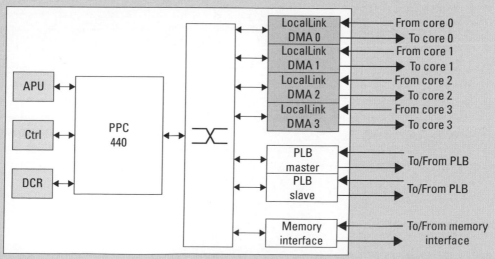

Figure 2.23. PowerPC 440 processor block diagram showing the available interfaces between processor and FPGA fabric.

Another unique feature of the PowerPC is the auxiliary processing unit (APU). The APU connects to the fabric coprocessor bus (FCB) to support custom instructions implemented in the FPGA fabric. For example, a double precision floating point unit (FPU) can be connected to the APU. Then, anytime the application needs to perform a floating point computation, the processor will offload the computation to hardware where the computation can be performed faster than a software-emulated FPU.

The PowerPC will be at the heart of our initial Platform FPGA designs. The PowerPC provides us with a "comfortable" environment (for those more familiar with C/C++ than HDLs) to begin to interact with the various components and, soon, our own custom soft cores.

Overall, we mention these components not because a designer is forced to implement Boolean logic or distributed RAMs at such a low level; ideally that is what the hardware description languages and synthesis tools are for. Instead, we mention them to help clarify for the reader the flexibility available with FPGA designs and to illustrate a point. The designer should not place too much dependency on low-level tools to create resource-efficient and high-performing designs. For example, being able to identify when a design is better suited for BRAM FIFOs instead of LUTs and specifying that information within the HDL context make the synthesis tool's job easier. Finally, we mention these different components to help a designer understand what their design is being mapped to. It is a gratifying feeling when a designer can look at a synthesis report and understand where and why all of the resources are being used and, if necessary, modify their design to take better advantage of the remaining resources.

2.B. Xilinx Integrated Software Environment

2.B.1. Overview of Commands

The Integrated Software Environment (ISE) contains a suite of commands that can turn FPGA designs described in hardware description languages and netlists into bitstream configuration files. ISE is a comprehensive program and there are plenty of tutorials and FAQs available. Instead, we aim to cover the underlying commands (Xilinx, Inc., 2009a) that are called by the ISE GUI. For those more comfortable with the ISE GUI, it is hoped that this section will at least clarify what each step is in the hardware synthesis process.

Xilinx Synthesis Tool At the core of the ISE tool chain is the Xilinx Synthesis Tool (Xilinx, Inc., 2009c) (XST). XST is used to synthesize hardware description languages into a netlist. XST is not the only synthesis tool available; however, as it is available within the ISE tool chain, it will be the synthesis tool used throughout this book.

To help in our understanding of how XST works, we begin by synthesizing the VHDL version of the 1-bit full adder. In addition to the VHDL file, two files are needed to run XST in command line mode. These two files are the project file and the synthesis script file. The project file specifies all of the HDL in the project to be synthesized. It is commonly named with the extension `.prj`. In the full adder example, only one VHDL file exists so the project file simply contains:

```
vhdl work fadder.vhd
```

The three columns denote the HDL type, library name, and VHDL filename. The `work` library is the default library to use.

The XST synthesis script is commonly named with the extension `.scr`. This script contains parameters used by XST during synthesis. It is possible to run XST without a script by entering each option on the command line; however, creating a single script is the preferred method, as it reduces redundant typing of the long series of inputs at the command line. For the full adder example, the synthesis script is shown in Figure 2.24.

```
run
-ifn fadder.prj
-ofn fadder.ngc
-ofmt NGC
-top fadder
-opt_mode Speed
-opt_level 1
-iobuf NO
-p xc5vfx130t-ff1738
```

Figure 2.24. Example of a synthesis script for XST.

The run keyword will indicate to XST to execute synthesis with the following attributes. XST requires an input project filename (ifn), which is the project file we mentioned earlier, containing the HDL to be synthesized. The output filename (ofn) is the name the synthesized netlist will be given after successful synthesis. The output file format (ofmt) is set to NGC, Xilinx's proprietary netlist format.

Next is the top-level entity name (top) attribute, which in our full adder example is fadder. Two synthesis optimization options follow, one that specifies whether XST should synthesize for speed (provide the highest operational frequency possible) or area (pack the logic as tightly as possible). The second option is for the synthesis effort level. A trade-off between synthesis effort and synthesis time is made; higher levels may provide more resource efficient or frequency efficient designs at the expense of longer synthesis times. The iobuf attribute adds I/O Buffers to the top-level module. In the full adder example, we choose not to insert I/O Buffers. The last attribute in this example is the FPGA part type (p). Here we specify to synthesize for the Virtex 5 FX 130T FPGA.

Additional command line arguments can be added to the synthesis script file; however, for this first XST example these are sufficient to produce a netlist. The final step is to run XST on the command line, passing the synthesis script file as an input file.

```
xst -ifn fadder.scr
```

After XST completes, a report is written to a synthesis report file (srp). If there are any syntax errors, XST will report approximately where in the file the line exists that failed synthesis. This is similar to compiling a C binary. Understanding the report is an important tool for the designer when trying to identify resource consumption and timing analysis. Under the Final Report heading, in Figure 2.26 is a list of the basic elements (BELS) needed to represent the digital circuit. In our design, two 3-LUTs are needed for the two outputs, sum and carry-out, in the full adder.

Another important section in the synthesis report file is the device utilization summary section, in Figure 2.27. For the specific FPGA device the number of slices, LUTs, flip-flops, BRAMs, and so on are listed, giving the designer an approximation to the amount of resources the design requires.

The synthesis report also provides some rough timing information, in Figure 2.28. These numbers are not accurate because they do not consider the actual placement of the circuit on the FPGA. It does provide an approximation to the minimum period and maximum combination delay the circuit can obtain. Looking at the report, our full adder example was purely combinational, there is no clock and as a result no minimum period. For larger designs when trying to meet timing, the synthesis report can help identify which component is causing the timing error.

```
===============================================================================
*                             Final Report                                   *
===============================================================================

Final Results
Top Level Output File Name         : fadder.ngc
Output Format                      : NGC
Optimization Goal                  : Speed
Keep Hierarchy                     : no


Design Statistics
# IOs                              : 5


Cell Usage :
# BELS                             : 2
#       LUT3                       : 2
===============================================================================
```

Figure 2.25. XST synthesis final report section.

```
Device utilization summary:
----------------------------
Selected Device : 5vfx130tff1738-3

Slice Logic Utilization:
  Number of Slice LUTs:                    2  out of  81920     0%
      Number used as Logic:                2  out of  81920     0%

Slice Logic Distribution:
  Number of LUT Flip Flop pairs used:      2
      Number with an unused Flip Flop:     2  out of      2   100%
      Number with an unused LUT:           0  out of      2     0%
      Number of fully used LUT-FF pairs:   0  out of      2     0%
      Number of unique control sets:       0

IO Utilization:
  Number of IOs:                           5
  Number of bonded IOBs:                   0  out of    840     0%
```

Figure 2.26. XST synthesis report device summary section.

For large systems with many components and subcomponents, XST can be used to synthesize the entire design into a single netlist or it can be used to synthesize individual components in a hierarchical fashion. Figure 2.25 depicts this synthesis flow for a design. While each approach has their merits, the second approach is used more often as it provides more immediate and useful feedback for each component. This approach stems from the philosophy of bottom-up design where components are created (typically in HDL) from the ground up. It is only natural to synthesize in the same

```
Timing Summary:
----------------
Speed Grade: -3

    Minimum period: No path found
    Minimum input arrival time before clock: No path found
    Maximum output required time after clock: No path found
    Maximum combinational path delay: 0.400ns
```

Figure 2.27. XST synthesis timing report.

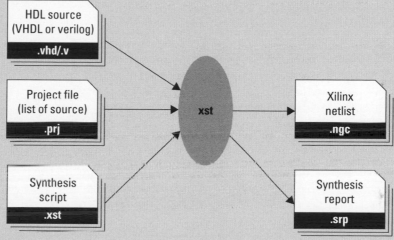

Figure 2.28. Synthesis flow.

fashion, generating intermediate netlists and producing individual synthesis reports. The designer can view these reports to determine which of the components need to be redesigned in the event of resource or timing constraints.

For example, consider a design consisting of three components, a top-level component and two low-level components. XST would be used to synthesize the top-level and two low-level components individually. The top-level component synthesis would treat the two low-level components as `black boxes`. The immediate advantage of such an approach is to minimize the number of times each component needs to be resynthesized. If one of the low-level components of HDL is modified, only that low-level component requires being synthesized again, unless the modification changes the entity's generic or port list, forcing the top-level component to change its instantiation of the design. These are akin to how libraries are in C/C++ during compile time.

Netlist Builder After XST is used to generate the system's netlist(s), the next step is to combine the netlists and any specific design constraints to produce a single netlist file. This is typically accomplished through the use of the following two commands, NGCBuild and NGDBuild.

NGCBuild

If the designer has chosen to synthesize each component individually in a hierarchical manner, the next command needed is called `ngcbuild`, which combines multiple netlist files into a single `ngc` netlist. NGCBuild opens the top-level netlist and matches any subcomponents with existing netlists. NGCBuild does not produce errors or warning when one or more subcomponent's netlists are not found. In the previous example, the top-level component netlist may be combined with one of its two subcomponents in the event that only one of the two components has been synthesized. NGCBuild can be called again to combine the missing subcomponent netlist when it is available. This is similar to the component level synthesis approach mentioned earlier in that a component's netlist is constructed as its subcomponents become available. An example of the NGCBuild command line execution is:

```
% ngcbuild input/system.ngc output/system.ngc -sd implementation
```

The input to NGCBuild is the top-level netlist (in this example it is `input.ngc`). The output netlist generated by NGCBuild is the second parameter (`output.ngc`). The flag `-sd` specifies the search directory to look for any component's netlist to be combined to generate the output netlist.

NGDBuild

Once all of the components have been synthesized, `ngdbuild` is used to generate a Xilinx Native Generic Database (NGD) file. The NGD file is used to map the logic contained within the netlist to the specific FPGA device. NGDBuild takes a single NGC file (created by NGCBuild) and any device, netlist or user constraints file. User constraints would include specific FPGA pins to be used for components such as RS232 UARTs and off-chip memory like DDR2 modules. NGDBuild will produce errors during run time if any of the netlists are missing in the input NGC file or if there are unbound user constraints. An example of NGDBuild is:

```
% ngdbuild -p xc5vfx130tff1738-2 -bm system.bmm -uc system.ucf system.ngc system.ngd
```

The command line flags used by NGDBuild are `-p` to specify which FPGA device to generate the ngd file for, `-bm` to specify the block memory file, which is a listing of all of the BRAMs in the system, and `-uc` to specify the constraints file listing all of the design's constraints (timing, IO, IO Standards). The final input is the NGC file generated by NGCBuild. The single output is the NGD file that will be passed on to Map, the next stage in the tool flow.

Map The Xilinx program MAP takes an NGD file and maps the logic to the specified FPGA device. MAP performs a design rule check (DRC) to uncover physical or (if possible) logical errors in the design. After passing the DRC, the logic described by the netlist is mapped to the FPGA device's components (such as BRAM, IOBs, and LUTs). MAP also trims any unused logic or netlists. Unlike synthesis, which can only trim internal signals if they are not used, MAP analyzes the entire system to determine if any logic is unnecessary. As a result, after MAP completes, the resource utilization reported is a more accurate representation than the synthesis report. We previously used the synthesis report's resource utilization to give us an approximate first-order estimation of the resources needed by the design; now we can rely upon MAP's report. When finished, MAP produces a Native Circuit Description (NCD) file, which can be used by the next tool to place and route the design to the target FPGA. An example of MAP's command line is:

```
% map -o system_map.ncd -pr b -ol high system.ngd system.pcf
```

The MAP flags are used to give the designer more control over what MAP placements are or are not run. Typically, a designer will be able to use the default options; however, there are cases when a nearly fully occupied FPGA design may require additional constants to MAP successfully. The `-o` flag sets the output file name; in this case the resulting NCD file is named `system_map.ncd`. The `-pr b` option specifies to pack flip-flops and latches in both input and output registers, while the `-ol high` flag is the overall effort level for the placement algorithm; high is used to achieve the best placement at the expense of longer MAP run times. The input file, `system.ngd`, is the NGD file generated by NGDBuild.

We already stated that `system_map.ncd` is the output from MAP, but in addition it is `system.pcf`, the physical constraints file, which are constraints placed during the design creation.

Place and Route   The Xilinx program PAR actually consists of two programs, Place and Route. These two programs are commonly run in series, so Xilinx has combined them into a single command. PAR takes the NCD file generated by MAP and runs both placement and routing algorithms to generate a routed NCD file. To begin, Place tries to assign components into sites (LUTs, BRAMs, DSP48E slices, etc.) based on any specific constraints (i.e., use pin location P38 to output the transmit (TX) line of the RS232 UART component) to maximize resource utilization while minimizing component distances (which will make routing and meeting timing requirements easier). Placement occurs through multiple phases (passes) and produces a placed NCD file.

After Place, Route is run to connect all of the signals for the components based on the timing constraints. Routing is also a multiphase operation, resulting in a routed NCD file. After PAR completes, a timing analysis is performed to verify that the design has met timing and, if not, produce a short log to help the designer identify and fix the errors. As the design size and clock rates increase, timing becomes more constrained and difficult to meet. It is possible to run PAR with different command line arguments in order to meet timing. PAR consumes the most time during the execution of all of the tools (as the design increases). Identifying fast *build* computers to run PAR on will reduce the amount of "idle" time before a design can be tested on the FPGA. An example of an PAR command line is:

```
% par -ol high system_map.ncd system.ncd system.pcf
```

PAR takes as input the NCD file and PCF files from MAP and generates an output `system.ncd` file, which is the placed and routed native circuit description.

Figure 2.29 illustrates the flow from netlists to a fully placed and routed design. This flow picks up after the Figure 2.25 that is run for each component in the design. These flows are used to help picture the process from source files (HDL) to just before generation of a bitstream, which is covered next.

Configuration Bitstream Generation   The final command in generating a configuration file for the FPGA is called `bit-gen`. BitGen produces a configuration `bitstream` for the specific Xilinx FPGA device. The bitstream (.bit) file contains

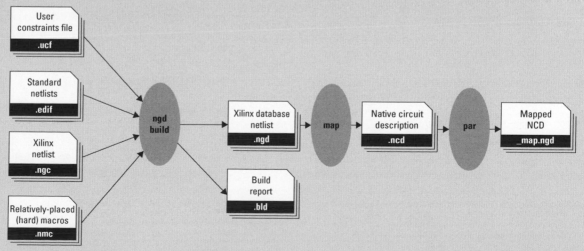

Figure 2.29. Netlist builder, map, and place-and-route.

configuration information (proprietary to Xilinx), which is downloaded and stored into the SRAM cells of the FPGA device. The concept of *programming* an FPGA is when the bitstream is downloaded to the FPGA, at which point the digital circuit is realized on the FPGA fabric. An example of the BitGen command line is:

```
% bitgen -f bitgen.ut system.ncd system.bit
```

The BitGen command can take as input a parameters file (`-f bitgen.ut`), which can be used to specify configuration information. The output `system.bit` is the configuration bitstream that can be used to "program" the FPGA.

Additional commands can be used during the synthesis and bitstream generation process; however, their uses are more application specific. We will hold off on covering those designs until later chapters where their use is better understood.

These short sections have covered the process from synthesis to bitstream in an effort to better understand the entire ISE flow and, when the need arises, to fix problems and to know where to look to identify the problems. Initially, it is easiest to use the XPS GUI to create complex designs and then use the GUI to synthesize, map, place and route, and generate the bitstream. However, as you gain more experience you may become more interested in the low-level commands that are producing the various log and report files. Going forward we encourage you to at least look at the logs and reports generated by these commands and then, if you are comfortable, try to execute these commands on your own.

2.C. Creating and Generating Custom IP

Up until now we have been laying a foundation of terms and technology, and with the addition of hardware description languages we have begun to learn how to build upon this foundation. We have also covered the commands used by Xilinx's ISE, which provides us with the tools to turn the HDL into a bitstream used to program the FPGA.

In addition to ISE, there are tools that can aid in the rapid development of components and cores. In this section two of these tools will be presented along with a detailed example of how to use them to generate custom components and cores.

2.C.1. Xilinx Core Generator

Xilinx Core Generator (CoreGen) is a tool that designers can use to quickly create hardware components. These components can range in complexity from FIFOs and math operators to DSP filters and transforms to network and memory controllers. While CoreGen unfortunately does not generate every type of component, it does provide a sufficient number to help most designers get their projects started. Earlier we mentioned FIFOs as being one of the more useful components, here we show how to use CoreGen to quickly generate a customized FIFO.

To start, from the command line run the command `coregen` and create a new project, specifying which FPGA device will be used when generating the FIFO. This is necessary because of the different FPGA devices and packages, and their supporting FPGA resources. If you are using the Xilinx ML-510 development board, then you can specify the Virtex 5 xc5vfx130t-ff1738 part. Figure 2.30 shows the main CoreGen GUI with the FIFO Generator component selected, ready to be run for the next step. The left half of the GUI is a list of the IP that can be generated with CoreGen. The upper right half lists information and actions relevant to the selected IP component, while the lower right half is a console to show output information, warnings, and errors. Now we are able to pick from a list of available components. Under the `Memories & Storage Elements` category are CAMs, FIFOs, Memory Interface Generators, and RAMs and ROMS. For this example we will select `FIFOs` and double click on `FIFO Generator`.

The first page of the FIFO Generator wizard, Figure 2.31, lets you specify the component name and FIFO implementation type. We have named our component `fifo_comp` and set it to be a Common Clock Block RAM implementation. From our

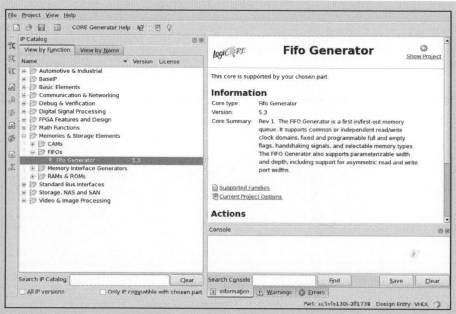

Figure 2.30. Main Xilinx CoreGen GUI.

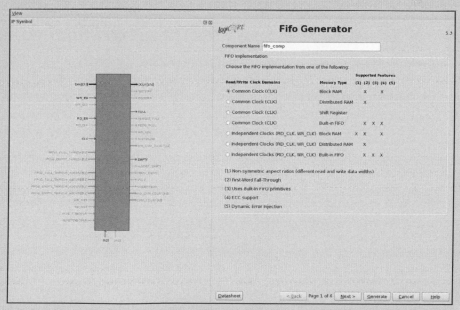

Figure 2.31. CoreGen's FIFO Generator initial configuration page.

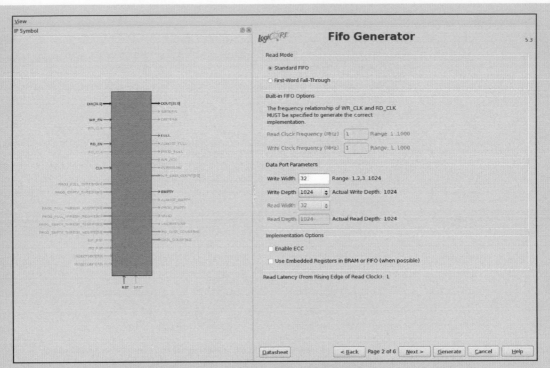

Figure 2.32. CoreGen's FIFO Generator read mode and FIFO options page.

previous discussions on RAM we can start to see how CoreGen gives us the ability to specify constraints (such as common vs independent clocks) and we are able to see what type of memory options are supported.

In the second page (Figure 2.32) we can specify the read mode, port width, and depth of the FIFO. In a standard FIFO read mode, when data are read from the FIFO via the read request (dequeue) signal, read data will be valid the next clock cycle. The first-word fall-through FIFO eliminates the single cycle read latency by storing the head of the FIFO in a register. Each subsequent read request pops the next element off the FIFO and stores it in the head register. These design decisions are less important in this example, but for latency sensitive designs it can play an important role in the system. For now, select the *standard FIFO* read mode.

At this point we can also specify the depth and width of the FIFO. Because we have selected a common clock BRAM, the width and depth are fixed for both the write and the read sides of the FIFO. If we had selected an independent clock implementation we could specify independent port parameters. For this example we use a write *width of 32 bits* and a *depth of 1024 data elements*. Recall that the 1K × 32 configuration requires a single 36-Kb BRAM. CoreGen gives us the ability to increase the depth without requiring us to write the HDL to cascade the BRAMs.

Figure 2.33 shows the third page, which provides the design with the ability to specify additional flags and handshaking ports. Depending on the amount of control needed in the design, these ports may or may not need to be added. The *valid flag* is a signal that will be assert high ('1') when valid data are on the dout port of the FIFO.

In the fourth page the initialization/reset options are set, as can be seen in Figure 2.34. Here it is possible to have either a synchronous reset or an asynchronous reset and what the default output should be after a reset. An important

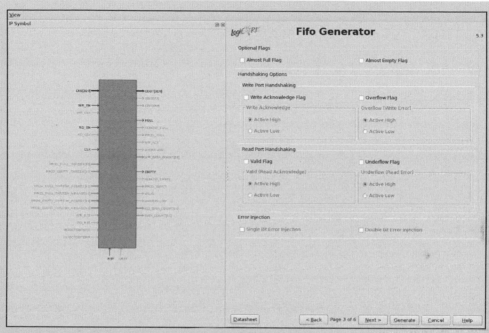

Figure 2.33. CoreGen's FIFO Generator handshaking signals page.

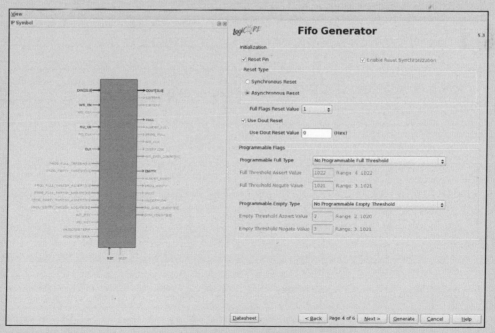

Figure 2.34. CoreGen's FIFO Generator initialization and programmability page.

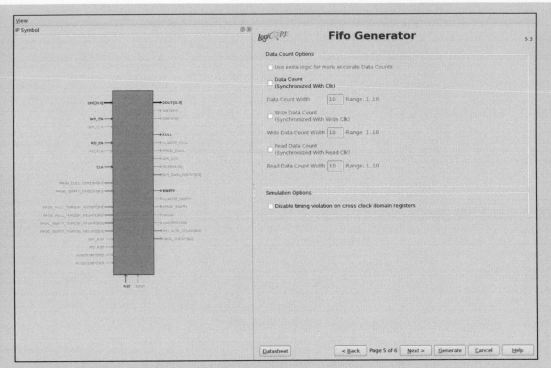

Figure 2.35. CoreGen's FIFO Generator data count page.

note, when a FIFO is reset, both `empty` and `full` flags will be asserted high, indicating data should not be read from or written into the FIFO. Once the reset is complete, the `full` signal will be deasserted, indicating the FIFO is ready for operation.

There is also an ability to add programmable full and empty flags. For data-flow purposes it might be important to have high and low watermarks for the FIFOs. When the FIFO reaches a predetermined threshold, the programmable full signal would be asserted high. Likewise a programmable empty signal can be used in the event a FIFO nears becoming empty. These programmable thresholds provide user-defined values for the FIFO compared to the `almost_full` and `almost_empty` signals, which are only asserted one data element before the respective signal is asserted.

Figure 2.35 shows page five of the FIFO generator, which provides the designer with the option to output a count of the number of data elements in the FIFO at any given time. This may be useful under conditions when a design needs to perform different operations based on the status of the FIFO.

The final page, Figure 2.36, provides a summary of the options taken by the designer and, most importantly, relays the resources based on those configuration settings. In our example, we have a 1K × 32-bit FIFO, which results in a single 36K BRAM being used.

At this point all that is left is to generate the FIFO. Once complete, a set of files are produced that provide simulation HDL (.vhd), HDL component declaration and instantiation templates to be used within calling HDL (.vho), and the proprietary Xilinx netlist file (.ngc), which will be used by the synthesis tools when the FIFO component is instantiated

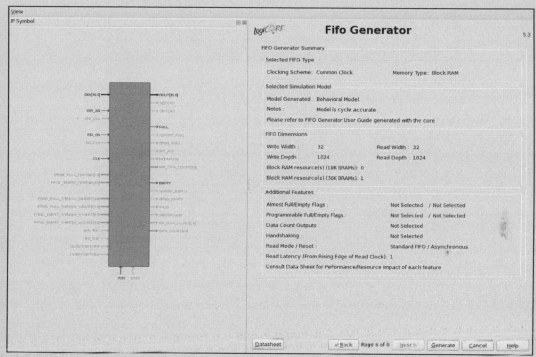

Figure 2.36. CoreGen's FIFO Generator summary page.

by the top-level component. A README is also provided, which details the remaining files generated automatically by CoreGen.

2.C.2. Create/Import Peripheral Wizard

In Section 1.A the Xilinx Platform Studio was introduced. With XPS we are given a rich set of GUI applications and wizards to expedite the base system building design process. We have already been introduced to the base system builder wizard. Using the Create/Import Peripheral (CIP) wizard we will generate a hardware core template. This template will give us the ability to read and write into software addressable registers. In Section 1.A we augmented this hardware core to support more functionality. In the meantime, we focus on the CIP wizard.

We can launch the CIP wizard from within the XPS GUI menu option:

Hardware → Create or Import Peripheral . . .

or by launching the wizard from the command line:

```
% createip
```

Once the wizard launches, you will walk through the following pages.

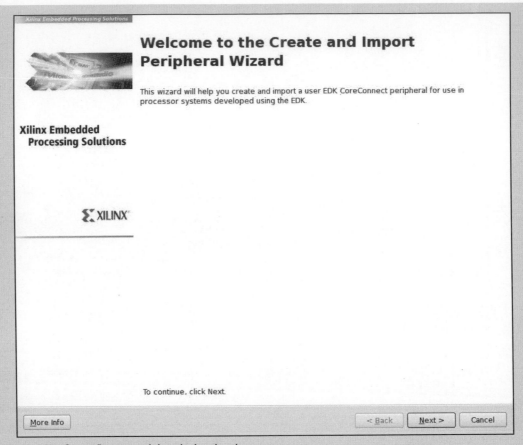

Figure 2.37. Create/import peripheral wizard welcome page.

The first page of the wizard is a simple welcome screen, which explains that the core you will be creating will be for an EDK CoreConnect-based project. In our case, this means we will be creating a core that will connect to the Processor Local Bus.

The second page is the peripheral flow; here we have the ability to create a new core (based on a template) or import and modify an existing peripheral. For now, we will continue with creating a template for a new peripheral.

The next page indicates where this project will be added, to an existing repository or directly into a project. Because this is the first core that we have created, this can be added to the existing project that was created previously at the beginning of this section. (Note: the path may differ based on where you saved your project.)

The fourth page lets us set the core's name and version number. There are characters that are not allowed in the name, the core contain any uppercase characters, start with a number or symbol, or end with a symbol. When naming the core, if the text turns red it indicates an invalid core name. We will name our core `my_test_core`.

On the next page we specify which bus our core will be able to attach. The PLB allows the PowerPC to communicate with our core. This will be the most common interface for PowerPC-based systems.

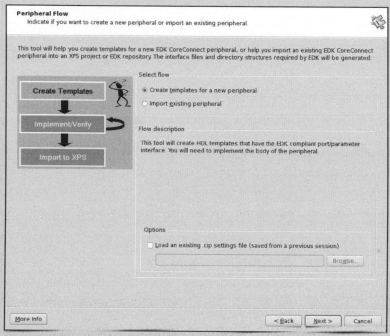

Peripheral Flow
Indicate if you want to create a new peripheral or import an existing peripheral.

This tool will help you create templates for a new EDK CoreConnect peripheral, or help you import an existing EDK CoreConnect peripheral into an XPS project or EDK repository. The interface files and directory structures required by EDK will be generated.

Create Templates → Implement/Verify → Import to XPS

Select flow
● Create templates for a new peripheral
○ Import existing peripheral

Flow description

This tool will create HDL templates that have the EDK compliant port/parameter interface. You will need to implement the body of the peripheral.

Options

☐ Load an existing .cip settings file (saved from a previous session)

[   ] Browse...

More Info   < Back   Next >   Cancel

Figure 2.38. Create/import peripheral wizard peripheral flow page.

Repository or Project
Indicate where you want to store the new peripheral.

A new peripheral can be stored in an EDK repository, or in an XPS project. When stored in an EDK repository, the peripheral can be accessed by multiple XPS projects.

○ To an EDK user repository (Any directory outside of your EDK installation path)
Repository: [▼] Browse...

● To an XPS project
Project: [/projects/examples/ch02/hw/system.xmp   ▼] Browse...

Peripheral will be placed under:

/projects/examples/ch02/hw/pcores

More Info   < Back   Next >   Cancel

Figure 2.39. Create/import peripheral wizard project page.

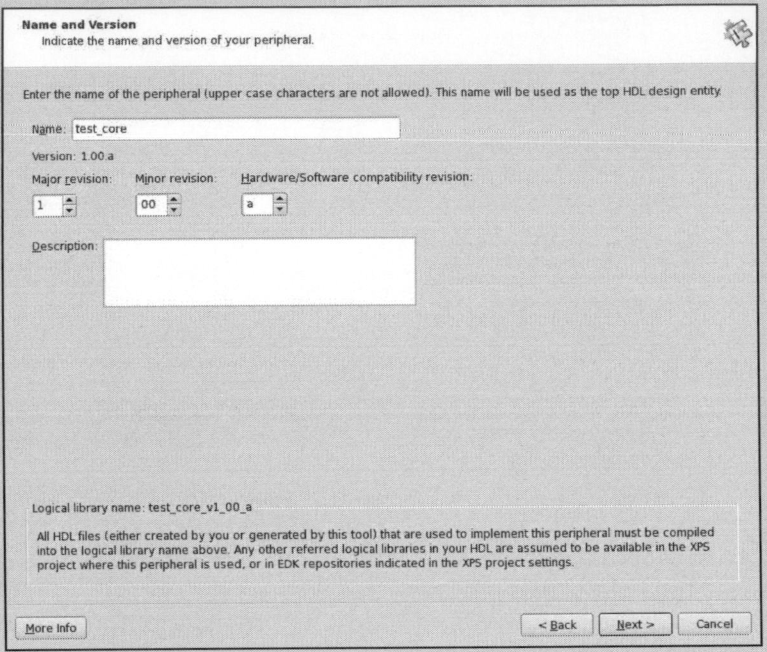

Name and Version
  Indicate the name and version of your peripheral.

Enter the name of the peripheral (upper case characters are not allowed). This name will be used as the top HDL design entity.

Name: test_core

Version: 1.00.a

Major revision: Minor revision: Hardware/Software compatibility revision:

1   00   a

Description:

Logical library name: test_core_v1_00_a

All HDL files (either created by you or generated by this tool) that are used to implement this peripheral must be compiled into the logical library name above. Any other referred logical libraries in your HDL are assumed to be available in the XPS project where this peripheral is used, or in EDK repositories indicated in the XPS project settings.

More Info   < Back   Next >   Cancel

Figure 2.40. Create/import peripheral wizard name and version page.

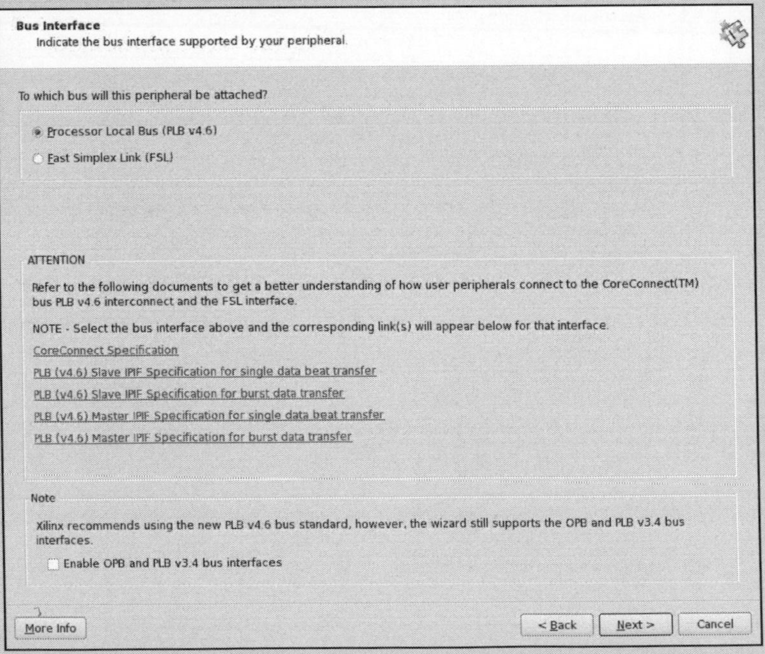

Bus Interface
  Indicate the bus interface supported by your peripheral.

To which bus will this peripheral be attached?

◉ Processor Local Bus (PLB v4.6)

○ Fast Simplex Link (FSL)

ATTENTION

Refer to the following documents to get a better understanding of how user peripherals connect to the CoreConnect(TM) bus PLB v4.6 interconnect and the FSL interface.

NOTE - Select the bus interface above and the corresponding link(s) will appear below for that interface.

CoreConnect Specification
PLB (v4.6) Slave IPIF Specification for single data beat transfer
PLB (v4.6) Slave IPIF Specification for burst data transfer
PLB (v4.6) Master IPIF Specification for single data beat transfer
PLB (v4.6) Master IPIF Specification for burst data transfer

Note

Xilinx recommends using the new PLB v4.6 bus standard, however, the wizard still supports the OPB and PLB v3.4 bus interfaces.
  ☐ Enable OPB and PLB v3.4 bus interfaces

More Info   < Back   Next >   Cancel

Figure 2.41. Create/import peripheral wizard bus interface page.

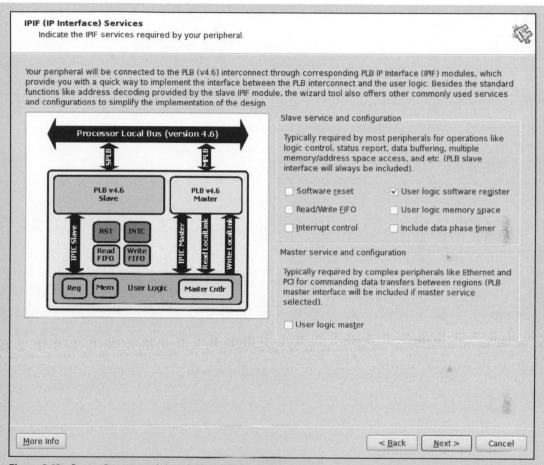

IPIF (IP Interface) Services
Indicate the IPIF services required by your peripheral.

Your peripheral will be connected to the PLB (v4.6) interconnect through corresponding PLB IP Interface (IPIF) modules, which provide you with a quick way to implement the interface between the PLB interconnect and the user logic. Besides the standard functions like address decoding provided by the slave IPIF module, the wizard tool also offers other commonly used services and configurations to simplify the implementation of the design.

Processor Local Bus (version 4.6)

PLB v4.6 Slave
PLB v4.6 Master
RST INTC
Read FIFO Write FIFO
Reg Mem User Logic Master Cntlr

Slave service and configuration

Typically required by most peripherals for operations like logic control, status report, data buffering, multiple memory/address space access, and etc. (PLB slave interface will always be included).

☐ Software reset   ☑ User logic software register
☐ Read/Write FIFO   ☐ User logic memory space
☐ Interrupt control   ☐ Include data phase timer

Master service and configuration

Typically required by complex peripherals like Ethernet and PCI for commanding data transfers between regions (PLB master interface will be included if master service selected).

☐ User logic master

More Info   < Back   Next >   Cancel

Figure 2.42. Create/import peripheral wizard IPIF page.

On the sixth page we can specify what interface our core will have to the PLB. This is known as the IPIF (Intellectual Property Interface). The PLB IPIF simplifies the number of signals and the complexity of the handshaking required to connect a custom core to the PLB. Here we will only add `User logic software registers`, allowing the processor to read and write to registers that will be created in our `my_test_core`. The registers provide a simple interface to exchange data and control signals between the processor and our core.

The slave interface page provides the core with the ability to support burst and cache-line transfers and nonnative data widths (beyond 32 bit). The PLB can support up to 128-bit transfers; however, the PowerPC is a 32-bit processor so for this first example we will continue with the default 32 bits.

The next page lets us set the number of software-addressable registers within the core. When the wizard generates the core's template, these registers and all of the necessary read-and-write logic will be included. We begin with three registers to help us demonstrate how to read and write to our core from the PowerPC.

The ninth page is the IP Interconnect (IPIC). These are the signals that our `my_test_core`'s user logic HDL will interact with. The IPIF mentioned earlier is responsible for interfacing with the PLB and generating the appropriate IPIC

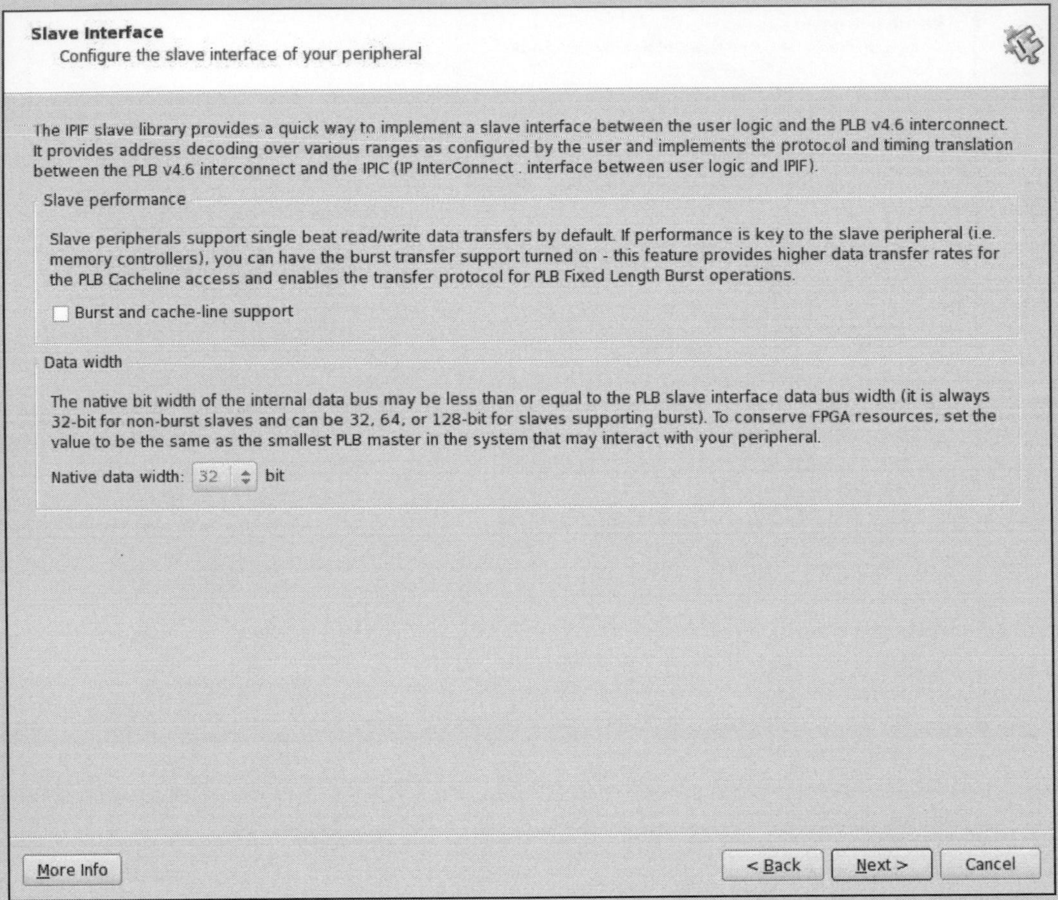

Slave Interface
Configure the slave interface of your peripheral

The IPIF slave library provides a quick way to implement a slave interface between the user logic and the PLB v4.6 interconnect. It provides address decoding over various ranges as configured by the user and implements the protocol and timing translation between the PLB v4.6 interconnect and the IPIC (IP InterConnect . interface between user logic and IPIF).

Slave performance

Slave peripherals support single beat read/write data transfers by default. If performance is key to the slave peripheral (i.e. memory controllers), you can have the burst transfer support turned on - this feature provides higher data transfer rates for the PLB Cacheline access and enables the transfer protocol for PLB Fixed Length Burst operations.

☐ Burst and cache-line support

Data width

The native bit width of the internal data bus may be less than or equal to the PLB slave interface data bus width (it is always 32-bit for non-burst slaves and can be 32, 64, or 128-bit for slaves supporting burst). To conserve FPGA resources, set the value to be the same as the smallest PLB master in the system that may interact with your peripheral.

Native data width: 32 ⬍ bit

| More Info | | < Back | Next > | Cancel |

Figure 2.43. Create/import peripheral wizard slave interface page.

signals. The IPIC signals are a much simpler set of bus interface signals, abstracting the complex bus signals away from the designer. For now we will stick with the default IPIC signals. Signals beginning with `Bus2IP` are input ports to the core from the PLB, and signals beginning with `IP2Bus` are output ports from the core to the PLB.

The next page provides an optional simulation support through use of a bus functional model (BFM). The BFM provides a cycle accurate simulator of the PLB, aiding in the design process by providing a quick simulation test bench for the custom core as if it were connected to the PLB. At this time we will skip the BFM; however, it will become very advantageous to use simulation when designing as it provides a quicker and easier debugging method than debugging a custom core running in hardware.

The next page explains the HDL structure of the created core. There will be two HDL files created, a top-level component (in our case named `my_test_core.vhd`) and the `User Logic` (`user_logic.vhd`). The user logic will contain

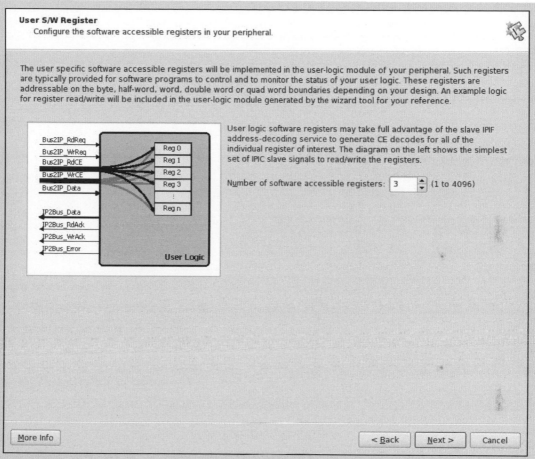

Figure 2.44. Create/import peripheral wizard slave registers page.

the initial HDL to provide the PowerPC access to the core. At this page we can also specify the user logic's hardware description language as Verilog instead of the default, which is VHDL.

Finally, the last page is the summary of the core that will be created. Once the IP core is generated we can look to see how the core's project directory is organized.

2.C.3. Hardware Core Project Directory

The create and import peripheral wizard generates a new hardware core's project directory inside the pcores directory. In our example we named the pcore my_test_core_v1_00_a. Within this directory are the subdirectories and files shown in Figure 2.49.

Within the data directory are two files, my_test_core_v2_1_0.mpd, which is the Microprocessor Peripheral Description (MPD), and my_test_core_v2_1_0.pao, which is the Peripheral Analysis Order (PAO). These files are mentioned because in order to augment the hardware core, we will need to modify these files.

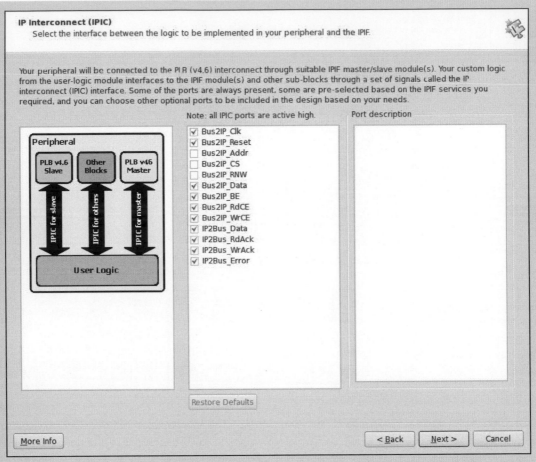

Figure 2.45. Create/import peripheral wizard IPIC page.

Microprocessor Peripheral Description File The MPD file contains project options, bus interfaces, generics, and ports for VHDI (or Verilog). *Peripheral Options* are used by the Xilinx EDK during Platform Generation (PlatGen), the stage before synthesis, to set up the synthesis script for XST. *Bus Interfaces* are a list of bus standards available in the EDK, such as the PLBV46 (Processor Local Bus v4.6). The bus interface is used to aggregate all of the bus signals (address, data, etc.) together to make connectivity of components to buses in the EDK easier. For example:

```
BUS_INTERFACE BUS = SPLB, BUS_STD = PLBV46, BUS_TYPE = SLAVE
```

Generics/Parameters are equivalent to the generics and parameters in a VHDL entity or Verilog module. These are the generics located in the top-level VHDL file in our hardware core, my_test_core.vhd. Assigning a default value in the MPD file for a generic will overwrite the default value given in the VHDL file. By placing a generic in the MPD file, it exposes the generic to the MHS file to be modified by the designer in the EDK. The parameters are specified with a name, default value, data type, and additional parameters depending on the data type. For example, C_SPLB_DWIDTH:

```
PARAMETER C_SPLB_DWIDTH = 128, DT = INTEGER, BUS = SPLB, RANGE = (32, 64, 128)
```

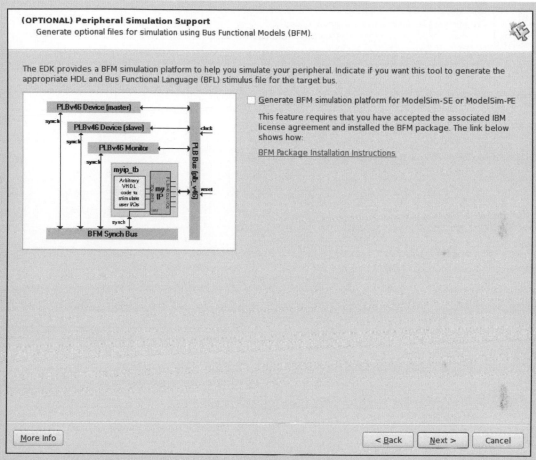

(OPTIONAL) Peripheral Simulation Support
Generate optional files for simulation using Bus Functional Models (BFM).

The EDK provides a BFM simulation platform to help you simulate your peripheral. Indicate if you want this tool to generate the appropriate HDL and Bus Functional Language (BFL) stimulus file for the target bus.

☐ Generate BFM simulation platform for ModelSim-SE or ModelSim-PE

This feature requires that you have accepted the associated IBM license agreement and installed the BFM package. The link below shows how:

BFM Package Installation Instructions

[More Info]   [< Back] [Next >] [Cancel]

Figure 2.46. Create/import peripheral wizard simulation page.

The priority of setting a component's generics is as follows:

1. set in the MHS File (highest priority)
2. set in the MPD File
3. set during instantiation of the component's generic map
4. set as the default value in the entity's generic listing

Ports, like generics, are used to connect the top-level entity to the rest of the system. In order to add an additional port in the top-level entity in a hardware core, it must also be added to the MPD file so that it will be visible to the EDK. The ports are specified with the port name, a default value (or connection to a bus interface), a direction, a vector size (if necessary), and a bus connection (if necessary). For example, PLB_wrDBus is the write data bus for the hardware core that is connected to the PLB.

```
PORT PLB_wrDBus = PLB_wrDBus, DIR = I, VEC = [0:(C_SPLB_DWIDTH-1)], BUS = SPLB
```

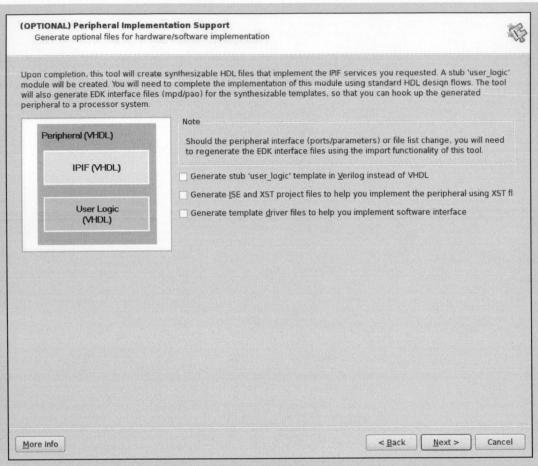

(OPTIONAL) Peripheral Implementation Support
Generate optional files for hardware/software implementation

Upon completion, this tool will create synthesizable HDL files that implement the IPIF services you requested. A stub 'user_logic' module will be created. You will need to complete the implementation of this module using standard HDL design flows. The tool will also generate EDK interface files (mpd/pao) for the synthesizable templates, so that you can hook up the generated peripheral to a processor system.

Peripheral (VHDL)

IPIF (VHDL)

User Logic (VHDL)

Note

Should the peripheral interface (ports/parameters) or file list change, you will need to regenerate the EDK interface files using the import functionality of this tool.

☐ Generate stub 'user_logic' template in Verilog instead of VHDL

☐ Generate ISE and XST project files to help you implement the peripheral using XST fl

☐ Generate template driver files to help you implement software interface

More Info   < Back   Next >   Cancel

Figure 2.47. Create/import peripheral wizard support page.

Peripheral Analysis Order File The PAO file contains a listing of all of the source HDL files, corresponding library, HDL type, and their synthesis order for the hardware core. If you notice the last two files listed are:

```
lib my_test_core_v1_00_a user_logic vhdl
lib my_test_core_v1_00_a my_test_core vhdl
```

These are the two main HDL files of our hardware core. The user_logic.vhd file is where the specific hardware core functionality will go. The my_test_core.vhd file is the top-level entity of the hardware core, hence it has the same name as the hardware core. The user_logic component is instantiated within the my_test_core component. For hardware cores connected to the Processor Local Bus (PLB), additional components may be instantiated within the my_test_core.vhd file to help connect the hardware core to the PLB. These components are generally known as the Intellectual Property Interface (IPIF), which coordinates the PLB signals to a simpler interface used by the user_logic known as the Intellectual Property Interconnect (IPIC). The IPIC signals are the

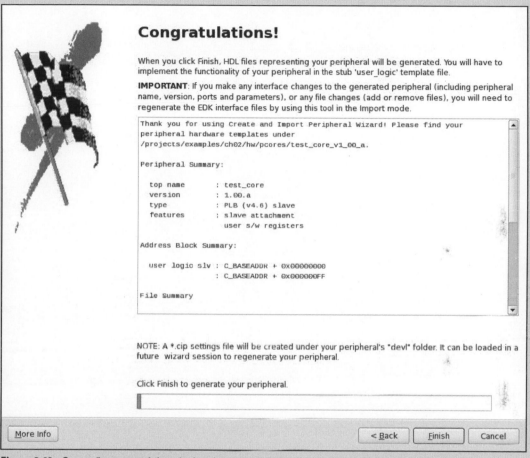

Figure 2.48. Create/import peripheral wizard summary page.

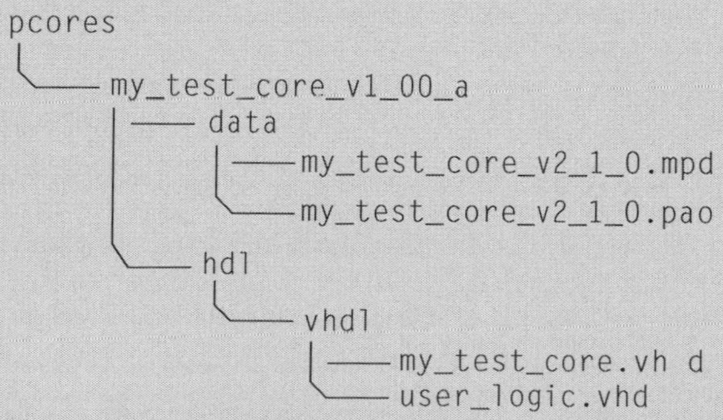

```
pcores
  └── my_test_core_v1_00_a
        ├── data
        │     ├── my_test_core_v2_1_0.mpd
        │     └── my_test_core_v2_1_0.pao
        └── hdl
              └── vhdl
                    ├── my_test_core.vh d
                    └── user_logic.vhd
```

Figure 2.49. Xilinx hardware core's pcore directory structure.

`Bus2IP_` or `IP2Bus_`, which denote the direction of the signal as either from the bus to the hardware core (an input) or from the hardware core to the bus (an output).

For now we will not modify the PAO file; however, if future hardware cores need to include additional HDL sources, the PAO file is how to tell the Xilinx synthesis tools to synthesize these files as well. We will revisit the `user_logic.vhd` file shortly to discuss modifying it to meet our design requirements.

Exercises

P2.1. If a designer exclusively used NMOS transistors in an integrated circuit NAND gate, what requires the most energy?

- maintaining a 0 on the output
- maintaining a 1 on the output
- transitioning from a 0 to 1 or a 1 to 0 on the output?

P2.2. If a designer used NMOS and PMOS transistors in a CMOS NAND gate, what requires the most energy?

- maintaining a 0 on the output
- maintaining a 1 on the output
- transitioning from a 0 to 1 or a 1 to 0 on the output?

P2.3. How does a fused (or antifuse) programmable logic device differ from an SRAM-based programmable logic device?

P2.4. What is the difference between using an array of AND and OR gates in devices such as PLAs or PALs and function generators in FPGA devices?

P2.5. How do the roles of SRAM cells and flip-flops differ in an FPGA device?

P2.6. Given the Boolean function

$$f_1(x, y, z) = xy + yz + xz$$

(a) draw the components of a 3-LUT, and
(b) show how the SRAM cells should be set.

Be sure to label all of the components including least significant bit and most significant bit of the MUX select lines and number its inputs.

P2.7. Given the Boolean function

$$f_2(x, y, z) = x + yz$$

(a) draw the components of a 3-LUT, and
(b) show how the SRAM cells should be set.

Be sure to label all of the components including least significant bit and most significant bit of the MUX select lines and number its inputs.

P2.8. How many different functions can be generated by a 3-LUT function generator?

P2.9. Are there any 3-input Boolean functions that are too complicated to be implemented in a 3-LUT?

P2.10. Can the function

$$f_3(w, x, y, z) = xy + yz$$

be implemented in a 3-LUT? Explain.

P2.11. Can the function

$$f_4(w, x, y, z) = xw + xw' + yz$$

be implemented in a 3-LUT? If so, how?

P2.12. Do CLBs and IOBs have flip-flops? Explain?

P2.13. How do CLBs and IOBs differ?

P2.14. Name three special-purpose cores found in Platform FPGA devices.

P2.15. Design a VHDL component that will compute the "next day of the week." Include in the comments the binary encoding for each day, "Monday," "Tuesday," and so on.

P2.16. Design a VHDL component that will repeatedly cycle through the Red → Green → Yellow pattern that one direction in a traffic intersection observes. The Green phase should last for three cycles, the Red phase should last for two cycles, and the Yellow should last for one phase.

P2.17. After synthesis, what is the difference between the ".ngc" file and the ".srp" file?

P2.18. In tool flow, what is the difference between the ".ngc" netlist and the ".ngd" netlist?

P2.19. After the netlist is mapped to LUTs and FFs, why is the PAR phase needed?

References

1394 Trade Association. (2010, January). *1394 TA specifications.* Also available at http://www.1394ta.org/developers/Specifications.html.

Agility. (2010, January). *Handel-C C for rapid FPGA implementation.* http://www.agilityds.com/products/c_based_products/dk_design_suite/handel-c.aspx, last accessed May 2010.

Ashenden, P. J. (2007). *Digital design: An embedded systems approach using verilog.* San Francisco, CA, USA: Morgan Kaufmann Publishers Inc.

Ashenden, P. J. (2008). *The designer's guide to VHDL* (3rd ed.). San Francisco, CA, USA: Morgan Kaufmann Publishers Inc.

Chow, P., Seo, S. O., Au, D., Choy, T., Fallah, B., Lewis, D., et al. (1991). *A 1.2 μm CMOS FPGA using cascaded logic blocks and segmented routing.* (Chap. 3.2, pp. 91–102). England: Abingdon EE&CS Books.

Compton, K., & Hauck, S. (2002). Reconfigurable computing: A survey of systems and software. *ACM Computing Surveys, 34*(2), 171–210.

Futral, W. T. (2001). *InfiniBand architecture: Development and deployment—a Strategic guide to server I/O solutions.* Hillsboro, OR: Intel Press

Grimsrud, K., & Smith, H. (2003). *Serial ATA storage architecture and applications: Designing high-performance, cost-effective I/O solutions.* Hillsboro, OR: Intel Press

IBM. (2009, December). *IBM CoreConnect.* Also available at `http://www-03.ibm.com/ chips/products/coreconnect/`.

Impulse Accelerated Technologies. (2010, January). *Impulse Accelerated Technologies.* `http://www.impulseaccelerated.com`, last accessed May 2010.

Open SystemC Initiative (OSCI). (2010, January). *Open SystemC Initiative (OSCI).* `http://www.systemc.org`, last accessed May 2010.

USB Implementers Forum (USB-IF). (2010a, January). *USB 2.0 Specification.* Also available at `http://www.usb.org/ developers/docs/`.

USB Implementers Forum (USB-IF). (2010b, January). *USB 3.0 Specification.* Also available at `http://www.usb.org/ developers/docs/`.

Xilinx, Inc. (2009a, December). *Command line tools user guide (UG628) v11.4.*

Xilinx, Inc. (2009b, October). *Embedded processor block in virtex-5 FPGAs (UG200) v1.7.*

Xilinx, Inc. (2009c, September). *XST user guide (UG627) v11.3.*

3

SYSTEM DESIGN

It is a distinguishing mark of a very good name that the plant should offer its hand to the name and the name should grasp the plant by the hand. . .

Carolus Linnaeus
Preface to Critica Botanica (1737)

Thirty years ago, when people thought of embedded systems, they primarily thought of custom computing machines built from electronic (discrete and integrated) circuits. In other words, most of the value of the work was in the hardware and hardware design; software was just some glue used to hold the finished product together. Yet with miniaturization of hardware, increases in speed, and increases in volume, commodity hardware has become very inexpensive. So much so that only high-volume products can justify the cost of building elaborate custom hardware. Instead, in many situations, it is best to leverage fast, inexpensive commodity components to provide a semicustom computing platform. In these situations, the final solution relies heavily on software to provide the application-specific behavior. Consequently, software development costs now often dominate the total product development cost. This is true despite the fact that software is so easily reproduced and reused from project to project.

Enter Platform FPGA devices. The embedded systems designer has a blank slate and can implement custom computing hardware almost as easily as software. The trade-off (there is always a trade-off) is that while configuring an FPGA is easy, creating the initial hardware design (from gates up) is not. For this reason, we do not want to have to design every hardware project from the ground up. Rather we want to leverage existing hardware cores and design new cores for reuse. This means we have to know what cores are already available, how these cores are typically used, how they perform, and how to build custom hardware cores that will be (ideally) reusable.

Another consequence of the shift from hardware to software is that system designers also have to understand much more about system software than the typical applications programmer, especially when designing for Platform FPGAs. Rather than simply writing applications that execute in the safe confines of a virtual address space found in a desktop system, we need to understand enough about system software (device drivers and operating system internals) to bridge the application code to our custom hardware. It is also necessary to understand what happens before the operating system begins along with the idea of using one platform (a workstation) to develop code for another platform (the embedded system). With commodity hardware, much of this complexity could be hidden from the average embedded systems programmer because the hardware vendor provided the tools and low-level system software. Not so when the hardware and software are completely programmable.

Therefore the aim of this chapter is to describe system design on a Platform FPGA. First we must discuss the principles of system design. Specifically, we address the metrics of quality design and concepts such as abstraction, cohesion, and reuse. Of course — as with many of the chapters in the text — whole books could be (and have been) written about each of these subsections. Our goal here is to provide enough of an introduction to address Platform FPGA issues. With a better understanding of design principles, we next consider the hardware design aspects, including how to leverage existing resources (base systems, components, libraries, and applications). Finally, the chapter concludes with the software aspects of system design. This includes the concepts of cross-development tools, bootloaders, root filesystem, and operating systems for embedded systems.

After completing the white pages of this chapter, the reader will have an abstract understanding of several central ideas in embedded systems design. Specifically:

- the principles of system design, including how to assemble Platform FPGA systems to be used in embedded system designs,
- the general classes of hardware components available to a Platform FPGA designer (and how to create custom hardware),
- the software resources, conventions, and tools available to develop the software components of an embedded system.

The gray pages of this chapter build on this knowledge with specific, detailed examples of these concepts in practice.

3.1. Principles of System Design

To manage the complexity of designing large computing systems we use a number of concepts. Abstraction, classification, and generalization are used to give meaning to components of a design. Hierarchy, repetition, and rules for enumeration are used to add meaning to assembled components. In this way, humans can develop software programs with tens of millions of lines of code, manage billion-dollar budgets, and develop multimillion gate hardware designs. This section focuses on some of the principles that guide good system design. This is far from an exact science and at times it is very subjective. The best way to read this section is to simply internalize the concepts, observe how they are applied to simple examples, and then consciously make decisions when you are building your own designs. In practice, it is difficult to learn good design skills from a textbook. It comes from experience and learning from others. Our goal here is to try to accelerate that learning by setting the stage and providing a common vocabulary.

3.1.1. Design Quality

To start we ask, what is "good" design? What is "bad" design? In short, system designs can be judged by many criteria and these criteria fall into one of two broad classes. ***External criteria*** are characteristics of a design that an end user can observe. For example, a malfunctioning design is often directly observable by the user. If the person turns up the volume on a Digital Video Recorder (DVR) and the volume goes down, then presence of the design flaw is obvious. However, there are many ***internal criteria*** that we also use to judge the quality of a design. These characteristics are inherent in the structure or organization of the design, but not necessarily *directly* observable by the user. For example, a user may not be able to observe the coding style used in their DVR but others (the manufacturer or a government procurement office) may be very interested in the quality of the design because it impacts the ability to fix and maintain the design. Clearly, some of these qualities can be measured quantitatively but many are very subjective.

A number of concepts and terms have been invented and reinvented in different domains and at different times. Hence, few of the terms that follow have universally accepted definitions. So when one author may use the term verification to casually mean that a system works with a set of test data, another author might call that validation (reserving that the term verification for designs has been rigorously proved to be correct).

The first set of terms is related to a system performing its intended function. The term *correctness* usually means the system has been (mathematically) shown to meet a formal specification. This can be very time-consuming for the developer but, in some cases, portions of the system must be formally verified (for example, if a mistake in the embedded system would put a human life at risk). Two other terms related to correctness are reliability and resilience. (In some domains, resilience is known as robustness.) The definition of *reliability* depends on whether it is applied to the hardware or software of the system. Reliable hardware usually means that the system behaves correctly in the presence of physical failures (such as memory corruption due to cosmic radiation). This is accomplished by introducing redundant hardware so that the error can be corrected "on the fly." Reliable software usually means that the system behaves correctly even though the formal specification is incomplete. For example, a specification may inadvertently omit what to do if a disk drive fills to its capacity. A reliable implementation might stop recording even though the specification does not state it explicitly. Because most complete systems are too large to formally specify every behavior, a reliable system results in designers making correct assumptions throughout the design. The last term, *resilience* (or robustness), is closely related to reliability. However, whereas reliability focuses on detecting and correcting corruptions, resilience accepts the fact that errors and faults will occur and the design "works around" problems even if it means working in a degraded fashion. In terms of software, one can think of reliability as doing something reasonable even though it wasn't specified. In contrast, resilience is doing something reasonable even though this should never had happened. Finally, there is dependability. This can be thought of as a spectrum, on one end is protection against natural phenomenon and on the other end malicious attacks. A dependable system shields the system from both. To help clarify the difference, consider the following three scenarios.

Correctness Example

As an example of correctness, consider the following example. Embedded systems are used in numerous medical systems and it is absolutely critical that the machine does not have any human errors in the design. This is usually accomplished by incorporating additional safety interlocks and formally proving the correctness of software-controlled, dangerous components. With the many different interacting components, one would describe all valid states mathematically and then prove that for all possible inputs, the software will never enter an invalid state. (An invalid state being defined as a state that couldharm the patient.)

Reliability Example

For most applications, formally describing all valid states can be enormously taxing (and itself error prone) so frequently designers fall back on informal specifications. Informal specifications often unintentionally omit direction for some situations. This can occur, for example, when a product gets used in a perfectly reasonable, but unexpected way. The specification might state that a camera needs to work with USB 1.1, 1.2, or 2.0. Assuming future versions of USB remain backwards compatible, a reliable design would not stop working if it was plugged into a version 3.0 USB hub. Likewise, if a Platform FPGA was intended to fly on a spacecraft, one would expect that the system will be more vulnerable to cosmic radiation. A reliable (hardware) design might put Triple Modular Redundancy on critical system hardware and periodically check/update the configuration memory to detect corruption.

Resilience Example

Resilience and robustness are different from reliability. These become very important in an embedded system because the computing machines interact with the physical world and the physical world is not as orderly as simple discrete zeroes and ones. For example, many sensors change as they age. Actuators are often connected to mechanical machines that can wear out. A resilient design behaves correctly even when something that is not supposed to happen, happens. For example, a thermometer connected to an embedded system may be expected to be in an environment that will never exceed 100° Celsius. If, because the sensor has become uncalibrated, the sensor begins to report temperatures such as 101, 102, or 103, for example, then the system should behave sensibly. A reliable system would try to fix the result coming from the sensor; a resilient system might treat 103 the same as 100 and continue.

In addition to these three system design characteristics, many other terms are used to judge the quality of a system design. For example, *verifiability* would be the degree to which parts of the system can be formally verified, i.e., proven correct. The term *maintainability* refers to the ability to fix problems that arose from unspecified behavior, whereas *repairability* refers to fixing situations where the behavior was specified but the implementation was incorrect. We tend to think of maintainability as the ability to adapt a product over its lifetime (version 1 followed by version 2 and so on). A repairable system design allows for easy bug fixes — especially once the product is in the field (upgrading a version 1 product to version 1.1). Along the lines of maintainability is the idea of *evolvability*; the subtle difference is the changes are due to new features (evolvability) versus previous changes to existing requirements (maintainability). Of course, ***portability*** (a system design that can move to new hardware or software platforms) and ***interoperability*** (a system design that works well with other devices) are important measures of design quality as well.

By themselves, these "-ability" terms do not have quantities associated with them. Nonetheless, being conscious of them during the development of an embedded system can be constructive. When the system developer needs to make a decision, these terms provide a common vocabulary to discuss, document, and teach design. For example, given two options, the system designer can record their reasoning — i.e., this option will be more portable and maintainable. This is critical in design reviews and helps teach less experienced designers why a decision was made. Lacking a written justification, a beginner might easily assume the decision was arbitrary. Or worse, without a concise way of describing decisions, the option not taken ends up not even being documented. The less experienced designer might not even realize a decision was made.

3.1.2. Modules and Interfaces

As suggested earlier, we are not going to be able to build very large systems by directly connecting millions of simple components. Rather, we use simple components to build small systems and use those systems to build bigger systems and so on. This is more commonly referred to as the ***bottom-up*** approach. We may also want to consider the design from a top level and work our way down defining each subcomponent in more detail, which is referred to as the ***top-down*** approach. Of course, these approaches are widely used in both hardware *and* software designs. (If one were to replace the last few sentences with subroutine or function in place of component, we could have just as easily been discussing the modularity of software designs.) Overall, these two approaches, or design philosophies, will be used throughout Platform FPGA design. The next few subsections will dwell on these in more detail.

First, we will use the general concept of a ***module*** to mean any self-contained collection of operations that has (1) a name, (2) an interface (to be defined next), and (3) some functional description. Note that a module could be hardware, software, or even something less concrete. However, for the moment, one can think of it as a subroutine in software or a VHDL component. We will expand on two key aspects interface and functional description, next.

There are two meanings to the term *interface*: usually one is talking about a formal, syntactical description, but for system design we have to also consider a broader definition. The term ***formal interface*** is the module's name and an enumeration of its operations, including, for each operation, its inputs (if any), outputs (one or more), and name. It may also include any generic compile-time parameters and type information for the inputs and outputs. The ***general interface*** includes the formal interface *and* any additional protocol or implied communication (through shared, global memory, for example).

Broadly speaking, the formal interface is something that can be inspected mechanically. So if two modules should interact, then their interfaces must be compatible and a compiler or some other automated process can check the modules' interactions. However, the general interface is not so carefully codified. Someone may think "how a module is intended to be used" and this cannot, in general, be checked automatically.

To make these concepts more clear, consider a pseudorandom number generator. (For those not familiar with the pseudorandom number generators, there are two main operations. The first "seeds" the sequence by setting the first number in the sequence. The second operation generates the next number in the sequence.) The formal interface might include two operations and a name drand48:

```
void srand48(long int seedval)
double drand48(void)
```

The general interface includes the way to use these operations; for example, the function called srand48 is used in order to seed the pseudorandom number generator and drand48 produces a stream of random numbers. How to interact with the module is part of its general interface, but is not formally expressed in a way that, for example, a compiler could enforce.

Other cases of general interface occur when some of the inputs or outputs of a module are stored somewhere in shared memory. For instance, a direct memory access (DMA) module will have a formal interface that includes starting addresses and lengths, but its general interface also includes the fact that there are blocks of data stored in RAM that will be referenced. To be more explicit, the DMA engine transfers a block of data, but is ignorant of its internal format.

A module can also include a functional description. The description can be ***implicit***, by which we mean the name is so universal that by convention we simply understand what the function is. For example, if a module is called a "Full Adder" (Figure 3.1), we do not need to say any more because the functionality of a full adder is well known.

The functional description can be ***informal***, which means its intended behavior is described in comments, exposition, or narrative. This is a very common way of describing a module: someone records the functionality in a manual page or some

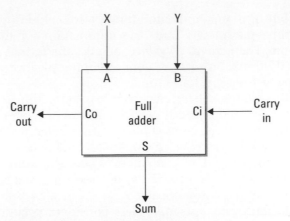

Figure 3.1. Implicit module description of a full adder.

document or in the implementation as comments. For example, when describing the full adder in narrative we could state:

```
The full adder component will add three bits together, X, Y and a
carry in bit. The addition will result in both sum and carry out
bits.
```

The functional description can also be ***formal*** where the behavior is either described mathematically (in terms of sets and functions) or otherwise codified, such as a C subroutine or a behavioral description in VHDL, for example:

```
-- Assign the Sum output signal
S <= A xor B xor Ci;
-- Assign the Carry Out output signal
Co <= (A and B) or (A and Ci) or (B and Ci);
```

Graphically, a module is very simple to denote, it can be as simple as just a box. If we wanted to be more specific, we can give a module a name, which, using a relatively new standard, is shown with a colon (:) followed by the name. In a design, we might want to distinguish between a module (a component in our toolbox) versus an instance (a component being used in a design). Instances (formally defined later) are shown with the module name underlined. Finally, we might want to have multiple instances in our design. If there is no ambiguity, we simply show multiple (underlined) boxes with the same module name, as is seen in Figure 3.2(a). However, if we want to make the distinction clear or if we need to refer to specific instances, we can give them names, illustrated in Figure 3.2(b). This is accomplished by putting a unique, identifying name to the left of the colon and module name.

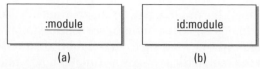

Figure 3.2. Two instances of a module system: (a) the default instance format and (b) the id to give the instance a unique name.

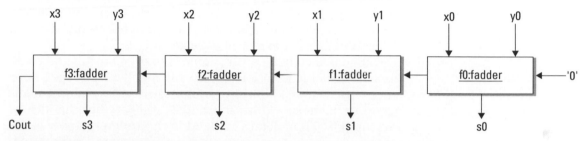

Figure 3.3. Four formally defined modules to generate a 4-bit full adder from 1-bit full adders.

Two more related terms are implementation and instance. An ***implementation*** (of a module) is some realization of the module's intended functionality. Of course, a module may have more than one implementation (Just like in VHDL where an entity may have more than one architecture). An ***instance*** is use of an implementation. In software, there is generally a one-to-one relationship between an implementation and an instance because the same instance is reused in time. However, in hardware it is common to use multiple copies of an implementation; each copy is an instance. (The verb ***instantiate*** means "to make a copy of an instance.")

Graphically, we distinguish an instance from a simple module by underlining the name in the box. If we want to highlight the fact that there are multiple instances of the same object, we can name the instances by labeling the module with an instance name, colon, module name. For example, Figure 3.2 show two instances of a module. Figure 3.3 is a simple example of four 1-bit full adders being instantiated to produce a 4-bit full adder.

3.1.3. Abstraction and State

Now we are ready to define two major concepts in system design: abstraction and state. These concepts are applied to the modules of a system and are described here. The dictionary defines abstract as an adjective to do with or existing in thought rather than matter (as in abstract art). The verb means to take out of, extract, or remove; in short, to summarize. An ***abstraction*** is the act or an artifact of abstracting or taking away; an abstract or visionary idea. Hence, a module is an abstraction of some functionality in a system. We will talk about wanting to make good abstractions — and we'll discuss the mechanics shortly — but, first, let's make sure we understand abstraction.

A module is a good abstraction if its interface and description provide some easily understood idea, but its implementation is significantly more complex. In other words, a good abstraction captures all of the salient features of an idea and cuts out all of the details unimportant to realizing the idea.

Our goal in creating abstractions is to overcome the fact that humans can only keep a relatively few number of things in our short-term memory. Typically, psychologists will say that we manage seven items at a time. Individuals will vary but the idea of keeping 100,000s of details on hand in short-term memory is not feasible. So what a good abstraction does is create an uncluttered, coherent picture of the module while shielding us from details that are not immediately relevant. If it is a bad abstraction, it forces us to think not about the module as a single entity, but rather makes us think about how it is implemented. We will see that a good abstraction is also important in reuse.

A great example of how a good abstraction can serve us comes from a subway map. A map of the London subway was first published in 1908. The map was rich in detail: it shows the exact route that the trains take, it is drawn to scale, it shows when the train is above ground, and it even shows rivers and other features.

However, take the perspective of a rider. If I walk up to the map, what am I looking for? Most likely, I want to know what train I need to get on. I am trying to get from point A to point B, and I need to know quickly (before my train departs!) which train I need to hop on. So information such as whether the train travels above ground or below ground doesn't matter to me. The number of turns a train takes or whether it travels east-west for a while is irrelevant to whether it gets me to point B.

In 1933, the London Underground map was changed. The new map created a bit of stir because a number of people felt like it was less informative. One might ask, what does it hurt to have *extra* information? The answer is subtle: by abstracting away much of the detail, the map could print the station names in a larger font. By not being true to the relative distances and physical locations of stations, the size of the map could be made smaller. Removing physical features, such as rivers, allowed the train lines to be drawn thicker. The results of these changes make the map much more readable. Functionally trumps literal correctness.

This is the goal of good abstraction: hide the details that do not serve a purpose. If the primary function is to be able to walk up to the map and make a fast decision about which train to get on, then readability is the most important feature. All of the extra information harms this function by distracting the user. So it is with reusable components.

Next, we want to consider another key concept: state. Hardware designers are very familiar with idea of state because it is explicit in the design of sequential machines and we can point to the memory devices (flip-flops) and say, "that's where the state is stored." However, in system design, it is a little less concrete because a module's state is stored in multiple places and in different kinds of memory devices (flip-flops, static RAM, register files, . . . even off-chip).

Formally, *state* is a condition of memory. We say something "has state" or is "stateful" if it has the ability to hold information over time. So an abacus and blackboard both have state. A sine wave does not have state. (Note: a sine wave is closely tied to the sense of time and it does not hold information over time.) In general, anything that is (strictly) functional does not have state. Most handheld calculators nowadays have state (they can keep a running total, for example). However, it is possible to build a calculator that does not have state.

Our interest in state has to do with identifying the state in a module. In the mechanics of system design, we will need to separate functionalities into modules, and abstraction and state are going to be major factors in how we derive a module. Because state in a module is not as explicit, the designer needs to consciously identify the states a module can be in and what operations might change that state. In short, good abstraction and careful management of state will lead to good modules and improve the design.

3.1.4. Cohesion and Coupling

The concepts are abstraction and state; the measures are cohesion and coupling. Cohesion is a way of measuring abstraction. If the details inside of a module come together to implement an easily understood function, then the module is said to have **cohesion**.

Coupling is a measure of how modules of a system are related to one another. A system's coupling is judged by the number of and types of dependencies between modules. Explicit dependence exists when one module communicates directly with another. For example, if the output signal of module A is the input to another module B, then we say that A and B depend on each other.

In software, if A might invoke a function in module B, then we say A depends on B (but B does not necessarily depend on A). The rule for determining dependence is "if a change to module A requires a designer to inspect module B (to see if the change impacts B), then B depends on A.

However, and this is where state comes into play, dependence is not always explicit. Two modules can be dependent in a number of ways. If two modules share state, then there is dependence. For example, if one module uses a large table in a Block RAM to keep track of some number of events and another module will occasionally inspect that table, the latter is dependent of the former. If someone wants to make a change to the format of the table, that change will impact the latter module. This is where explicitly identifying state becomes critical to the formation of modules within a system.

Dependence can crop up in even more subtle ways. Two modules may not have any shared state and may not explicitly communicate via signals or subroutine calls but are dependent. For example, they may be dependent because the system may rely on them completing their tasks at the same time. Hence, the system is coupled in time. Another more esoteric example might be in an embedded system that has sensors and actuators. Two modules may not communicate directly, but if one module is changing an actuator that another module senses, there may be a dependence. Dependence in itself is not bad. Indeed, some dependence is necessary between modules because, we assume, the modules are designed to work together to form the system. What we are interested in is the degree of coupling in the system — that is, the number and type of dependencies.

In general, explicit dependencies that arise from formal interfaces are the best forms of dependence, and a system composed of modules with a unidirectional sequence of dependencies will generally lead to good quality designs. However, if there are many implicit dependencies, circular dependencies (where one module A depends on module B and B depends on module A), or large numbers of dependencies, then chances are the design can be improved.

One way of reducing coupling in a system is through encapsulation. Encapsulation involves manipulating state and introducing formal interfaces. The idea is to move state into a module and make it exclusive (not shared). Often called "information hiding," it sounds overly secretive, but it is a very effective technique. One consequence is that if one wants to change the module, then there is much more freedom to do things like change the format of the state without the risk of introducing a bug into another

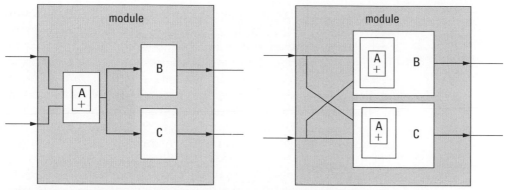

Figure 3.4. (a) Original design and (b) modified design with lower coupling.

module. If the module has good abstraction, then information hiding also allows the module be reimplemented in isolation. All that is necessary is to keep the interface the same.

Coupling is the result of dependencies between modules. Some coupling is inevitable. What kind of system is composed of modules that do not interact at all? The goal here is to avoid coupling when it is unnecessary. A number of techniques will manipulate the degree of coupling. For example, consider the two (very simple) designs in Figure 3.4.

In the first design of Figure 3.4(a), the two inputs to the module, x and y, are summed together in submodule A and the results are passed to the submodules B and C. In the second design, the summation is duplicated *inside* each of the submodules. In the first design, we had two dependencies — submodule B depends on submodule A and submodule C depends on submodule A. In the second design there are no submodule dependencies, so we have clearly reduced the coupling in the design.

What is the advantage? It may be hard to see the advantage in this example because submodule A is so simple and is unlikely to change. However, suppose submodule A was originally designed to work only on unsigned numbers. Over time, it was determined that submodule C also needed to be able to work with signed numbers. So a designer that is looking at modifying submodule C would necessarily have to change both C and A. However, if one changes A, then one *must* consider the effect on submodule B. Perhaps B will work fine with the change to A. However, the point is that systems with a high degree of coupling have this cascading effect where a simple change cannot be made in isolation. Rather, coupling forces the designer to consider the whole module and understand everything in order to ensure that a change does not break something.

There are two disadvantages to this change. One might argue that we have traded an explicit measure of quality (design size) for a subjective improvement in another measure of quality (maintainability). However, this change does not necessarily increase the design size! In this case, it is entirely possible that submodule A's functionality will simply merge into the CLBs already allocated by the submodules B and C. Hence, it is possible that there is no net gain in CLBs allocated. Of course, this is not always true and one has to weigh the costs — would the extra CLBs require a larger FPGA?

A second disadvantage to duplicating submodules in order to decrease coupling is that the designer now has to maintain the same component in two places. So if a bug is discovered in submodule A, then it is fixed once in the first design. However, in the alternative design, one has to fix the bug in both submodules B and C. (A word from the trenches: one person's bug might be another person's feature. So A may be performing in a way that doesn't match its description and B requires that it be fixed; however, module C may actually depend on the incorrect behavior to work!)

Wrapping up this discussion, it should be clear that many factors go into applying the design principles described. As this example shows, there are a number of trade-offs for even very simple designs!

3.1.5. Designing for Reuse

In addition to improving design quality, another use of these design principles is to make reusable components. With the increasing complexity of designs, it is in our favor to construct designs with the intention of their reuse. To get started, we must first understand what is necessary to create and identify reusable designs. One indicator is with high cohesion and low coupling, which leads to reusable design components. Note the hidden costs though:

- RCR — relative cost of reuse
- RCWR — relative cost of writing for reuse

Essentially, RCR says that you have to read the documentation and understand how to use the module before you can reuse it and RCWR says that someone has to put extra effort into designing a module for others to reuse (Poulin *et al.*, 1993). For example, when writing a C program to copy data, say 32 bytes of data, we could write our own `for-loop` to exactly copy the data. We could reuse this loop and possibly generalize it over time (copy words versus bytes, a variable size, or forward versus backward). In contrast, one could learn the "string.h" module in the standard C library. This module provides a rich set of data movement operations. These include operations such as strcpy, strncpy, memcpy, and memmove. The trade-off is learning how to use `strcpy` versus the time it takes to create your own copying function. In some cases, it could be easier to generate your own in place of learning a potentially complex component. This would suggest that the RCR of the module is fairly high and this discourages reuse. Another cost associated with reuse is RCWR, which is the cost associated with making your custom-created component fully reusable.

One way of managing RCWR is to take an incremental approach: design a specific component for the current design. If it is needed again, copy-and-generalize it. Over several designs, add the generality needed to make it a reusable component.

In VHDL, this can be done through introducing generics into the design. Moreover, one point of doing custom computing machines is to take advantage of specialization! So simply adding generality without leaving the option of generating application-specific versions through generics is counterproductive.

Refactoring is the task of looking at an existing design and rearranging the groupings and hierarchy without changing its functionality. Figure 3.4 illustrates refactoring. Often, it is done to make reusable components. The common use of refactoring is to improve some of the implicit and explicit quality measures mentioned in subsection 3.1.1.

One final word about testing. The value of reusable components is clear. But, of course, there is the danger that components might be refactored and accidentally change their functionality. *Regression testing* is used to prevent this. It usually is automated and might be simulation driven (à la test benches) or it may be a set of systems that wraps around the component and exercises its functionality. (Multiple systems are needed because one wants to also test all of the generics that are set at compile-time.)

3.2. Control Flow Graph

Throughout the text we use the idea of a software reference design. There are many ways to represent a system — from formal specifications to informal (but common) requirements, specification, and design documents to modeling languages such as UML. In addition to these representations, it is often common for a designer to build a rapid prototype — a piece of software that functionally mimics the behavior of the whole system, even hardware modules that have not been implemented yet. We refer to this software prototype as the ***software reference design***. The major drawback of software reference design is the cost associated with creating it but, as a specification of the system, it has a number of advantages. The first is that it is generally a complete specification. (If there is any question about how the future system is to behave, one can observe how it behaves in the reference design.) Another advantage is that it is executable — a designer can gather valuable performance data from running the software reference design with specific input data sets. Finally, because the software reference design is computer readable, the specification can be analyzed by existing software tools.

Through the next several chapters we will assume that a software reference design exists. Here we show how computation in software reference design can be represented mathematically. The next chapter uses this notation to help make decisions about what parts of the system should be implemented in hardware versus software.

We do this by borrowing some concepts from compiler technology — primarily the control flow graph. The control flow graph summarizes all possible paths of program from start to finish in a single, static data structure. Formally, a ***Control Flow Graph (CFG)*** is graph $G = (V, E)$ where the vertices (or nodes) V of the graph are basic blocks and the directed edges indicate all possible paths that program could take at run time. A ***basic block*** is a maximal sequence of sequential instructions with a Single Entry and Single Exit (SESE) point. Figure 3.5 illustrates this definition. The first group of instructions (A) is not a basic block because it is not maximal — the first instruction (store word with update) should be included. The second group of instructions (B) is a basic block. The last group (C) is not a basic block because there are two places to enter the block (at the store word instruction after the add immediate, or by branching to label .L2).

An edge (b_1, b_2) in a control flow graph indicates that after executing basic block b_1 the processor may immediately proceed to execute basic block b_2. If the basic block ends with a conditional branch, there are two edges leaving the basic block. If it does not end with a branch or if the last instruction is an unconditional branch, there will be a single edge out. Two special vertices in the graph, called *Entry* and *Exit*, are always added to the set of basic blocks. They designate the initial starting and stopping points, respectively.

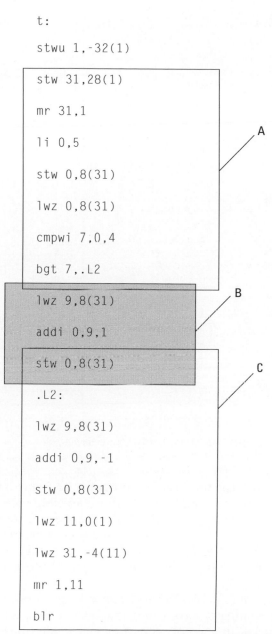

```
t:

stwu 1,-32(1)

stw 31,28(1)

mr 31,1

li 0,5

stw 0,8(31)

lwz 0,8(31)

cmpwi 7,0,4

bgt 7,.L2

lwz 9,8(31)

addi 0,9,1

stw 0,8(31)

.L2:

lwz 9,8(31)

addi 0,9,-1

stw 0,8(31)

lwz 11,0(1)

lwz 31,-4(11)

mr 1,11

blr
```

A

B

C

Figure 3.5. Groups of instructions; (A) and (C) are not basic blocks, (B) is a basic block.

Informally, a basic block sequence of instructions that we know, by definition, will be executed as a unit. The edges in the CFG show all the potential sequences in which these units are executed. For example, the simple subroutine shown in Figure 3.6(a) has

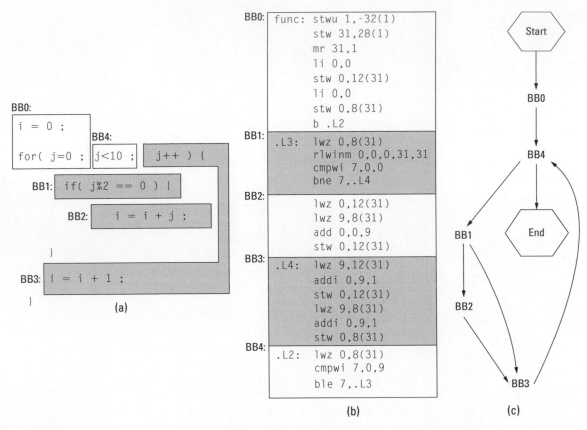

Figure 3.6. Basic blocks in (a) C source code (b) translated to assembly (c) a control flow graph.

identified the basic blocks in a C program. In Figure 3.6(b), the C program has been compiled to PowerPC assembler code and the basic blocks have been identified. Finally, the control flow graph is illustrated in Figure 3.6(c). Note that it is possible to identify the basic blocks in a C file if one knows how the compiler emits assembly code. Unless it is obvious, we use assembly code to illustrate basic blocks.

Compiler researchers and developers use the control flow graph in a number of ways. Often, graph algorithms applied to the CFG will result in a set of properties that guide transformations (optimizations) designed to improve the performance of the program. When combined with data dependence (see Chapter 5) the CFG can be used to determine what parts of the program can be executed in parallel (and thus implemented spatially in hardware). For now, our immediate need is to visualize the software reference design. The next chapter uses the basic blocks as the atomic unit that can be partitioned between hardware and software.

3.3. Hardware Design

Thus far, the discussion of system design has been high level and general. Next we turn toward hardware design and, very specifically, the hardware components available to the Platform FPGA designer. We begin with a brief description about how these common architectural components evolved and then describe several of the broad classes of hardware modules available. This section ends with a general description of how the designer can expand their toolbox with custom hardware modules.

3.3.1. Origins of Platform FPGA Designs

Simply put, designers rarely want to build an embedded system from scratch. To be productive, an embedded systems designer will typically begin with an existing architecture, remove the unneeded components, and then add cores to meet the project requirements. The processor-memory model, seen in Figure 3.7, which is basic desktop PC architecture, has worked well as a starting point. To begin, we briefly review some key computer architecture components so we are able to understand and use them in our Platform FPGA designs.

The introduction of the IBM Personal Computer (PC) in 1980 had an enormous impact on the practice of building computing machines. The intent was to make a system that would appeal to consumers and hobbyists alike; therefore, low-level details of the system were readily available. This spurred third-party development of peripherals and (probably unintentionally) compatible machines from other manufacturers. As the speed of the microprocessor, volume of machines, and competition increased, the cost actually decreased. It became possible for manufacturers and vendors to try different computer architecture designs. Ultimately, the architecture that has evolved is what is common in today's desktop computers.

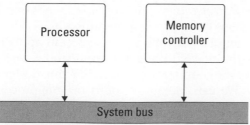

Figure 3.7. The fundamental process-memory model to be used as a base system in Platform FPGA designs.

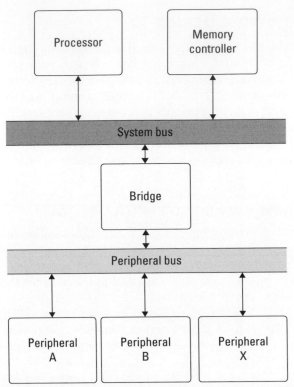

Figure 3.8. The two bus process-memory model used to support parallel, independent high-speed and low-speed communication.

Later computers use a two-bus system where the processor and memory reside on a processor-specific system bus and the lower speed peripherals (serial ports, printers, video monitor) reside on a generic, standard peripheral bus. Figure 3.8 shows this arrangement, which allows the system components to evolve rapidly in terms of clock frequencies, voltages, and so on while maintaining compatibility with the third-party peripherals, which do not change as quickly.

Embedded computing architectures have not changed much from this basic arrangement; in fact, this foundation has allowed designers to focus on improving the individual components. The bus model is arguably insufficient for certain application needs, but it serves the needs of general applications well. This organization makes a good starting point for our designs with Platform FPGAs. We can utilize the hard or soft processor core(s), on-chip memory and off-chip memory controllers, and peripherals to support system input and output to build a base system that resembles these computer organizations.

Platform FPGAs have adopted these two basic processor-memory model from the desktop computer architecture because it provides an established framework that can be built upon for custom designs. With the addition of existing components and cores, more complex systems can be constructed, often within a considerably shorter time frame than traditional embedded systems designs. In fact, in section 3.A you already assembled a simple Platform FPGA system when building the "hello world" FPGA example. Within this section we aim to go into more detail on what comprises a Platform FPGA base design (system). We use the the basic computer organization processor-memory model and expand on it to work up to a useful base system.

A valid question to ask at this point is "why create a generic base system when we are using FPGAs?" FPGAs by their very nature are programmable and application specific. In an ideal world where a designer's time could be infinitely spent on the project, there were no deadlines, and money was no object, creating completely custom designs would make sense. Unfortunately, the ever-growing demand for first-to-market solutions requires designs to be up and running and brought to market quickly. FPGAs offer an additional advantage of providing field programmability to the mix, allowing a potentially less than ideal solution being initially offered, then updated in later revisions.

3.3.2. Platform FPGA Components

Chapter 2 discussed the components that exist in an FPGA, such as logic cells, blocks, and on-chip memory. While these components are useful in all FPGA designs, they will be the building blocks for much larger systems. With the ideas of modularity, cohesion, and coupling of components and designs, we want to begin to build *base systems* that can be used and reused as a starting point for embedded systems design. We have already discussed the strengths of this approach and introduced the prevalent organization with the processor-memory model. Because each design is different, it is obvious that the design will require modifications, but the processor is a good place to start.

Processor

Generally speaking, the processor offers the designer control and a familiar design environment. Even if the final design will require little or no involvement from the processor, its use within the rapid prototyping or early development stages can help rapidly evolve the design. For us, the processor is an obvious starting point when describing and building a processor-memory model design.

In a Platform FPGA, two types of processors can exist, hard and soft core processors. Chapter 2 discussed the hard processor cores in detail and even gave examples of PowerPC 440 integrated in the FPGA fabric on the Xilinx Virtex 5 FX series FPGAs.

Other Platform FPGAs provide sufficient reconfigurable resources that a soft processor core can be implemented in the logic blocks and surrounding resources of the FPGA. Soft processors offer a great deal of flexibility, as they are by their very nature configurable. Unlike hard processors, where functions have been fixed since fabrication, incremental improvements to a soft processor (such as the more recent addition of the memory management unit to the Xilinx MicroBlaze processor) provide the designer with a more flexible design approach.

While this could quickly turn into a long discussion on the advantages and disadvantages of processors in hardware versus software, we focus instead on the processor's capabilities. For instance, even the most basic processors, requiring a minimal amount of resources (i.e., PicoBlaze), can operate in what is called *stand-alone* mode, offering only the most basic functionality (such as stdin/stdout). More advanced processors may include memory management units (MMU) to support more verbose operating systems (such as Linux). There are even processors that offer coherency with shared memory between multiple processors to create a multicore processor (similar to what exists in commodity desktop PCs).

Overall, knowing the processor's role in the application can help dictate which processors can and cannot be used. For instance, the PicoBlaze is well suited for more complex state machines, but not for running Linux. Likewise, soft processors may offer wider flexibility when migrating from one FPGA device family to another (or even to a different FPGA vendor). Before choosing which processor will be the cornerstone of the design, consider the following questions:

- Does the FPGA include hard processors?
- Are there sufficient resources to implement a soft processor?
- What role will the processor play in your design?
- What type of software will be used on the processor?
- How much time will the processor be used versus hard cores?
- Are there future plans to using a different FPGA?

Some of these questions may be easier to answer than others, but being able to address them before moving too far along the design process is important. Chapter 4 helps address questions regarding identifying suitable functions to be implemented in hardware versus software. This chapter is more concerned with construction of the platform and augmenting it to meet the design's needs.

If the FPGA does include a hard processor core(s), the initial design might be best suited to include its use rather than expend additional resources on a soft processor. If, however future generations of the system will include different FPGAs, it might make more sense to use a soft processor core that can be moved between FPGAs with as little effort as possible. Vendors and developers of soft cores should be able to provide enough information for a design to determine if it is feasible to include in the chosen FPGA.

Memory

In order for the processor to do any useful work, memory must be included to store operations and data. Different computer organizations and memory hierarchies could be discussed at this point. Whether the processor follows the Von Neumann or Harvard architecture or contains levels 1, 2, and 3 cache is arguably too low level for embedded systems designers. Instead, we focus on the following questions:

- What type of memory is available?
- Is there on-chip and/or off-chip memory?
- How much on-chip/off-chip memory is available?
- Is the memory volatile or nonvolatile?
- How does the processor interface with the memory?
- How does the system interface with memory?
- How do specific cores interface with memory?
- Is the memory being utilized efficiently?

As mentioned in Chapter 2, modern FPGAs include varying amounts of on-chip memory (often referred to as block RAM). The uses of this memory are wide and varied based on the application. The memory can be included within a component, core, or as part of the base system. The location of the memory dictates its interface and accessibility. For example, say a custom core includes a FIFO built from on-chip memory, it may have a standard FIFO interface (enqueue/dequeue), which only the custom core can access, or it may be accessible to a processor as a way of loading data into the custom core to be operated on (such as a single precision floating point addition core). When designing systems with on-chip memory needs, it is important to identify how it will be used within the system.

In the event the design requires more memory than is available on-chip, off-chip memory is required. There are many different forms of off-chip memory and knowing which type to use is a difficult decision that goes beyond the scope of this chapter. However, interfacing with the particular memory is important to address now. A *memory controller* is required to handle *memory transactions*, requests for access to memory in order to read or write data from a specific address. The memory controller is a soft core that can be configured to meet the specific needs of the memory it is interfacing with. For example, each DDR2 DIMM has specific operational characteristics that require complex state machines to interface with it. Fortunately, for the designer, many of these memory controllers have already been designed with generic parameters to allow for quick integration with different memory manufacturers. Within FPGA designs it is possible to include processor-centric memory access, where the processor issues all requests on behalf of the system, or to include Direct Memory Access (DMA), where cores within the system can request memory directly.

For both on-chip and off-chip memory it is difficult to provide strict design rules, as they can be used in such a variety of ways. However, utilizing them efficiently is of critical importance because the rate at which memory is increasing lags behind the processor (Wulf & McKee, 1995), and with the addition of multiple sources contending for a single resource, this demand for memory only further exacerbates the problem. Chapter 6 covers how to tackle memory bandwidth management questions more efficiently. Of key importance is configuring the system to tightly integrate the components needing low-latency access to memory and separating them from components that may not access memory as frequently or at all.

Buses

Now that we have described the two main components in a processor-memory model design, we must start to address the various ways to connect them. The simplest approach (beyond a strict point-to-point interface) is to provide a *bus*. The processor(s) and memory controller(s) connect to the bus via a standard bus interface. The bus interface is specific to the particular bus, but at the simplest level consists of address, data, read/write requests, and acknowledgment signals. The bus also includes a bus arbiter, which controls access to the bus. When a core needs to communicate with another core on the bus, it issues its request for access to the bus. Based on the current state of the bus, the arbiter will either grant access or deny access, causing the core to reissue its request. A core that can request access to the bus is considered a *bus master*. Not all cores need to be bus masters; in fact, many custom cores are created as *bus slaves*, which only respond to bus transactions.

A bus on an FPGA is implemented within the configurable logic, making it a soft core. For example, Xilinx uses IBM's CoreConnect library of buses, arbiters, and bridges. More details regarding the CoreConnect library are presented in section 3.A. Some important design considerations need to be addressed when using buses.

- What cores will need to directly communicate?
- Do certain cores communicate more often than others?
- Do specific cores require a higher bandwidth between them?
- Can any cores function on a slower bus?

As mentioned earlier, it is common to find a two-bus system in desktop computers. This is done to isolate the lower speed peripheral devices from higher speed devices (such as the processor and memory). In system design, it may be advantageous to put certain cores on one bus and others on a separate bus. By adding a bridge between the two buses, it is possible for cores to still communicate, although at the cost of additional latency.

System Bus

In multiple bus designs the bus with the highest bandwidth that connects the processor, memory controller, and remaining high-speed devices (such as a network controller) together is often referred to as the *system bus*. Xilinx uses IBM CoreConnect's Processor Local Bus (PLB) as its equivalent system bus. When the number of cores needing access to the bus is relatively small, connecting all of the cores on a single bus is a logical, resource-efficient decision. In Platform FPGA designs, the system bus is the fundamental mechanism to communicate between the processor and custom hardware core. As the number of hardware cores grows, a decision must be made as how to most efficiently support these additional cores in combination with the already existing system. One solution is to introduce a second bus.

Peripheral Bus

A second bus may be added to separate the design into different domains. In some cases this is done for high-speed and low-speed designs. In others, it may be to provide a subset of the cores with a dedicated bandwidth for communication. In either event, addition of a second bus, often known as the *peripheral bus*, allows two arbiters to control communication across the two buses. With a single bus, if the processor was accessing data from the memory controller, any other cores needing to communicate would be required to wait for the memory transaction to complete. In a two-bus system, those cores could be migrated to the peripheral bus and allowed to communicate in parallel.

Bridges

In some cases it is necessary for a core on the system bus to communicate with a core on the peripheral bus. This requires the addition of a *bridge*. A bridge is a special core that resides on both buses and propagates requests from one bus to another. A bridge functions by interfacing as a bus master on one bus and a bus slave on the other. The slave side responds to requests that must be passed along to the other bus. The master side issues those requests on behalf of the original sender. Sometimes only a single bridge is required; if the peripheral bus only will respond to requests from the system bus, a system-to-peripheral bridge is required. However, if cores on the peripheral bus need access to the system bus (say for access to the off-chip memory controller), then a second bridge, a peripheral-to-system bridge, is required. The common nomenclature for describing the interfaces of a bus in terms of master side and slave side is as follows. The system-to-peripheral bridge means that the bridge is a slave on the system bus and a master on the peripheral bus. This may seem backward, but the reason is quite simple. In order to communicate from the system bus to cores on the peripheral bus, the bridge must respond to system bus requests (making it a slave on the system bus) and issue the request on the peripheral bus (making it a master on the peripheral bus).

Peripherals

Now that we have established a mechanism to connect cores together we should address some of the peripherals a system designer may add to a design. When we talk about peripheral cores, we usually are referring to hardware cores that interface to peripheral devices, such as printers, GPS devices, and LCD displays. Peripherals themselves are traditionally the components around the central processing unit. In our case, some peripherals (such as a video graphics adapter) may be entirely implemented in the FPGA, but often the peripheral is external to the FPGA and the hardware core provides the interface.

In Chapter 7, we dedicate the whole chapter to interfacing with the outside world. Here we simply mention common peripherals found in Platform FPGA designs.

A number of high-speed communication cores have been implemented as FPGA cores. There is a PCI Bridge and a PCI Arbiter — the former is needed to connect the FPGA's system bus to an existing full-function PCI bus, whereas the latter includes the logic to create a full-function PCI bus. A variety of Ethernet cores are available for connecting the Platform FPGA to a (wired)

Ethernet network. Likewise, a variety of USB cores provide support for different versions and capabilities. Many of the older (low-speed) communication cores have been implemented as well, including UARTs, I2C, and SPI.

As part of the principles of system design, building cores for reuse leads to the eventual accumulation of a library or repository of cores. These cores may provide functionality that many of the base systems need. For example, more and more cores are requiring Internet access whether it be for debugging purposes or to update an embedded system's database; having a core that can be integrated quickly into a design that has been tested previously reduces design time significantly.

Building Base Systems

Enough talk, it is time to put these concepts together and build a simple base system, consisting of a process, on-chip memory, off-chip memory, and a UART for serial communication with a host PC. We are being a little generic still in terms of the actual core, but in section 3.A we will be more specific with respect to Xilinx FPGAs. Still, we have a processor, two types of memory, and a UART. We have yet to mention what would be used to connect these components together because that is a little more application specific. In some cases it makes sense to separate the high-speed and low-speed devices onto different buses.

In this example, there would be no immediate need for such a separation. Remember that this is the base system from which larger designs will be built. As a result we want to be flexible enough in our design to allow custom cores to be added without requiring significant changes to this base system. For that fact, we will include both a system bus and a peripheral bus. Figure 3.9 depicts this initial base system. Notice that with the addition of buses, we need to include a bridge. Because the UART is a slave on the peripheral bus, there is no requirement for adding a second bridge to allow it to master the system bus. If future designs require this, we can go back and add the peripheral-to-system bridge.

While drawing boxes suffices for an academic approach, it is insufficient for practical implementations. We would not build this base system with schematic capture software because the number of signals and wires to connect would be enormous. Instead, we use hardware description languages. Using the bottom-up design approach, we could create a single HDL file and instantiate each of the components within it.

Not only do we need to connect all of the components input/output signals, we need to connect the input/output signals

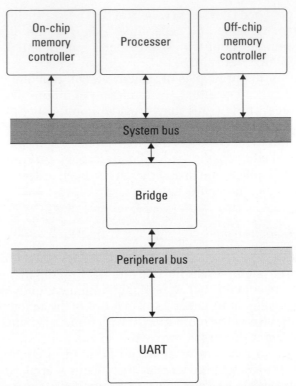

Figure 3.9. Block diagram of the base system consisting of a processor, on-chip and off-chip memory, and a UART.

that go off-chip. For example, the UART includes transmit and receive pins that are routed off-chip to a RS232 IC to provide serial communication between the FPGA and the outside world. This requires additional constraints to be set to notify the synthesis tool that a signal is to be routed to a specific pad and pin.

In practice, even this amount of work is inefficient. FPGA and software tool vendors provide GUIs or wizards to help automate this process. For the beginning designer, the tools are a great starting point because they help the designer identify key components and how they are connected. For the more experienced designer, the tools and wizards may prove to be less useful.

3.3.3. Adding to Platform FPGA Systems

Now that we have a base system, let's go ahead and add some custom compute cores. Many embedded systems devices are now including some form of network interface, whether it be wired or wireless. For demonstration purposes, we will add a TCP/IP 10/100/1000 Mbit Ethernet network core. The network core will

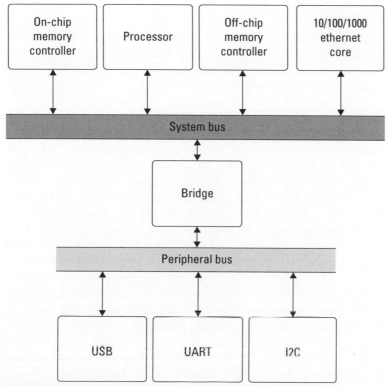

Figure 3.10. Block diagram of the base system with the additional cores: networking, USB, and I2C.

provide us with access to the FPGA from anywhere in the globe via a Web interface. Adding this core to the base system can be as simple as adding the instance to the top-level HDL file and updating the pin constraints file. We will add the network core to the system bus because as data are transferred to and from the FPGA, we will use the off-chip memory as a temporary storage buffer. Figure 3.10 is the modified block diagram of this base system. In addition to networking, we have also connected a USB and I2C interface to the peripheral bus just to start to round out the theoretical design.

The benefit of using the bus (or two-bus) within the design is it allows the system to be modified (adding or removing cores) with ease. As long as the new core has the same bus interface, typically a standard such as the PLB that is published and widely available, a systems designer just needs to connect up the signals. In the event that the new core is a slave on the bus, the new core will need to be assigned a unique address range from the system's address map. The address range is how other cores on the bus will communicate with the new core.

Address Space

From a hardware perspective, the "location" of data is straightforward and easily identifiable. Off-chip memory is physically located outside of the FPGA's packaging, typically as a separate memory module or, more commonly, as a Dual In-line Memory Module (DIMM). From a design perspective, accessing the memory is nothing more than through the memory's address space. Off-chip memory may be addressable from the address range 0x30000000–0x3FFFFFFF, for a total of 256 MB of addressable data. Globally addressable on-chip memory, located on the same bus as the off-chip memory controller, may have an address range 0xFFFF0000–0xFFFFFFFF, for a total of 64 KB of addressable data. In Platform FPGA designs it is possible to automatically generate these address ranges or to set specific address ranges.

For on-chip memory that is only local to a compute core, the address range is user defined and commonly word addressable (instead of byte addressable). The address range is also important for any compute cores that must communicate with the processor. We mention this information here because up until now we have not interacted with a range of compute cores or memory. Figure 3.11 shows that address map for the two-bus base system mentioned previously. Each core that has an address range is at least a slave on the bus. The processor is a bus master only and therefore does not include an address range within the address map.

In the two-bus system, the bridge acts as a middleman between requests issued from the system bus to the peripheral bus. The bridge must be assigned an address range that will span all of the cores on the peripheral bus. If the bridge is given the incorrect address range, requests may never make it to the peripheral bus or to the destination hardware core.

3.3.4. Assembling Custom Compute Cores

In Chapter 2, hardware description languages were introduced with a few examples to help the reader grasp some of the concepts. We also covered how to use existing tools and wizards to create components and custom core templates. Now it is time to cover how to design and assemble custom compute cores. While there is a large body of writing on how to design digital computing systems, both manually and automatically, we look at the process from a more systematic engineering approach. To start, we want to answer the age-old question of "why build custom compute cores?" Once answered we discuss design approaches, consider design rules and guidelines, look at how to test and

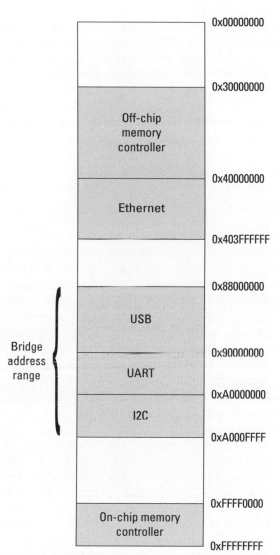

Figure 3.11. One possible address map for the theoretical two-bus system with networking, USB, I2C, and UART.

debug hardware, and finally culminate with a functional custom core.

Why Build Custom Cores?

We begin with the most important question, "why build custom cores?" It is widely believed that hardware development is difficult and because there are more software professionals in the workforce than there are hardware engineers, some might argue to use processors and build software. Furthermore, processors are inexpensive and cost-effective, and processor manufacturers put an enormous amount of design effort into a piece of hardware that will ship many millions of units over its lifetime.

Often, the immediate response to the question "why design hardware?" is "for performance" and by performance the speaker typically means "speed." This is true; however, there are other compelling reasons to implement hardware as well. These include computational efficiency and predictability. We'll look at each of these reasons in detail because it is important for a hardware designer to know when a hardware solution is and is not justified.

Advantage: Speed

Because custom hardware designs are often used to speed up applications, some designers will occasionally make the mistake of generalizing that "hardware is faster than software." However, the idea that hardware is faster than software is a fallacy. In fact, if naively implemented, hardware is often *slower* than software. Moreover, it is true that any hardware design implemented in an FPGA will perform $5\times$ to $10\times$ slower (and consumes more area on a chip) than the same circuit implemented directly in silicon (using the equivalent process technology). If the design we happen to implement is similar to a processor, then we gain nothing and lose much in speed (and area). So how does FPGA hardware outperform a processor?

Practically speaking, there are two reasons why some FPGA designs have a performance advantage plus a couple of minor reasons. The first practical advantage is rooted in the execution model. The sequential computing model of the standard von Neumann processor creates an artificial barrier to performance by expressing the task as a set of serial operations. With hardware, the task's inherent parallelism can be expressed directly. To compensate for its inherently serial operation, modern processors commit a significant portion of the hardware resources to extracting the instruction-level parallelism to increase their speed, which we will revisit shortly when we discuss efficiency. Although less significant overall, another way that the execution model can impede performance is in instruction processing. The processor model has to commit resources to fetching, decoding, and dispatching instructions, all of which are functionality that is commonly part of the hardware design. For some applications, memory bandwidth limits the performance and part of the bandwidth is consumed by instructions being fetched from memory. In hardware designs, the instructions are implicitly part of the design.

The second practical reason FPGAs have been able to outperform standard processors has to do with specialization. In general-purpose processors, data paths and operation sizes are organized around general requirements. This typically means 32- or 64-bit buses and functional units. Thus, to multiply an integer by some constant c requires a full-sized multiplier in a processor. However, if additional information about the application is known, an FPGA-based implementation can be created with customized function units to meet the exact needs of the application. For example, a constant multiplier can be orders of magnitude faster than a general-purpose multiplier.

Although, run-time reconfiguration is possible, it is currently not in widespread use. Nonetheless, FPGAs can use this technique to outperform a general-purpose processor by using information only available at run time to produce specialized circuits. For example, a circuit that computes Triple-DES (an encryption algorithm) can be several orders of magnitude faster if the key is known in advance. This particular example has been demonstrated elsewhere (Hennessy & Patterson, 2002). Unfortunately, building run-time reconfigurable designs is a challenging, time-consuming process. Until design tools and methodologies mature and become easier to use, it is unlikely that this important source of improved performance will become common. The final chapter discusses run time reconfiguration in more detail.

Advantage: Efficiency

Suppose a hardware implementation of task A takes exactly the same amount of time to complete as the equivalent task executed on a processor. We can assume that the software implementation is easier to develop. Is there any reason to build a hardware implementation? The answer is yes when the hardware solution is more efficient. By efficiency, what we are concerned with is how to accomplish a fixed task with a variable amount of resources. By resources, we could be talking about area on a chip, the number of discrete chips, or the cost of the solution. While speed is a predominant reason to commit to a hardware design, efficiency is still a valid reason. A hardware design plus a processor is often more efficient than two processors.

For example, suppose we have a network interface that implements a standard protocol (such as TCP/IP over Ethernet). If we needed to augment an existing computer (that is already loaded to its capacity) to handle network events, then the two options might include adding another processor dedicated to network traffic and building a custom network interface that offloads the network tasks. If both approaches meet the minimum criteria, then we say the more efficient solution is the one with a lower cost. If the system is being deployed on a single chip, then the more efficient solution is the one that uses less area.

Advantage: Predictability

While efficiency is an important consideration, it is often the case that a processor is not being used to its full capacity. Thus, someone might argue that the processor can simply multitask in new functionality. Even if this does not overload the processor, there is another compelling reason to use a hardware implementation. This case arises in embedded systems where timing constraints are very important. When there are real-time demands on the system, scheduling becomes important. Hardware designs have the benefit of being very predictable, which in turn makes scheduling easier.

So, there are cases when it makes sense to move a task to hardware if it makes that task more predictable or if it makes scheduling tasks on the processor easier. For real-time systems, where the goal is to satisfy all of the constraints, predictability is often more valuable than simply making the system faster.

Disadvantages

Perhaps the biggest disadvantage to building hardware solutions is one already mentioned, the development effort required. Compared to the numbers of professionals that can code software, there is a small number of hardware designers. Moreover, most people will assert that designing hardware is more difficult than coding software. Certainly, for large designers, there are more details that one has to attend to. So, unless there is a compelling reason (in terms of the performance metrics from Chapter 1 or the advantages just mentioned), then it may not be worth the extra effort. A second disadvantage is the loss of generality. It is simply the nature of our modern world that product requirements will evolve over time. The loss of generality has the potential of negatively impacting the design over time.

In summary, Platform FPGAs offer speed, efficiency, and predictability advantages over software-only solutions — compelling advantages for many emerging embedded computing system projects. However, there is no universal answer to the question. As a Platform FPGA designer, part of your task includes determining when a simple microcontroller is appropriate.

Design Composition

In general, there are two ways of building digital computing machines. The first is the one that has been traditionally covered in most sophomore-level digital logic courses, which begins with logic gates. The second, which is sometimes covered in later courses, starts with higher level logic blocks from which complex systems are composed.

In the first approach, the designer begins with requirements that are translated into a specification (expressed in various forms such as Boolean expressions, truth tables, and finite state machines). From there, all of the various formal techniques are used to reduce the number of states, minimize the Boolean functions, and realize the machine in the fewest number of components. When we say "built from gates" this is the approach we are talking about.

In the second approach, the designer starts with logic blocks that have a predetermined functionality — such as decoders, n-to-1 multiplexers, and flip-flops. These components are selected by the designer and are arranged creatively to meet the requirements. Logic gates may be used but their need is diminished by the functionality provided by the other components. The second approach is what we will focus on for the remainder of this book. There are many reasons to use the first approach; however, for practical designs utilizing millions of gates, maintaining the design becomes a daunting task that is simplified by a more *modular design approach*.

Generally speaking, there are three steps when designing modular custom cores. The first step is to identify the inputs and outputs of the core. In some designs the inputs and outputs are already set based on the functionality of the system. For example, in a bus-based system the inputs and outputs are initially fixed to at least the bus signals. Additional signals may be added based on the design (i.e., connecting an interrupt signal from the core to the processor). These signals may change through the design process, but establishing a solid interface to a top-level component will greatly assist in not only the component's composition, but aid in the design of any components that eventually will use this core.

The second step is to identify the operations and compose a data path, usually a collection of multistage computations (i.e., a pipeline). Each component is designed with a particular function in mind. The exact operations needed may not be as clear during the beginning of the design phase, but determining the necessary low-level functionality (or subcomponents) allows for the construction of a data path. The data path represents the flow of data through the component. Once the flow has been established it becomes possible to construct a computation pipeline. A pipeline in hardware contributes to the performance and efficiency of the design. Capturing the stages of the pipeline may initially be difficult, but starting a design with the concept of supporting pipeline operations makes the process much more manageable.

The third step is to develop a controlling circuit that sequences the operations, usually a finite state machine. We often think of hardware in terms of parallelism, that is, independent operations that can be executed at the same time. Parallelism is one of the keys to achieving speedup over processor-based designs. However, many designs still require computations in some sequential flow. Consider a simple equation:

$$f(x, y, z) = x * y + (4 * z)$$

It is possible to build hardware to multiply $x * y$ in parallel with the multiplication of $4 * z$, but addition of the two results must wait for completion of the multiplication. A finite state machine can be used to control the computation by first performing the two independent (parallel) multiplications and then performing the addition.

Bottom-Up and Top-Down Design

Earlier we mentioned two design approaches, bottom up and top down. In many FPGA designs, the bottom-up approach is used when assembling systems, as described previously. This same approach can be used for assembling custom compute cores. For example, when using the structural HDL method, each component is built by instantiating subcomponents. Before the top-level design can be completed, each of the subcomponents must be designed and tested. In this approach, modularity and designing for reuse are very important.

In the *bottom-up* approach, each subcomponent can be treated as a *black box*, where only inputs and outputs are known to the designer. The underlying functionality may be represented in a data sheet with definitions of latency, throughput, or expected outputs. In fact, designs are often completed by more than one person. As a result, each designer relies on a black box that will be filled in later by another designer.

Alternatively, starting from the top-level design and working down to low-level components is known as *top-down* design. When designing custom compute cores, the designer would begin with the core's interface (inputs and outputs). This creates a black box representation of the core. Once the interface is set, the designer can begin to systematically decompose the design into its subcomponents. This process is repeated for each subcomponent until low-level components are identified to be simple enough to design. The top-down approach does not necessarily associate with behavioral or structural HDL, but a designer may use more behavioral HDL.

The end product of either a top-down or a bottom-up approach should result in the same functionality. Internally, the design may look drastically different, but the top-level interface and operation should perform identically based on the specification.

Let's work with an example to illustrate the different ways components can be combined to form large modules. Consider the simple system illustrated in Figure 3.12. The desired functionality is to add four numbers together.

Beginning with a familiar approach, we consider the *temporal* implementation shown in Figure 3.13. While this implementation may not be immediately comparable to a software solution, consider how sequential addition is performed. In this solution, four numbers (a, b, c, d) are connected to a multiplexer, which

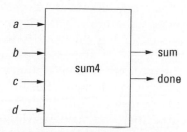

Figure 3.12. The top-level component to add four numbers.

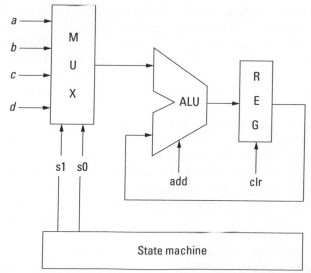

Figure 3.13. A temporal implementation.

is controlled from a simple state machine. The state machine increments from 0 to 3 to select each of the four inputs. The multiplexer feeds each input to the arithmetic logic unit (ALU), which is set to `add` mode. The ALU stores the result in a register, and the state machine increments the state bits `s1 s0` to add the next input. In this approach, the addition would take four additions:

1. $reg_{t0} = 0 + a$
2. $reg_{t1} = reg_{t0} + b$
3. $reg_{t2} = reg_{t1} + c$
4. $reg_{t3} = reg_{t2} + d$

In a system were only one ALU and register exist, this would be a sufficient minimal resource solution. Furthermore, augmenting this design to add eight numbers instead of four would only require a large multiplexer and addition state bit `s2`.

Clearly this is not the fastest approach. In terms of speed it is desirable to perform as many independent operations in parallel as possible. Unfortunately, there is a cost with parallel approaches, that is, added resources. Using three ALUs we could perform `temp1 = a + b` and `temp2 = b + c` in parallel and then add `temp1 + temp2`. The trade-off between latency and resources is ultimately at the hands of the designer, but it is wise to consider both low latency and low resource utilization approaches early in the design phase in case there is a need to switch between the two in a later phase.

Spatial Composition

Most programmers are familiar with the typical sequential composition rules of a von Neumann compute model. The simplest rule — two operations are sequenced — implies that there is a thread of control where the second operation takes place after the first computes its result. This is a degenerate case of spatial composition, which relaxes the strict ordering of operations. Hardware designs are not limited to sequential execution (unless dictated by the design). Thus, when a hardware designer specifies two operations, they will execute simultaneously unless the designer further specifies an ordering. Figure 3.14 shows a spatial implementation of the four adder example. In this case, the additions are pipelined such that results are fed forward to the next adder.

The loose ordering of operations in time is both a boon and a bane for Platform FPGA design. Concurrency is what gives system designers speed, and control of timing is what gives system designers predictability — both primary motivations for using hardware. However, simply expressing timing relationships between operations is a challenge, let alone determining what the correct timing

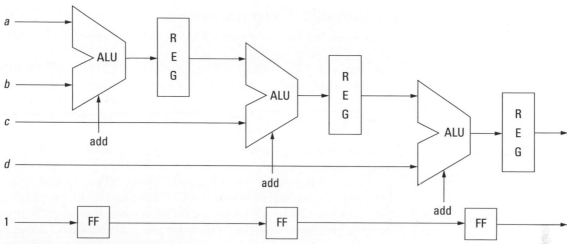

Figure 3.14. A spatial implementation.

relationships are. Consequently, this is a frequent source of system design errors. Chapter 5 goes into more detail regarding spatial design.

3.4. Software Design

Embedded system products have rapidly become more sophisticated in recent years. In the past, software was simple, had a single task, and played a relatively minor role compared to the design of the hardware. If two tasks were needed, they were often kept independent — logically and physically. As microcontrollers increased in speed, embedded systems added system software to manage multiple threads of control. This allowed a single microcontroller to time-multiplex separate tasks. Nowadays, some embedded system processors have memory management units, support virtual memory, and are fast enough to support full, modern operating systems. This has been a boon for users because it resulted in an explosion of product features. With a full-featured operating system, embedded systems designers can incorporate or adapt large software applications that were originally written for general-purpose desktop or server machines. This section aims to cover the various background information to help embedded systems designers understand and implement complex system software in Platform FPGA-based systems. We cover specific design and tool flows in section 3.A with the end result being running a complete Linux system on the Xilinx ML-510 development board. Here we address the concepts and definitions.

3.4.1. System Software Options

Just as with hardware, an embedded systems designer has a wide range of choices when it comes to system software. By *system software*, we are referring to any software that assists the application — usually by adding a software interface to access the hardware. This ranges from a simple library of routines to a full-fledged operating system that virtualizes the hardware for individual processes.

In the simplest situations, almost no system software is needed at all. In this case, the C start-up files (subroutines that the compiler/linker adds to every C program) are modified. At run time, these subroutines execute before calling the `main` function of the designer's application. With no operating system, these initial subroutines are responsible for setting up the processor and peripherals. (There is a collection of files with names such as `crt1.o`, `crti.o`, and `gcrt1.o`. The CRT part stands for C run time and, depending on the compiler options, different variants of the start-up files are used. Also, different processors will have different start-up files and the names may vary as well.) Even if the processor has a memory management unit, simple cases such as this execute the application in "real" or "privileged" mode and no memory protection is used. This is called a *standalone C* program because it runs without the support of any additional system software. In addition to being simple, an advantage of this approach is that there is essentially no overhead.

For Platform FPGAs, this is often a first step when testing new hardware cores because the C program typically has complete access to the hardware, it is very simple to compile a small test program, and there are fewer steps to test a live system. Often, this solution produces a small enough executable that the entire software system (application and system software) can fit within the block RAMs of the Platform FPGA. Avoiding off-chip RAM can be a significant advantage for some embedded systems. The disadvantage, of course, is that it offers little to the developer. There is often no protection against mistakes in the software. Perhaps the biggest drawback today is that it is very difficult to take advantage of existing software that assumes a full C library and a workstation- or server-type operating system. Examples of a stand-alone C system include those provided by Xilinx's Standalone Software Development kit, μlibc-only, and newlib.

Sometimes additional functionality from the system software is useful — such as supporting multiple threads — but the overhead of a full-featured operating system is undesirable. Numerous products and Open Source solutions are available that specifically target embedded systems to meet this need. They range from simply adding timer interrupt service routine and the ability to switch between different threads of control to full-featured operating systems that only lack a memory management unit. In some cases, the system software is combined with the application when the application is compiled.

One step up from "stand-alone" is a simple threading library. This solution includes the ability to create, schedule, and destroy multiple threads of control. The simplest of these just provide library calls so that the developer does not have to manage context switches and the program has to explicitly yield the processor. More advanced threading libraries include preemption (a thread does not have to explicitly yield the processor) and have the ability to schedule the frequency, priority, and deadlines of various tasks. Examples of this include eCos, XilKernel, Nucleous, and μC/OS-II — there are many others.

Somewhere between lightweight threading system services and a full-fledged operating system is the μC Linux project. This project grew out of the Linux 2.0 kernel and was intended to provide support for processors that lacked a Memory Management Unit (MMU). Without an MMU, there is no virtual memory. This means that an operating system cannot create true processes (since a process has its own virtual address space). So even though μClinux does not support all of the usual Linux system calls (`fork`, `sbrk`, for example) and any "process" can crash another process by overwriting its memory, a large degree of compatibility is maintained.

Operating systems provide a number of services to an application developer but they also have a cost. The obvious cost is that they add overhead or, conversely, use hardware resources (processor cycles, memory, power) that would otherwise be available to the application. There is also a cost associated with using the system software. Often, embedded systems use OS software that is different from what is found on a desktop or server. This means that the developer has to learn new interfaces, conventions, and what is or is not available. The type of services that the system software can provide ranges from simply time-sharing of the processor among multiple threads to simple protection of resources to complete virtualization of the hardware platform. A natural consequence of such a wide range of costs and benefits is a spectrum of system software choices. Some of the advantages and disadvantages of these choices are highlighted here.

At the far end of the spectrum, we have a full-featured operating system. These are the operating systems that one would find on desktop PCs, workstations, and servers. The chief disadvantage of using an ordinary operating system in an embedded system is that it requires a substantial overhead — the processor has to have an MMU, the OS generally has a large memory footprint (almost always requiring external RAM), and the operating system will include a number of extra processes running concurrently with the embedded application. Moreover, there are additional things that a developer has to do. Most of the system software thus described can run without a secondary storage subsystem (i.e., a filesystem). However, most full-fledged operating systems need, at minimum, a root filesystem. This doesn't have to be in the form of a hard drive but the developer has to create and store it somewhere on the embedded system.

Until recently, it simply was not feasible to consider using a full-fledged operating system in embedded systems because the required resources far exceeded what was found in embedded system hardware. However, with newer devices — such as Platform FPGAs — it is possible and becoming more common. Having a full-fledged operating system offers some enormous benefits to the embedded systems developer. First, it breaks with the trend thus far. A stand-alone C system is simple to work with. As we added services, there was more and more burden put on the developer to know what is provided by the system software and how to use it. With a full-featured OS, this is no longer an issue — it is the OS most programmers are intimately familiar with. Second, because it is the common OS, an enormous catalog of software becomes available. As embedded systems become more ubiquitous and connected to the Internet, they need to support more interfaces and more communication protocols. With a full OS, it becomes much much easier to leverage existing software.

3.4.2. Root Filesystem

UNIX and its variants (Linux, BSD, Solaris, and many, many others) share the concept of a root filesystem. A filesystem is a data structure implemented with a collection of equal-sized memory blocks that provides the application with the capability of creating, reading, and writing variable-sized files. Most filesystems provide the ability to organize the files hierarchically. In UNIX, files and subdirectories are grouped in directories. That is, there is one special directory called *root* that contains files and subdirectories; the subdirectories can contain files and other subdirectories. The filesystem data structure is implemented most often on secondary, nonvolatile storage such as disk drives or, more recently, solid-state drives. However, the underlying blocks of memory can be copied sequentially to other forms of memory, including RAM, ROM, or even a file of another filesystem! When the filesystem is being manipulated this way — being copied as sequential blocks of memory — it is typically referred to as a *filesystem image*. When the filesystem is being used (to manipulate files) it is called a *mounted filesystem*.

The simplest embedded designs, such as stand-alone C systems, usually do not require a formal filesystem. Nonvolatile storage is organized specifically to hold the application's data and often is customized for the problem at hand. However, as embedded systems become more complex, they use full-featured operating systems. In the case of UNIX, it means that the designer must create some initial filesystem called the *root filesystem*. Unlike some operating systems that place all of their start-up code in a single executable, the boot process for UNIX-like operating systems has the kernel interacting with the filesystem very early. In some cases, the kernel is actually stored on the filesystem and the bootloader (see later). After the kernel is running, it will look in prescribed directories for start-up files, system configuration files, and a special application called init, which is the first process to run. The init process is then going to use configuration files stored on the root filesystem to start other processes in the system and finish booting from the system configuration files. What this means is that the embedded systems designer has to how to create a filesystem and how to populate it.

Later we talk about the specifics for Linux, but the universal answer to the first question "where do we get started?" is that we need to create a filesystem image. There are two main ways of doing this. In both cases, one creates a subdirectory that will become the root filesystem in the embedded system. This directory is populated with the files and subdirectories required for the embedded system. This includes configuration files such as what commands to run at start-up, required system and application executables, kernel modules, and run-time libraries (shared objects for dynamic linking). The first way is to create a filesystem on a spare partition of a disk drive or use a loop-back device, which allows you to treat a file as if it were a partition. Once the filesystem is created, then you can mount the filesystem by simply copying your root filesystem to the newly mounted location. The only significant drawback to this approach is that on most operating systems, several of the steps require superuser privileges. The alternative approach does not require root; instead it uses a special-purpose application to generate a filesystem image directly. Examples of this include genisoimage, genext2fs, and mkfs.jffs2, which are programs that create a filesystem image for ISO9660/Joliet/HFS, Ext2, and JFFS2 filesystems. The first is intended primarily for things such as CompactDisk storage, and the latter works with Memory Technology Devices (MTD, i.e., flash memory). The middle option works well for conventional disk drives.

In both cases, the resulting filesystem image can be directly written to some media (a drive partition on the embedded system, an EEPROM, an MTD flash device) or combined with the operating system and loaded into RAM when booted. Since it is common to copy a filesystem image to RAM and use it as the root filesystem, the image is often called a "RAM Disk" and 'ramdisk.image.gz' is a common file name for a compressed root filesystem image. Several well-known distributions of GNU/Linux-based systems will use a RAM Disk as the root filesystem. This allows a single kernel to be first booted with the RAM Disk, then probe the hardware and install the required kernel modules, and finally mount the "real" root filesystem. This allows the system to finish booting using the "real" root filesystem and the RAM Disk's RAM is reused. Because this use is so common, many places refer to it as the "initial ramdisk" or "initrd." The name refers to how it is used but it is no different from the filesystem images we create.

3.4.3. Cross-Development Tools

Regardless of which operating system choice, an embedded systems developer will need to compile one or more applications. As it is often the case that the developer's workstation has a different processor and/or operating system, the designer will need to use a different set of compiler tools to create an executable.

A compiler translates a High-Level Language (HLL) to efficient assembly code. An assembler translates one-for-one mnemonic instructions and assembler directives into machine code in an object file format. A linker/loader combines multiple object files and archives of object files (library files) into a single object file and assigns memory addresses to all of the symbols in the object file. A *cross-compiler* is a high-level language translator that runs on one platform but produces executables for another platform. By platform, we mean (1) a specific processor, (2) a C library, and (3) an operating system. By default, most compilers now dynamically link to a C library, so the *version* of the C library is important as well as the specific version of the operating system. (In the case of Standalone C systems, the platform is just the processor as there is no operating system and any libraries are statically linked into the executable.)

Along with the cross-compiler, there is a matching set of "cross-tools." This includes what are typically called "bin tools," which is a reference to Unix object files and executable files (called binaries) stored in subdirectories such as /bin and /usr/bin. Bin tools include a cross-assembler, a cross-linker, and other tools to read and manipulate object files. The debugger is typically included in the cross-development tools as well.

3.4.4. Monitors and Bootloader

In the earliest days of microprocessor-based embedded systems, simple 8-bit microprocessors migrated from hobby computers and games into other commercial products (what we now call embedded systems). Vendors of these microprocessors typically made developer kits that included fabricated boards that highlighted the chips capabilities) and a Board Support Package (BSP) that included compilers, power-on-self-test (POST) software, libraries of Basic Input/Output System (BIOS) software, and a built-in debugger. The POST did exactly what its name says and often was executed before any other software simply to verify that nothing had worn out since the last time the system was turned on. POST software typically relied on the BIOS software to provide functionality, such as "read a character from a UART" or "write a disk sector." Because the POST (and by extension the BIOS) had to be stored in nonvolatile memory (ROM), this meant that embedded systems designers could use those subroutines "for free." That is, by using the subroutines in the BIOS, the size of the embedded application execution code size was kept small. The other software component that was typically included was a simple debugger called a monitor.

A *monitor* is a primitive type of a debugger. Modern debuggers typically run as a separate process (hence require an operating system), have access to the compiler's symbol table, and give the developer a rich, flexible interface. In contrast, a monitor is interrupt-driven — either the processor is interrupted or the application being debugged traps to the debugger. Also, a monitor usually only supports the most basic functionality — reading/writing absolute addresses, setting breakpoints, and manipulating registers. Some were able to disassemble (convert machine code back to assembly) but, again, they only showed absolute addresses (not symbol names). Monitors typically had one functionality not found in debuggers today — monitors support the transfer of memory over the serial communication channel used to interact. Because the communication channel usually transmitted ASCII (seven significant bits per byte) and executables use all

8 bits of a byte, blocks of memory were transmitted and encoded. Two popular formats were common: Intel Hex files and Motorola S-Records. Thus while developing the application, the designer could typically start the monitor and then copy their application to RAM. This helped shorten the test/debug their software.

We mention these historical notes because the vestiges of this approach remain today. For example, the GNU debugger (or gdb) is a popular debugger. It has a configuration where a small "gdb-server" code is cross-compiled and its role is to interface to a monitor. The gdbserver then uses a serial line to talk to the full gdb client. The client, running on a workstation, has access to the compiler's symbol table, a graphical display, and a full-featured operating system. This provides the developer with a rich user interface in which to debug.

Modern systems have moved one step beyond. The modern replacement of a monitor might be a JTAG interface.[1] JTAG controllers take over the processor and perform arbitrary read and writes to any physical address, including main memory. This provides an alternative approach to the same end. In this case, the debugger talks to an interface to the JTAG controller.

Likewise the POST/BIOS functionality has morphed into desktop PC's BIOS software. This code begins right after the power is turned on. There may be a little message "Press F10 for BIOS Setup," which gives the user a chance to change the main board's configuration. (Some computers say "CMOS Setup," which is the same thing — CMOS refers to a battery-backed memory that the BIOS uses to store parameter parameters between power cycles.) For some operating systems, it is critical that the BIOS puts the computer and its peripherals into a known state. For others, such as Linux, the early boot code assumes nothing and initializes the hardware itself.

Partially concurrent with development of the PC, workstations emerged with a slightly different approach. These machines used a small software program called a bootloader (or sometimes called PROM). In its earliest form, it was simply a program that read the first sector of a hard drive (which contained a more advanced start-up program) into main memory and then jumped/branched to the first address of the loaded sector. This program then proceeded to load the operating system. This multistage start-up sequence was called booting the system, which is short for "boot strap." The name comes from the expression, "pulling yourself up by your bootstraps" and was a way of addressing the question of

---

[1] JTAG is an acronym for Joint Test Action Group. However, its use here and in practice is so different from its intended purpose that we just refer to it as JTAG.

"how do you start an operating system that exists on secondary storage when there is no operating system to manage secondary storage?" Well-known bootloaders from the past include Sparc Improved Boot LOader (SILO), LInux LOader (LILO), and MIPS PROM.

Bootloaders have emerged in the PC world as well. The BIOS still runs first, then a bootloader is launched, and then the bootloader starts the operating system. Popular bootloaders today include GRUB (GNU Project, 2007), U-Boot (Denk, 2007), and RedBoot (eCos, 2004). Newer bootloaders are significantly more sophisticated as well. A modern bootloader has the ability to communicate over various networking protocols, provide graphical interfaces, and support booting multiple operating systems from different media, as well as know how to read a disk sector from secondary storage.

For embedded systems, the BIOS/monitor approach still dominates very small (8-bit microcontrollers) and legacy systems while the bootloader approach is gaining ground as full-featured operating systems become necessary to support widely used Internet protocols.

Chapter in Review

This chapter focused on the principles of system design and the hardware and software background necessary to be able to construct embedded system designs on a Platform FPGA running with a full-fledged operating system. In addition, we also emphasized important design concepts to support the ability for base systems, custom hardware cores, and low-level components to be reused within a system. From a hardware design point of view, the processor-memory model plays a key role in the rapid assembly and reuse of existing code. Likewise, by including Linux into the software design, we can quickly incorporate an already well-established code base that works well for both general-purpose and embedded systems.

Certainly there has been much information presented, and the reader may find that the gray pages of these chapters help tie everything together with some practical examples. Because we are still concerned with assembling base systems, we must spend time understanding the additional tools, wizards, and GUIs that can help expedite this process. Finally, in the last section of the gray pages is a comprehensive Linux example, covering everything necessary for acquiring, compiling, and running Linux on an FPGA.

Practical Expansion: System Design

The three major learning objectives for these gray pages are roughly grouped into the following categories: hardware tools, software tools, and configuration tools. In the first section, the reader will learn more about the Xilinx-provided components in the Embedded Development Kit and Xilinx Platform Studio (XPS), the tool used to assemble these hardware components into systems. The second section is concerned with software aspects of such system. This includes everything from building a customized cross-compiler from source code to compiling a Linux kernel and GNU operating system. The last section describes the tools and different options for combining the hardware and software for a Platform FPGA system.

3.A. Platform FPGA Architecture Design

While the strengths of building custom computing machines lie in the ability to implement custom hardware cores, these cores are often used in conjunction with typical computer organizations. This practical expansion describes how to assemble simple processor-memory systems and then how to add customized cores to these systems. There are three main concepts to be covered.

The first concept is to learn the predefined (Xilinx and third-party) cores provided by the Xilinx Embedded Development Kit (EDK). Many of the system cores come from a collection of System-on-Chip cores called CoreConnect from IBM, Inc (IBM, 2009). Some of the CoreConnect cores are not available for FPGA designs, and some have been reimplemented so that they map well to an FPGA. Hence, some cores only support a subset of the original features found in the IBM CoreConnect cores. Also, the interfaces may change slightly with each version of the Xilinx tools. So when trying to use these cores, it is best to start with a quick look at the data sheets provided with the version of the tools you are using.

The second concept is to learn how to assemble systems from these cores. Xilinx provides a tool — Xilinx Platform Studio (XPS) — that helps in this regard. We have already seen this approach in section 3.A with the Base System Builder (BSB) wizard and the XPS tool. However, it is possible to start with a top-level VHDL file that names all of the components and necessary signals and provides an architecture that instantiates all of the cores. But the number of signals, generics, and ports on individual cores can make this approach tedious and error-prone.

The final concept is to learn how to customize and extend a base system. Again, this could be done by editing the top-level VHDL file and other files directly, but it is usually easier to use the graphical user interface provided by XPS. We end this section with a practical example, building a custom hardware core to perform single precision floating point addition and run the design (and sample application) and the FPGA.

3.A.1. Xilinx EDK and IBM CoreConnect

The Xilinx EDK includes a large repository of prebuilt, customizable soft cores and wrappers that provide an interface for instantiating hard cores. These cores range from UARTs to buses to memory controllers and even soft processor cores. IBM CoreConnect adds to this repository by providing soft core versions of the already widely used buses, bridges, and arbiters that have been configured for use with the Xilinx FPGA devices. Using these cores it is possible to build a processor-memory system and then add custom cores to implement specific functionality.

Soft IP Cores To get a first-cut idea of what sort of peripherals are in a computer system, consider the back panel of a standard PC. A standard desktop PC five years ago might have one or two serial ports, a mouse and keyboard connector, a parallel port, a video port, an Ethernet port, and some audio jacks. In addition to these peripherals, there are on-board peripherals that are unseen until the case is opened. This includes obvious peripherals (such as disk drives), as well as subtle peripherals, such as the temperature sensors and little read-only memories located on the DIMM memory sticks.

First, we discuss the various backbone peripherals (processors, buses), next we discuss memory, and finally we finish with a plethora of practical peripherals that might reasonably be implemented as cores in a Platform FPGA system. In the peripherals section, we describe the characteristics of these components.

Processors The Virtex 5 FPGA can implement both hard core processors (PowerPC 440 (Xilinx, Inc., 2009a)) and a wide range of soft processor cores, such as MicroBlaze, PicoBlaze, and OpenSPARC. The main difference between these processor choices is their performance versus area trade-off. The PowerPC 440 is a 32-bit processor that implements the PowerPC architecture and adds some features for embedded systems. The MicroBlaze (Xilinx, Inc., 2009c) soft processor core is also a 32-bit processor, and current versions now support virtual memory, making it a desirable alternative to the PowerPC. One benefit of the MicroBlaze is that it can have multiple instances, limited by the size of the device. Compared to the PowerPC, which on the Virtex 5 it has zero, one, or two hard processor cores. The PicoBlaze (Xilinx, Inc., 2010) is a very small 8-bit processor, which is appropriate for a number of control applications — especially those with large, complex state machines. Depending on the number of states and transitions, a PicoBlaze can implement the same functionality with fewer resources. However, it is unlikely that one will ever run a full-featured OS on one. The LEON Open Source Sparc David Weaver (2008) processor is gaining more momentum as its implementation in FPGA configurable resources is becoming more efficient.

Many other soft processors could be discussed here as well; however, for more rapid development the obvious processor to use is the PowerPC, as there are two hard PowerPC 440 cores on the chip used throughout the text to demonstrate the book's concepts. To use these cores, all that is needed is to instantiate a black box component, which will expose the PowerPC signals to the design. The component is called `ppc440` and, using EDK nomenclature, instances are named `ppc440_0` and `ppc440_1`.

Buses A system designer has a number of choices for buses. They include the PLB, LMB, OPB, FSL, DCR, and two kinds of OCM! Why so many? Two reasons. Some of these buses are designed for very specific situations and provide the required functionality with the minimal amount of resources. Second, each general-purpose bus provides different performance characteristics, features, and resource requirements. It is up to the system designer to find the right bus structure to meet performance goals given the fixed amount of resources available.

The *Processor Local Bus* (Xilinx, Inc., 2008, 2009d), or PLB, comes from the IBM CoreConnect library. The PLB is considered the system bus, connecting the processor (in our case the PowerPC 440) to high-speed peripherals such as off-chip memory. It is a high-performance 128-bit data (32-bit address) bus designed to interface to the PowerPC 440 core. The PLB offers the highest bandwidth of the other available buses from CoreConnect, along with other features, such as concurrent reads and writes, address pipelining, fixed and variable burst, and cache fill features. Consequently, the resources needed to implement master and/or slave interfaces to this bus can be high. Also, implementing a PLB interface is more complex, as there are additional protocols to go along with the additional features.

The *Local Memory Bus*, or LMB, is a 32-bit bus specifically designed to be a system bus for the MicroBlaze (the MicroBlaze can also connect to the PLB). Unlike the PLB, it is designed to have only one master (the MicroBlaze) and one slave (a memory controller). With these restrictions, many of the features of the PLB are unneeded. It has a reasonably high bandwidth and requires a modest amount of resources.

A number of peripherals (such as serial ports or timers) require very little bandwidth and few features. The *On-chip Peripheral Bus*, or OPB, is a 32-bit bus and is an excellent compromise between performance and resource requirements.

It requires a moderate number of logic blocks, yet still offers a substantial amount of bandwidth. It is also has a simpler protocol, which makes it an easier interface to implement in cores. It can be used with either PowerPC or MicroBlaze, and a number of bridges allow a number of effective bus structures to be developed.

The *Fast Simplex Link* bus, or FSL bus, is a special-purpose bus designed to push (or pull) a stream of data out of (or into) a MicroBlaze. In essence, it is a first-in, first-out (FIFO) that directly ties the processor to a peripheral. It uses little in terms of resources and by construction there is no bus contention. Because the FSL is essentially point-point, the bandwidth is considerably high between the two components.

One of the key distinctions between the buses discussed thus far is how much bandwidth is needed. However, in some cases, a peripheral core will need high bandwidth for some types of access and low bandwidth for others. A simple response would be to use a high-bandwidth bus to share both kinds of access. However, mixing these two types of access can quickly degrade the effective bandwidth of a high-performance bus. (By arbitrating for simple, one-byte transfer, low-bandwidth access actually steals many productive cycles from a high-bandwidth bus.) The ideal solution is to use two buses.

The *Device Control Register* bus (or DCR bus) is designed specifically to do this. It requires very little resources and, in fact, is fairly limited in its functionality (only one master and only a 10-bit address bus). However, it allows a processor to directly communicate small bits of information (such as configuration information) to a peripheral without interfering with the effective bandwidth of a bus like the PLB.

Memory The last set of components that form the backbone of a typical computer organization is related to the memory hierarchy. This includes off-chip memory and on-chip memory (formed from the BRAMs). On-chip memory is limited based on the available resources; however, it provides high bandwidth and low latency access. This is not to be confused with cache on a conventional CPU. Instead, this is more along the lines of local storage, requiring some control mechanism to populate the memory and provide access to its contents. Off-chip memory is conventionally in the form of either static or dynamic random access memory (SRAM or DRAM). Off-chip memory can provide a significantly greater storage capacity while still providing high bandwidth and respectably low latency access.

On-chip RAM is, by default, built from the diffused Block RAM resources distributed throughout the Virtex family chips. These RAMs are usually instantiated as a single core that allocates the proper number of BRAMs and associated decode logic to match the controller's data bus width and capacity (specified by the number of address lines). Using the tools, when instantiating a BRAM core, an on-chip memory controller must also be instantiated to connect the BRAM to the bus. It is also possible to create your own hardware core with an interface to the BRAM. Each BRAM core includes two ports for dual read/write access, which can give a lot of flexibility to the designer with accessing memory. More of these memory access concepts are discussed in Chapter 6.

The interface between off-chip memory and an on-chip bus is handled by a *memory controller*. The memory controller is chosen to match a bus (typically PLB, OPB, or LMB) with various memory technologies. The variety of off-chip memory technologies warrants some discussion because, in many ways, its importance in computer system design is growing.

More recently, systems using an external memory controller use the Multi-Ported Memory Controller (MPMC). The MPMC provides one or more connection ports to off-chip memory, which means that two cores could be directly connected to the MPMC and independently issue requests to and from memory. This is useful for components such as the processor, which need frequent access to memory and also provide an additional benefit of alleviating the system bus (in our case the common PLB) from all of the memory traffic. Which cores are connected is an important design consideration that we will get into more in Chapter 6.

Peripherals Finally, the group of cores one might find in an embedded system are the peripherals. These cores are typically associated with the various sensors and actuators used in the enclosing product plus all of the basic cores used to communicate with subcomponents in the basic computing system. Needless to say, the number and the variety of peripherals that may appear in a Platform FPGA are immense. Moreover, newer technology is always emerging and some older technology will mature and fall out of favor (but not completely). So at best, we can hope to characterize some of the major players here and provide enough information to help a designer sort out which cores might serve them and which ones have nothing to do with their project's goals.

(Also, it is worth noting that *devices* is a synonym for peripheral in CoreConnect terminology. Peripheral is predominant in the Xilinx literature and appears in the IBM documentation, but "device" occasionally appears in some literature and when it does, it usually means a peripheral core.)

To organize this large set of cores, we can begin by grouping them into categories by their function. Communication (low-speed and high-speed) is associated with cores that interface to off-chip components (from PROMs to terminals to other computer systems). Digital and Analog I/O also transmit information from the FPGA to external devices (and the converse) but these tend to be less sophisticated. For example, a digital I/O line tends to be a TTL signal (either 0 or +5 volts) whereas communication includes an encoding and semantics. There are also cores that are responsible for distributing and controlling clock signals. Other specialized cores are responsible for managing interrupts, creating programmable timers, and such. Of course, there are function units used to off-load processor computation. We briefly describe several possible cores later. Note that many of the communication cores are detailed later in Chapter 7.

A number of low-speed communication cores are used. For many decades, the king of low-speed cores was the industry standard RS-232C. A Universal Asynchronous Receiver/Transmitter (UART) core implements this standard and it typically transmits 10- to 12-bit messages serially at a relatively slow rate. In a standard desktop PC, this is the COM1/COM2 port that one used to connect to dial-up modems. In embedded systems, these cores are often used for debugging because the PC's port can be interfaced easily to a UART on the embedded system. In a Platform FPGA design, a UART is often included for this very purpose. Other low-speed protocols include the Inter-Integrated Circuit (IIC or I^{2}C), Serial Peripheral Interface (SPI), and Dallas 1-wire. The subtle differences between these protocols and the cores that implement them are explained later. Often the choice of one over another is dictated by a third-part sensor, display, or actuator that is to be integrated into the embedded system. (Note Intelligent I/O, or I^{2}O, is completely different and was designed to work with PCI to off-load some I/O processing on hardware RAID cards, for example.)

High-speed communication protocols can be grouped into two broad categories: parallel and serial. The older protocols used a relatively slow clock rate, high voltages, and wide parallel buses to increase bandwidth. As newer techniques, such as Low-Voltage Differential Signaling (LVDS), became popular, newer high-speed serial protocols have become extremely popular. The highest speed communication protocols today aggregate multiple high-speed serial paths, or lanes, into one high-speed channel. (Again, details are covered later.)

Examples of parallel high-speed protocols include Small Computer System Interconnect (SCSI) and the Advanced Technology Attachment (ATA) buses implemented by cores of the same name. The SCSI protocol used a parallel bus, originally designed for high-speed peripherals (such as disk drives). The ATA protocol was specifically designed to talk to disk drives, and the actual chip-to-chip protocol, called Integrated Drive Electronics or IDE, was ubiquitous for years in the standard PC desktop. Both of these standards have been updated recently to have high-speed serial counterparts — iSCSI and SATA. The Serial ATA (SATA) (Grimsrud, Knut, and Smith, Hubbert, 2003) protocol subsumes both ATA and IDE; to distinguish SATA from the older parallel ATA/IDE protocol, the latter has been retroactively called PATA for Parallel ATA (InterNational Committee for Information Technology Standards T13, 2010).

Other well-known high-speed serial protocols include the Universal Serial Bus (USB) (USB Implementers Forum (USB-IF), 2010a,b), which has almost completely replaced RS-232C (Electronics Industries Association, 1969) in commodity PC hardware today, FireWire (IEEE 1394) (1394 Trade Association, 2010), InfiniBand (Futral, 2001), HyperTransport (Holden, Brian, Anderson, Don, Trodden, Jay, and Daves, Maryanne, 2008; HyperTransport Consortium, 2010), and PCI Express (or PCIe) (PCIIG, 2010).

Of special note is a number of networking protocols specifically designed to allow multiple, independent computing systems to communicate. The Ethernet family of protocols is the most common of all of these and is the dominate protocol for wired networks. Others, such as the Controller Area Network (CAN), are used in a number of embedded systems situations.

3.A.2. Building Base Systems

Our goal is to assemble systems that can be reused through a variety of designs and easily expanded to meet each new design need. Using the bottom-up approach to assemble the components in a piecewise fashion is a good starting point. Instead of building up from the lowest levels, we focus more on a more course grain, compute core, granularity. These were discussed briefly earlier, but now we will be more Xilinx and Virtex 5 specific.

Our key building blocks will be the PowerPC 440, PLB, and PLB BRAM (which is technically composed of a BRAM component and a BRAM interface, which connects PLB requests to the BRAM). The PLB will act as the system bus to connect the processor to the on-chip memory. We start with this base system because it provides us with the greatest flexibility; however, it does seem to lack much functionality. Strictly speaking, if we are starting with just these components we would not have a system clock or reset capability.

The ML-510 development board comes with an oscillator to provide a 100-MHz clock that can be used to drive additional clocks through one or more DCMs. In Xilinx's 11.x EDK tools, the DCMs are wrapped by a new component called the `clock_generator`, which can generate up to 16 output clocks. If we wanted to operate the PowerPC 440 at 400 MHz and the PLB at 200 MHz, we could instantiate the clock generator with the two additional clocks. This will be useful later when interfacing with components running at nonstandard frequencies. The upper limit of the Virtex 5 FPGA fabric is 550 MHz and care should be taken when trying to operate large amounts of the fabric at high frequencies as it becomes difficult to meet timing conditions and can produce greater amounts of heat. Finally, we want to include reset circuitry to allow the system to be reset. A reset differs from a power cycle in that the reset does not requiring reprogramming the FPGA, whereas the power cycle removes power from the fabric and the volatile memory cells lose their configuration. The EDK provides an additional component known as `proc_sys_reset`. This component connects an external reset switch to the system and can provide bus and peripheral component resets.

In addition to the component instantiations, we need to specify which pins on the FPGA are connected to the off-chip oscillator (for the clock) and reset switch. If using a Xilinx development board (such as the Xilinx ML-510), this information can be found from the board's schematic documentation. Otherwise consult the vendor for the necessary documentation. On the ML-510 the pin location for the system clock is `L29` and the reset pin is `J15`. Using the user constraints file (UCF) we can specify the pin constraints that will connect to the top-level system as I/O pins.

We would like to take a moment to reflect on what we have discussed. Some may wonder why we jump into building a custom design when there is already a functional base system builder (BSB) wizard available. It is true, while we recommend use of the wizard as a beginning tool, the earlier you understand what the tool is building, the quicker you will be able to augment and even build your own custom designs. What we would like to do now is actually go into more detail regarding the files produced by the BSB wizard and used by XPS.

3.A.3. Augmenting Base Systems

The base system described so far lacks much functionality, at least that is visible by any I/O device. So writing a program to print "hello world" would not print to a console because no such peripheral device exists yet. We can augment the processor-memory base system to include a peripheral, a UART, to provide a serial console interface to the system. As discussed earlier, we added the UART to the peripheral bus; however, current Xilinx tools focus less on the separation of buses (system and peripheral) and instead rely on a higher bandwidth and lower latency system bus. It is still possible to use a two bus system (and there are many reasons to do so), but for components that have low utilization like a UART, adding them to the PLB does not impact the design negatively.

Because the UART is a slave on the bus, it requires an *address range* to which it will respond to bus requests. The address range must be unique and not encompass any existing addresses for other cores. For the UART, we will give the address range between 0x84000000 and 0x8400FFFF. Any requests crossing the bus between these addresses will be responded to by the UART. In addition to the address range, we must also specify the transmit and receive pins that the UART will connect to off-chip. These pins may be connected to an RS232 chip on the developer board. As with the clock and reset pins, we must identify the pins for transmit and receive for the UART and connect them in the user constraints file. The ML-510 developer board contains two RS232 chips (it is possible to add multiple UARTs; however, only two connect to the standard DB9 serial port).

```
Net fpga_0_RS232_Uart_1_RX_pin LOC = T11 |   IOSTANDARD=LVCMOS33;
Net fpga_0_RS232_Uart_1_TX_pin LOC = H9  |   IOSTANDARD=LVCMOS33;
```

At this point we have constructed the base system that was created during section 3.A. Before we build and add our own custom core we should take a brief moment to discuss some of the important Xilinx project files that can be used to more quickly build, augment, and implement working systems.

3.A.4. XPS Project Files

The BSB wizard can quickly generate base systems using the processor-memory model. In fact, more than just the processor, bus and memory can be added and modified with the BSB wizard. We can add peripherals such as UARTs, timers, interrupt controller, and even I/O for components such as LEDs and LCD. We can also add networking interfaces and off-chip memory interfaces. We encourage you to play around with the BSB wizard and to read the accompanying documentation to understand what adding each peripheral or modifying each parameter does.

Upon completion of using the BSB wizard, a Xilinx Microprocessor Project is created. We would like to take some time to identify the important project files that are quite useful when assembling and modifying a Platform FPGA system.

system.xmp Xilinx Microprocessor Project (XMP) file used for project management (it is recommended not to modify this file directly).

system.mhs Microprocessor Hardware Specification (MHS) file used by Platform Generator (platgen) to assemble base systems based on the configuration settings specified in this file. This file will be a common point of contact if you prefer not to use the XPS GUI.

system.mss Microprocessor Software Specification (MSS) file used by Library Generator (libgen) to assemble the operating system libraries and drivers.

system.ucf User Constraints File (UCF) located in the data directory, which specifies constraints for system. Here you can set physical pin constraints, I/O standards, clock constraints, etc.

The choices while using the BSB wizard customize each of these four files. Looking at the MHS file (from any standard text editor or from within the XPS GUI) shows the I/O ports and each component within the base system. For example, the PowerPC's component instance is specified by:

```
## Anything beginning with the # symbol is a comment
## PowerPC 405 MHS Instance
BEGIN ppc440_virtex5
 PARAMETER INSTANCE = ppc440_0
 PARAMETER C_IDCR_BASEADDR = 0b0000000000
 PARAMETER C_IDCR_HIGHADDR = 0b0011111111
 PARAMETER C_SPLB0_NUM_MPLB_ADDR_RNG = 0
 PARAMETER C_SPLB1_NUM_MPLB_ADDR_RNG = 0
 PARAMETER HW_VER = 1.01.a
 BUS_INTERFACE MPLB = plb_v46_0
 BUS_INTERFACE JTAGPPC = ppc440_0_jtagppc_bus
 BUS_INTERFACE RESETPPC = ppc_reset_bus
 PORT CPMC440CLK = clk_400_0000MHzPLL0
 PORT CPMINTERCONNECTCLK = clk_200_0000MHzPLL0
 PORT CPMINTERCONNECTCLKNT01 = net_vcc
 PORT CPMMCCLK = clk_100_0000MHzPLL0_ADJUST
 PORT CPMPPCMPLBCLK = clk_100_0000MHzPLL0_ADJUST
 PORT CPMPPCS0PLBCLK = clk_100_0000MHzPLL0_ADJUST
END
```

Microprocessor Hardware Specification   The MHS file can be used to quickly instantiate additional compute cores (custom or from the Xilinx IP library repository) or modify parameters and ports. The parameters can be generally associated with the generics of a VHDL entity while the ports are associated with the ports of the entity. The bus interface is a Xilinx specification, which aggregates all of the associated bus signals into a single structure. For example, the PowerPC is connected to the PLB as a bus master on the BUS_INTERFACE_MPLB. For cores connecting to the PLB via a slave interface, bus addresses are set in the MHS file through base and high address parameters.

The MHS file is used by the Platform Generator (PlatGen) tool. This is one of the many underlying tools (or commands) called by XPS. Generally speaking, PlatGen parses the MHS file and locates the HDL for each core to generate a top-level component (often called system.vhd for VHDL based systems, located in the hdl directory). PlatGen invokes the HDL synthesis of each of the cores, culminating with synthesis of the top-level entity. The system then continues through the NGCBuild, NGDBuild, MAP, PAR, and BitGen flow discussed in section 3.A.

Microprocessor Software Specification   The MSS file is used to specify parameters for the operating system, included libraries, and drivers. Initially, the BSB wizard generates the MSS file with a *stand-alone* OS and specifies standard in/out to a UART (assuming one is provided). The MSS file also specifies which cross-compiler to use, for example, powerpc-eabi-gcc. Each component may include a driver that specifies software drivers and libraries to be used to interface with the component, if necessary. The MSS file is used by the Library Generator (LibGen) to create the software platform for the embedded processor system.

User Constraints File   The UCF is located in the data directory and specifies the pin location constraints, attributes, timing constraints, and I/O standards, among many others. In embedded systems design it is common to include I/O in the design, whether to some display (LEDs or LCD) or some digital input (say from an analog-to-digital converter for a sensor). For these designs the general-purpose I/O is typically fixed at manufacturing of the PCB layout, but a developer board may include a variety of components to interface with, such as EEPROMs or temperature sensors. For our designs, we will need to include at least the clock:

```
TIMESPEC TS_sys_clk_pin = PERIOD sys_clk_pin 100000 kHz;
Net fpga_0_clk_1_sys_clk_pin LOC = L29  |  IOSTANDARD=LVCMOS25;
```

This initial introduction to these files is meant to help break the ice when it comes to working with Xilinx tools. Clearly, Xilinx documentation is going to provide a wealth of information that this book cannot include. What is important is that while beginning to use these tools, you spend the time to understand how all the files are used and interact.

3.A.5. Practical Example: Floating-Point Adder

To help tie all of these concepts together we will build a custom computing core to perform single precision addition:

$$F = A + B$$

We will use the available wizards and repository of cores already available to help expedite the build process. When building the base system and hardware core template, we will rely on the readers to have familiarized themselves with the material covered in the gray pages of Chapters 1 and 2.

The hardware core will use two input FIFOs to buffer the two operands (A and B) and one output FIFO to buffer the result (F). The operands will be written into the FIFOs by the processor, but only when both operands are available and when there is room in the result FIFO will the addition operation commence.

The FIFOs and single precision floating point adder will be generated through the Xilinx CoreGen GUI. We will then modify the hardware core to use these new components and create a simple software application to test the hardware. The goal of this example is to show how to integrate all of these tools, wizards, and components. We want to stress that there are more efficient ways to perform floating point computation, one of which is to directly use the APU/FPU supplied by Xilinx for the Virtex 5 FX130T designs.

3.A.6. Base System

Creating a base system through the base system builder wizard should be a fairly familiar process now. We will not show all of the wizard options, choosing instead to list the necessary capabilities of the system. Reader may choose to add more peripherals for their own testing, keeping in mind that the larger the system, the longer the tools will take to run and generate a bitstream. The base system should include the following:

- PowerPC 440 - 400 MHz
- PLB - 100 MHz
- UART Lite
- On-Chip Memory - 64 KB

Adding off-chip memory (either via ppc440mc or mpmc) is optional; our application will be stored within on-chip memory, but for further practice consider rebuilding the system to run out of off-chip memory once the on-chip memory system works. (Note: If you choose to include off-chip memory, you will still need on-chip memory. The processor looks to address 0xFFFFFFFC during boot up before jumping to the beginning of the program, which may or may not reside in off-chip memory.)

3.A.7. Create and Import Peripheral Wizard

Once the base system has been built, we will use the CIP wizard to generate a hardware core template. In the application's description we stated we would write operands A and B to the hardware core as well as retrieve result F from the hardware core. From this description we know we need at least three slave registers (software addressable registers) within our custom hardware core. The hardware core generated in section 3.A will suffice or you can choose to add additional registers to provide status information regarding the floating point computation (such as ready for data, result ready, FIFOs empty, and FIFOs full). Configure the hardware core by:

* naming the core *my_test_core*
* connecting the core to the *PLB version 4.6*
* checking *User Logic Software Registers*
* unchecking *Include data phase timer*
* setting the number of software addressable registers to *3*
* accepting the remaining default options

After the template has been generated, we will create the FIFOs and the single precision floating point adder before modifying the template.

3.A.8. Core Generator

Change directories into the newly created hardware core, which resides within the `pcores` directory of the XPS project. Create a directory within the `my_test_core` called `coregen`.

```
% cd pcores/my_test_core_v1_00_a
% mkdir coregen
```

Launch the CoreGen wizard and create a new project within the `coregen` directory for the Virtex 5 FX130T FPGA (unless you are targeting a different FPGA).

First we will generate a FIFO. Rather than generating three separate FIFOs (for each input and output) we will use one FIFO, but instantiate it three times with our hardware core. We will then generate the single precision floating point adder. This is followed by modifying the necessary project files for the hardware core to use the generated components. Finally, we include the hardware core in the previously generated base system, add some test C code, and generate a bitstream to test out the design.

Generate FIFO   With the CoreGen project created, let's begin by expanding the *Memories & Storage Elements* category. Then expand the *FIFOs* subcategory and run the *FIFO Generator*.

Set the component name to `fifo_32x1024`

Because single precision addition uses 32 bits of data, we must generate our FIFO to use both 32-bit inputs and outputs. Depending on how we want to clock the system, we can either use a common clock (where the read and write clocks are the same) or independent read and write clocks to allow us to use the FIFO to cross the clock domains. For now, let's use a common clock. The depth of the FIFO will dictate the number of resources used. We will set the depth to 1024 elements (again, with 32-bit data width) to use a single 36-Kb BRAM per each FIFO instance.

To support processor reading data from the result FIFO correctly, enable the *First-Word Fall-Through* read mode. Without this each read will be off by one data due to the 1 clock cycle latency to read from a standard read mode FIFO.

Finally, add the `valid` signal, which we will use to indicate when data are valid from the operand A and B FIFOs to perform the calculation by the single precision floating point adder.

Generate Adder   Next we generate the single precision floating point adder. Under the *Math Functions* category is the *Floating Point* subcategory. Expand these and run the *Floating-point* Generator (Xilinx, Inc., 2009b). Set the component name to `sp_fp_adder`.

Because we are only wanting to support addition, we will set the operation to *Add*. Because the Virtex 5 FX130T has 320 DSP slices, we can afford to use a few for this example. Select *High Speed* for the architecture optimizations and select *Full Usage* to use DSP slices. Finally, select all of the handshaking signals: *operation new data*, *operation read-for-data*, and *ready*.

We will use the handshaking signals along with the FIFO's handshaking signals to coordinate the computation, making sure not to process data unless both operands are valid and the resultant FIFO is not already full.

For this simple example only addition is supported; however, once complete the reader may wish to explore supporting additional floating point computations. We will leave this experimentation up to the reader, but will offer one point to consider. That is, how to select which operation the floating point unit is to perform.

3.A.9. User Logic

Now that we have generated the FIFOs and single precision floating point adder we have to modify the hardware core to actually instantiate these cores. The template hardware core contains a file called `user_logic.vhd` within the following directory:

```
pcores/my_test_core_v1_00_a/hdl/vhdl/user_logic.vhd
```

By default the user logic template supports reads and writes from software addressable registers, known as slave registers in the template. We must add our logic to the generated template in order to support the single precision floating point adder and FIFOs. If this is your first time creating and modifying a core generated by Xilinx Create and Import Peripheral wizard, spend some time exploring the template and reading the supplied comments to get a feel for the template.

Once comfortable, we will add the following signal declarations beneath the existing signal declarations. For those less familiar with VHDL, these signal declarations need to be inserted between the architecture and begin sections of the `user_logic.vhd` file.

```
architecture IMP of user_logic is
  -- Insert Signal and Component Declarations here
begin
```

These signals will be covered in more detail when they are instantiated within the design. For now, they are simply to be used to connect up the CoreGen components.

```
-- Internal Signals for SP FP Adder Example
signal adder_re    : std_logic;
signal slv_we_reg  : std_logic_vector(0 to 3);
-- Operand A FIFO Signals
signal fifo_A_din  : std_logic_vector(31 downto 0);
signal fifo_A_wr_en : std_logic;
signal fifo_A_rd_en : std_logic;
signal fifo_A_dout : std_logic_vector(31 downto 0);
```

```
signal fifo_A_full  : std_logic;
signal fifo_A_empty : std_logic;
signal fifo_A_valid : std_logic;
-- Operand B FIFO Signals
signal fifo_B_din   : std_logic_vector(31 downto 0);
signal fifo_B_wr_en : std_logic;
signal fifo_B_rd_en : std_logic;
signal fifo_B_dout  : std_logic_vector(31 downto 0);
signal fifo_B_full  : std_logic;
signal fifo_B_empty : std_logic;
signal fifo_B_valid : std_logic;
-- Result F FIFO Signals
signal fifo_F_din   : std_logic_vector(31 downto 0);
signal fifo_F_wr_en : std_logic;
signal fifo_F_rd_en : std_logic;
signal fifo_F_dout  : std_logic_vector(31 downto 0);
signal fifo_F_full  : std_logic;
signal fifo_F_empty : std_logic;
signal fifo_F_valid : std_logic;
-- Single Precision Floating Point Adder Signals
signal adder_a      : std_logic_vector(31 downto 0);
signal adder_b      : std_logic_vector(31 downto 0);
signal adder_result : std_logic_vector(31 downto 0);
signal adder_nd     : std_logic;
signal adder_rfd    : std_logic;
signal adder_rdy    : std_logic;
```

Directly after these signals will come the component declarations. These are the FIFO and adder components generated by CoreGen. These declarations can be found in each component's corresponding *.vho* file located within the `core-gen` directory created earlier. Even though we are using three instances of the FIFO, they all share the same component (`fifo_32x1024`) declaration.

```
-- FIFO Component Declaration - 32 bits, 1024 Elements Deep
component fifo_32x1024
  port (
    clk   : in std_logic;
    rst   : in std_logic;
    din   : in std_logic_vector(31 downto 0);
    wr_en : in std_logic;
    rd_en : in std_logic;
    dout  : out std_logic_vector(31 downto 0);
    full  : out std_logic;
    empty : out std_logic;
    valid : out std_logic);
end component;
-- Single Precision Floating Point Adder Component Declaration
component sp_fp_adder
  port (
    a             : in std_logic_vector(31 downto 0);
    b             : in std_logic_vector(31 downto 0);
    operation_nd  : in std_logic;
```

```
      operation_rfd : out std_logic;
      clk           : in std_logic;
      result        : out std_logic_vector(31 downto 0);
      rdy           : out std_logic);
end component;
```

Now that we have declared all of the signals and components to be used within our system, we can go ahead and instantiate each of the components. This code must go between the architecture's begin and end keywords.

```
begin
  -- Put signal and component instantiations here
end IMP;
```

Shortly we will explain how the intermediate signals connecting each component function.

```
-- FIFO A (Operand A) Component Instantiation
fifo_A_i : fifo_32x1024
  port map (
    clk   => Bus2IP_Clk,
    rst   => Bus2IP_Reset,
    din   => fifo_A_din,
    wr_en => fifo_A_wr_en,
    rd_en => fifo_A_rd_en,
    dout  => fifo_A_dout,
    full  => fifo_A_full,
    empty => fifo_A_empty,
    valid => fifo_A_valid);
-- FIFO B (Operand B) Component Instantiation
fifo_B_i : fifo_32x1024
  port map (
    clk   => Bus2IP_Clk,
    rst   => Bus2IP_Reset,
    din   => fifo_B_din,
    wr_en => fifo_B_wr_en,
    rd_en => fifo_B_rd_en,
    dout  => fifo_B_dout,
    full  => fifo_B_full,
    empty => fifo_B_empty,
    valid => fifo_B_valid);
-- FIFO F (Result) Component Instantiation
fifo_F_i : fifo_32x1024
  port map (
    clk   => Bus2IP_Clk,
    rst   => Bus2IP_Reset,
    din   => fifo_F_din,
    wr_en => fifo_F_wr_en,
    rd_en => fifo_F_rd_en,
    dout  => fifo_F_dout,
    full  => fifo_F_full,
    empty => fifo_F_empty,
    valid => fifo_F_valid);
-- Single Precision Floating Point Adder Component Instantiation
```

```
sp_fp_adder_i : sp_fp_adder
  port map (
    clk           => Bus2IP_Clk,
    a             => adder_a,
    b             => adder_b,
    operation_nd  => adder_nd,
    operation_rfd => adder_rfd,
    result        => adder_result,
    rdy           => adder_rdy);
```

With the components properly declared and instantiated we can begin to connect the intermediate signals that drive the inputs to our components. For the FIFOs these signals consist of data input, write enable, and read enable. For the adder these signals consist of operands A and B, and the new data operation.

FIFO A receives data from slave register 0. By this we mean when the processor writes data to the hardware core's base address with the offset of 0×00, data will then be written into FIFO A. Similarly, FIFO B receives data from slave register 1, which is the hardware core's base address with the offset of 0×04. When data are written to either of these slave registers, there is a single clock cycle delay between when data arrive from the bus and when it is valid in the register. To make sure we write the correct data into the FIFO, we must register the write select from the bus, using one of our intermediate signals we declared, slv_we_reg. This is done by adding the following process:

```
SLAVE_WE_REG_PROC : process( Bus2IP_Clk ) is
begin
  if Bus2IP_Clk'event and Bus2IP_Clk = '1' then
    if Bus2IP_Reset = '1' then
      slv_we_reg <= (others => '0');
    else
      slv_we_reg <= slv_reg_write_sel;
    end if;
  end if;
end process SLAVE_WE_REG_PROC;
```

Now we can drive the write enable signals for FIFO A and B, correctly storing the single precision operands A and B into the corresponding FIFO.

```
-- FIFO A Signals
fifo_A_din   <= slv_reg0;
fifo_A_wr_en <= slv_we_reg(0);
fifo_A_rd_en <= adder_re;
-- FIFO B Signals
fifo_B_din   <= slv_reg1;
fifo_B_wr_en <= slv_we_reg(1);
fifo_B_rd_en <= adder_re;
-- FIFO F Signals
slv_reg2     <= fifo_F_dout;
fifo_F_rd_en <= Bus2IP_RdCE(2);
fifo_F_wr_en <= adder_rdy;
fifo_F_din   <= adder_result;
-- FIFO A/B Read Enables
adder_re   <= not(fifo_A_empty) and not(fifo_B_empty) and
              not(fifo_F_full) and adder_rfd;
```

```
-- Adder Signals
adder_a     <= fifo_A_dout;
adder_b     <= fifo_B_dout;
adder_nd    <= fifo_A_valid and fifo_B_valid and not(fifo_F_full);
```

In addition to data and write enable signals, we are also driving the read-enable signals for FIFO A and FIFO B with the same signal, adder_re. This signal is asserted when both the operands are valid, the single precision adder is ready for data, and the resultant FIFO is not full. This condition assures that we do not consume data from FIFOs without the adder being able to process it or the result FIFO being able to store it.

Output data from the adder are written as input data to the FIFO. When data are *ready*, this indicates there are valid data to be written to the result FIFO. To support the processor reading data from the result FIFO, we connect the output from FIFO F to the third slave register (slv_reg_2) so that when the processor reads from the hardware core's base address with the offset *0x08*, the result will be returned to the processor. In order to support writing to slv_reg_2 we must comment out slv_reg_2 signal assignments in the SLAVE_REG_WRITE_PROC process. If we neglect this step we will receive an error during synthesis stating a *multisource* on the register.

3.A.10. Modify Hardware Core Project Files

Now that we have the core logic in place we can go ahead and modify the hardware core's project files to support the generated CoreGen netlists. First, we must create a *netlist* directory to store these netlists and allow the Xilinx tools to locate them during the NGCBuild process. Without these files we will be unable to generate a bitstream for testing.

```
% cd pcores/my_test_core v1_00_a
% mkdir netlist
% cp coregen/fifo_32x1024.ngc netlist/fifo_32x1024.ngc
% cp coregen/sp_fp_adder.ngc netlist/sp_fp_adder.ngc
```

Once the netlists have been properly copied into the newly created *netlist* directory, we must change directories to the *data* directory. Within the *data* directory we will create a new file with whatever text editor you are comfortable using called:

```
my_test_core_v2_1_0.bbd
```

This is known as the Black Box Description (BBD) file. The BBD file is used to identify the black box netlists of a hardware core. In our hardware core we have two black boxes, the FIFO and the single precision adder. To specify this within the BBD file we must add the following:

```
FILES
fifo_32x1024.ngc, sp_fp_adder.ngc
```

Finally, we must modify the Microprocessor Description File (MPD) within the same *data* directory so that the BBD file will be read and the netlists will be included in the final top-level netlist, generated by NGDBuild.

```
OPTION STYLE = MIX
```

This one line should go within the MPD file after the other peripheral *OPTION* flags. Forgetting this last step can result in an error during NGDBuild because it will be unable to find the netlists for the black boxes.

3.A.11. Connecting the Hardware Core to the Base System

Up until now the base system has remained unchanged because we have not yet connected the hardware core to the system. Within XPS we must locate and add the `my_test_core_v1_00_a` to the base system. Under the *IP Catalog* locate the *Project Local Pcores* category. Expanding this category should reveal a *User* category, which, when expanded, should reveal the *my_test_core* hardware core. In the event the local pcores category does not exist, it may be necessary to rescan the user repositories. This can be done through the

```
Project -> Rescan User Repositories menu option.
```

To add the hardware core to the base system, simply double-click (or right click and select Add IP). The hardware core should appear with the System Assembly View window within XPS.

Initially, the hardware core will be added to the design, but without a connection to any bus. In this project there is a single PLB, so we will expand the `my_test_core_0` instance and set the *SPLB* to *plb_v46_0* instance. This establishes the connection between the new hardware core and the existing base system. There is one last step needed to allow the processor to communicate with the hardware core: generate an address range for the hardware core. By selecting the *Address* tab within the System Assembly View window we can click on the *Generate Addresses* button to generate an address for our hardware core. Depending on the system (and version of the tools you are using) this may generate a similar address:

```
Instance          Base Name      Base Address   High Address   Size   Bus Interface
my_test_core_0    C_BASEADDR     0xC8200000     0x82C0FFFF     64 K   SPLB
```

Now we have a base system that is ready to be synthesized and tested. The final step is to synthesize the whole design and generate a bitstream. Select the

```
Hardware -> Generate Bitstream menu option.
```

The synthesis time depends on the build machine (and the additional peripherals added to the design), but should complete within approximately 5–15 minutes.

3.A.12. Testing the System

After the system synthesizes successfully, we must export the project to the software development kit (SDK). Following the direction we specified during section 3.A, make sure to launch the SDK after exporting the design. First we must create a new Software Platform:

- Project Name: standalone_platform
- Processor: ppc440_0 (ppc440_virtex5)
- Platform Type: standalone

Next, create a new *Managed Make C Application Project*:

- Project Name: adder_app
- Software Platform: standalone_platform
- Project Location: Use Default Location for Project
- Sample Applications: Empty Application

Once the empty application is generated we will need to add a new source file to the application. This can be done by right clicking on the *adder_app* and selecting *New* followed by clicking *Source File*. Name the source file *app.c*. To this source file we will add the following application.

```
#include <stdio.h>
#include ''xparameters.h''

// The Base Address of Custom Hardware Core
#define HW_BASEADDR XPAR_MY_TEST_CORE_0_BASEADDR

// A structure to make accessing the Slave Registers easier
typedef struct{
    float op_a;    // Operand A - Slave Register 0
    float op_b;    // Operand B - Slave Register 1
    float result;  // Result F  - Slave Register 2
}hw_reg;

int main() {
    // A pointer to the hardware core's slave registers
    // the address offsets are already handled by the struct
    volatile hw_reg *hw_core = (hw_reg*)(HW_BASEADDR);
    printf(''Single Precision Application Test\n'');
    printf(''Writing to Operand A\n'');
    hw_core->op_a = (float)123.45;
    printf(''Writing to Operand B\n'');
    hw_core->op_b = (float)678.90;
    printf(''Reading Results F\n'');
    printf(''%f = %f + %f\n'', hw_core->result, hw_core->op_a, hw_core->op_b);
    printf(''Test Complete\n!'');
return 0;
}
```

In this simple example we are writing a single set of operands A and B to the hardware core and reading the result back. Because we have FIFOs, we add a `for-loop` to populate multiple operands and read multiple results, but we will leave that exercise to the reader.

Programming the FPGA within the SDK   In section 3.A we generated an ACE file to program the FPGA. Alternatively, we could use a JTAG to program the device without needing to generate the ACE file. This process is only applicable for those who purchased a JTAG with the ml510 development board (or have a JTAG for their FPGA). Otherwise, follow the directions specified in section 3.A to create another ACE file.

Once the FPGA has been turned on and the terminal is opened to display output from the UART, we can use the SDK to program the FPGA's bitstream. To do this, select the

Tools -> Program FPGA menu option.

A new window will open to let you specify the bit and bmm files. Browse to the project's working directory's implementation directory, select the `.bit` and `_bd.bmm` files, and then select *Save and Program*. The FPGA will be programmed with the hardware bitstream, but the application will not yet be loaded into the on-chip memory.

```
Single Precision Application Test
Writing to Operand A
Writing to Operand B
Reading Results F
802.350037 = 123.449997 + 678.900024
Test Complete
```

Figure 3.15. Output for Single Precision test running on the Xilinx ML510 development board.

To load the program we need to right click on the *adder_app* and select the

Run As -> Run on Hardware menu option.

The application will be downloaded to the FPGA and the application will start running. If the terminal is open and configured to the correct baud rate you should see the output shown in Figure 3.15.

3.B. Embedded GNU/Linux System

Perhaps the greatest barrier to using free and open software in embedded systems is the learning curve. Over the next few chapters this text will introduce the bare minimum required to get started, including the tools and steps required to create a GNU/Linux-based system on a Platform FPGA. This section starts with the basic organization of a GNU/Linux filesystem, explains configuration and build tools, describes how to create a cross-compiler, builds a root filesystem, cross-compiles an application, and creates a cross-compiled kernel. That is a lot for one section but, fortunately, there are many supplemental resources available on the World Wide Web.

Specifically, we start with the organization of the UNIX filesystem, describe the two main configuration tools in use, and then how to build a cross-development environment. We then use these tools we just built to cross-compile Linux, build a root filesystem, and add a "Hello World!" application.

3.B.1. Organization of Unix Filesystem

If we are going to use a full-featured operating system, such as Linux, the first thing we have to concern ourselves with is the organization of the root filesystem. We need to decide where the required configuration files, system binaries (such as init), shared libraries, and application binaries will be located. We could put everything in the root directory, but as the number of files and supporting software packages increase, this becomes unwieldy. Likewise, we could come up with a completely novel scheme of directories and subdirectories — but again, this approach usually offers few advantages while potentially introducing considerable confusion. Instead, it makes sense to learn the conventional places to put things. Over the years a common hierarchy of directories has evolved and the Filesystem Hierarchy Standard (FHS) documents these *de facto* conventions. Using the convention minimizes the amount of configuration choices and reduces the chance of making a mistake. It is also much easier to ask for help from a colleague or the Internet if everything is where it normally is!

The first convention to follow is how to layout the directories and subdirectories of the root filesystem. (A detailed description of the entire tree for Linux can be found at www.pathname.com/fhs/.) A subset required for a medium-sized embedded system is described here. At the root, there are typically 15 directories. The directories bin, sbin, lib, and dev are used to hold general-purpose binaries, system binaries, shared/static library files, and device files, respectively. (Device files are used to provide applications with access to peripherals — we'll revisit them in the next chapter.) The `etc` directory is used to store configuration files that generally don't change frequently; the `var` directory is used to store run-time generated files that applications use to store state information or record events (such as log files). The directory `boot` is used by the bootloader to store configuration information about the different ways a system can be booted; embedded systems typically only use this directory during development since, in the field, the system only boots in one configuration. The `home` (or sometimes called `users`) directory has the home directories of all users except the super-user, root. The user root's home directory is just `root`. The `sys` and `proc` are special directories used for accessing kernel information, which are discussed later. The directories `opt` and `mnt` are for convenience — some developers install "optional" software in the former and the latter is a convenient place to mount new, temporary filesystems. The directory `tmp` should be obvious.

The remaining directory typically found in the root is called `usr`. Pronounced *user*, this directory has several subdirectories that may seem redundant: `bin`, `sbin`, and `lib`. Why two binary directories (`/bin` and `/usr/bin`)? There are two practical reasons. In the past, larger systems periodically checked a filesystem for inconsistencies but because one can't do a thorough check on a mounted filesystem, the system was booted without `usr` and a filesystem check could be performed. After the check, the root filesystem was extended by mounting the `usr` directory. Thus `/bin` and `/sbin` have the bare minimum to boot a system. The `/usr/bin` and other directories in `/usr` contain everything else needed to run the system.

Fortunately, one doesn't have to memorize these things. In practice, there are a number of simple shell scripts that create a template root filesystem (the book's Web site has one and its usage is explained shortly). Also, many Open Source packages already know the standard and automatically configure themselves to install in the proper locations. A typical layout is shown in Figure 3.16.

3.B.2. Configuration Software and Tools

There are two popular configuration tools used with Open Source software today. The first is often referred to as the "autotools" or "configure script" approach. The former name refers to a set of tools (autoconf, automake, and libtool) used to create a set of Makefiles and a `configure` script. The latter name refers to the output of running the tools. In either case, these software tools are used by package developers to make their software portable and easy to install. The second major configuration technology is called a "menuconfig" system. This software was introduced to replace a slow question-answer system that was originally used to configure the Linux kernel. Since then, it has been adopted by several other packages. In addition to these configuration tools, a handful of other software tools are useful for collecting and managing software packages.

GNU Release Process   Most GNU software and many (non-GNU) open source packages come with a `configure` script. This script, among other things, prepares the software to be compiled and installed. UNIX and UNIX-like systems have a long and interesting history. As a result, different libraries of subroutines have evolved over time. There have been "best practices" that have emerged, peaked, and are now rarely employed. This has led to great diversity in terms of what features and subroutines might be available. The autoconf part of autotools was intended to interrogate the local system and then provide system configuration information to the application. Although many of the portability issues of the past have been resolved, the tools continue to be extremely useful in cross-compiling and installing application files. (Éric Lévénez has an interesting Web page that summarizes how 100+ different versions of Unix have influenced each other, `http://www.levenez.com/unix/`.)

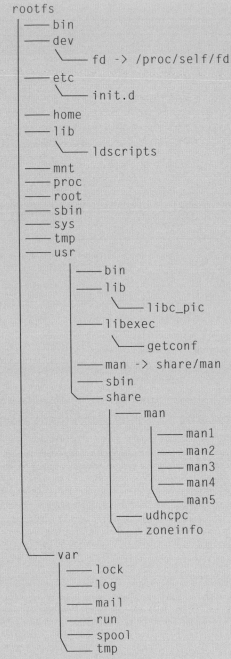

Figure 3.16. Typical bare-bones layout of FHS root filesystem.

The general process of installing a GNU package is:

1. Acquire the software (nowadays, this means downloading from the Internet) and unpack the software (uncompress and untar).
2. Go into the subdirectory created by unpacking and type:

```
./configure
```

3. Assuming that there were no incompatibilities that could not be worked around, the next step is to build the software package:

```
make
```

4. An optional step is to type "make check" — this will run regression tests and does a quick sanity check before the software is installed on your system
5. Assuming the software has compiled correctly and passed the optional regression check, the last step is to install the software. This is accomplished with

```
make install
```

This will not just copy the executables into the proper locations but also creates configuration files and anything else required to run the application.

By default, most software packages will install in a directory called /usr/local. This directory mirrors the /usr directory (in the sense that there is a /usr/local/bin, /usr/local/lib, /usr/local/include, etc.) and indicates that this is locally installed software that didn't ship with the workstation.

However, the location is very easy to change. Most configure scripts allow the user to specify a prefix that changes the install location. For example, the command:

```
./configure --prefix=/usr
```

will set up the software so that after it has been compiled, binaries will be installed in /usr/bin. In addition to —prefix there is another set of several options that let you further customize the specific directories that various types of files are installed. Most GNU packages have the option of adding arguments to the command that direct how the application is to be compiled and how it should behave.

Later it will become necessary to build the root filesystem in a subdirectory of another filesystem. Simply changing the —prefix option will not always work because some applications use the prefix to compile absolute paths inside the application. So instead of saying something like

```
./configure --prefix=/home/rsass/build/rootfs/usr
:
make install
```

one can use an environment variable, DESTDIR, to point at the subdirectory that will be the root filesystem in the embedded system. It is used on the make install command as shown below.

```
./configure --prefix=/usr
:
make DESTDIR=/home/rsass/build/rootfs install
```

Some of the most important of these configuration options (at least for embedded systems) are the options that direct how to cross-compile the application. These are `--host`, `--build`, and `--target`. Often in embedded systems, the "host" is the workstation where development and debugging take place and the "target" is the embedded system that is being developed. However, in GNU terminology, the terms *host* and *target* are used differently. So, for the moment, completely forget about target and we will just talk about the host machine and the build machine.

In GNU terminology, host refers to the machine that will ultimately execute the application. The configure script will assume the host is the machine the script is being run on, but in our case, the host is the processor and C library on our embedded system. So the host refers to what kind of binary executables we want to produce. Because these binaries will likely be dynamically linked to a library, the version of the C library is important as well. To tell the configure script to assume a different **host**, you use the `-host` argument. In GNU terminology, **build** refers to the machine we are going to use to compile the application. (Obviously, there has to be a cross-compiler on the build machine that knows how to produce the host's executables.) Often, the configure can guess the build machine by assuming it is the one that ran the configure script. So this is only strictly necessary when the person installing the software knows they are going to run "make" on another machine. To tell to configure that the package will be compiled on a machine, one uses the `--build` argument. (We will return to the `--target` option at the end of this discussion.)

These three configure options take a single argument called a machine triple. A **machine triple** is a string that uses hyphens to specify a machine's vendor, CPU model, and operating system. For example, `sparc-sun-solaris` indicates that the machine was manufactured by Sun, Inc., that the instruction set of the processor is SPARC, and that the OS is Solaris. This is in contrast to `sun-sparc-sunos`, which is the same hardware except that it is running the SunOS version of Unix instead of Solaris. And, like a lot of things, it complicated. For example, if part of the machine triple is implied or is generic (therefore not relevant to configuring), then it can be omitted. For example, `sparc-solaris` is a legal abbreviation because the vendor is implied. Further complicating the situation is that some parts of the triple need more detail. So, for example, we will be using the PowerPC in the Virtex 5 for many of the examples, but the PowerPC has many variants designed for different roles. The PowerPC 604 was intended for desktops and workstations, while the PowerPC 405 and PowerPC 440 are for embedded systems. To indicate the model, `powerpc-405` or `powerpc-440` could be used in the slot for the CPU (making the machine triple look like it has four components!). Similarly, many operating systems need to be clarified. For example, in common parlance, people refer to a popular open source operating system as "Linux." However, it is important to the configure script to know more. What most people mean by "Linux" is referred to more properly as `linux-gnu`. That is, the operating system kernel is Linux but the C library and the rest of the operating system come from GNU. In contrast, `mach-gnu` and `hurd-gnu` refer to the same library and OS utilities but are different OS kernels. So the machine triple for one of the authors' laptops is `i686-pc-linux-gnu`. The vendor has been omitted, the processor is `i686-pc`, and the operating system is `linux-gnu`. Likewise, the hard procesor core found on the Virtex 5 is `powerpc-440-linux-gnu`. The hard procesor core found on the Virtex 4 is `powerpc-405-linux-gnu`. Thus, to configure a software package to be compiled an Intel desktop for the embedded PowerPC in a Virtex 5, one might use the following command.

Thus, to configure a software package to be compiled an Intel desktop for the embedded PowerPC in a Virtex 5, one might use the following command

```
./configure --prefix=/usr --build=i686-pc-linux-gnu--host=powerpc-440-linux-gnu
```

Machine Triple	Type of Machine
i686-pc-linux-gnu	32-bit Intel processor Linux Distribution
x86_64-unknown-linux-gnu	64-bit Intel processor Linux Distribution
sparc-sun-solaris2.9	SPARC processor on Sun Inc. hardware running solaris
mips-sgi-irix6	MIPS processor on SGI Inc. hardware running Irix

Table 3.1 Various machine triples.

This will tell the software package that ultimately this system is going to be transferred to a PowerPC running GNU/Linux and that the cross-development tools to use have the prefix powerpc-440-linux-gnu-. For example, the name of the C compiler would be powerpc-440-linux-gnu-gcc.

What about --target? Well some applications are designed to use object files either as an input (a linker/loader) or as an output (an assembler). This introduces a third machine possibility: the package can be configured to be compiled on one machine (the build machine), executed on another (the host machine), and manipulate the object files of a third machine (the target machine). So, one might compile an assembler where the build and host machine are i686-pc-linux-gnu but the target is powerpc-440-linux-gnu. This would be a cross-assembler — it is an executable runs on Intel x86 machine but it takes PowerPC mnemonics and produce PowerPC object files. Clear? If not, you can use this rule of thumb: if the package has nothing to do with manipulating a machine's object files then the --target option is unneeded.

Menuconfig For software, such as Web server or an XML parser, the number of configure/install options is relatively small (less than 50). However, for other software that is highly configurable (such as Crosstool-NG) it is unwieldy to specify these on a command line. Moreover, it is difficult to keep track of a specific configuration. To handle this better, the Menuconfig system was developed. The most common version is a Curses-based application that presents the person installing the software with a menu of choices. In the case of the Linux kernel, every choice has three options: Yes, No, or Module (indicated by a [*], [ ], and [M], respectively). *Yes* means that the option should be compiled into the kernel. *No* means that the option should not be compiled and no support for incorporating the option at run-time should be included. The *Module* option indicates that the option should be compiled separately from the kernel and provisions made so that the option can be inserted and removed at run-time. Device drivers are often compiled as modules. There is no reason to have every device driver in the kernel when a system is booted because there are several hundred devices supported by Linux but any one system is only going to have a handful of devices installed. If a USB thumb drive is plugged in at run-time, the appropriate modules are inserted dynamically and the system can start using the device immediately.

These related options are grouped together into menus and then related menus are grouped to form submenus of the main menu in a hierarchy.

The program is accessed by unpacking the software and typing:

```
make menuconfig
```

If the system being configured is going to be cross-compiled, it is a good habit to specify the architecture and cross-compiler. (For some applications that use menuconfig, this doesn't matter, but for others it does.) So, for example, to configure the Linux kernel for the Virtex 5's hard processor, one would type

```
make ARCH=powerpc CROSS_COMPILE=powerpc-440-linux-gnu- menuconfig
```

The `ARCH` environment variable tells menuconfig to ask questions specific to building PowerPC kernels. The `CROSS_COMPILE` environment variable tells menuconfig the prefix of cross-development tools. Note the trailing hyphen after the machine triple; this string is the prefix of cross-development tools so the hyphen is required. In other words, this is a prefix of a command, not a machine triple.

The input to menuconfig is one or more `Kconfig` files that have all of the configuration options to present to the person compiling the software. The output is a `.config` file. Both are plain ASCII files that are human-readable. The `.config` file has one option per line in the form of:

```
CT_GMP_MPFR=y
```

or

```
CT_GMP_MPFR=m
```

or

```
# CT_GMP_MPFR is not set
```

for "compile into the kernel," "compile as a module," and "don't compile," respectively.

This means that one can save a configuration by simply copying the `.config` file to another place (or name). It also means that one can quickly determine the differences between two configurations with the utilities such as `diff`. The Linux kernel has several architecture-specific default configurations included. The command

```
make ARCH=powerpc CROSS_COMPILE=powerpc-440-linux-gnu-44x/virtex5_defconfig
```

will create a default `.config` that is a reasonable starting point for the Xilinx ML-510 boards.

Note that if you create a `.config` manually (by copying it from somewhere, for example), you still need to run through the menuconfig system at least once or by typing

```
make ARCH=powerpc CROSS_COMPILE=powerpc-440-linux-gnu- oldconfig
```

3.B.3. Cross-Development Tools and Libraries

The traditional Board Support Package found in microcontroller-based embedded systems is relatively unchanging and works well for other embedded system platforms because it can rely on a number of fixed characteristics. Either the processor has a floating point unit or it doesn't. Either there is RAM at a particular place in the address map or not. When there is variability in a traditional microcontroller-based system, it is fairly limited. Not so for Platform FPGAs! Although the hard processor core on the Virtex 5 FX device is fixed (a PowerPC 440) it is not the only processor option. In addition to the MicroBlaze that comes with XPS, there are a range of soft processor cores as SPARC and MIPS implementations. Furthermore, even the hard processor cores found on FPGAs can be customized. Thus, in addition to building hardware and creating the embedded system application, a Platform FPGA designer often has to build the cross-development tools and system software as well. Here we describe the steps used to build a GNU Compiler Collection (GCC) tool chain for the PowerPC 440 found on the Virtex 5 FX device. We assume that an FPU is not present. Only two changes are needed to build the tool chain for the Virtex 4's PowerPC 405.

There are four (or five if you want a debugger) major software packages central to building a cross-development environment. The first is called the binutils package. The binutils package includes a number of primitive utilities to create and manipulate binaries. This includes an assembler, a linker/loader, and a collection of utilities that operate on object files. The next package is the C library, which provides a collection of standard subroutines for C (and other languages). The C compiler needs to have a C library available when it is compiled, this presents an interesting chicken-and-egg problem. Because most of the C library is written in C, it needs a cross-compiler to compile it. The solution is a multistage build where a minimal C compiler is created to cross-compile the library, which is then used to compile the full C (and other languages) cross-compiler(s). It is also worth noting that several C library options are available. The newlib C library is the simplest and often used in embedded systems without an operating system. There is the GNU C Library (GLIBC), which is a full-featured library. Recently, there is a fork of the GLIBC project called Embedded GNU C Library (EGLIBC), which is the one we'll use here. The fourth essential software package is the Linux kernel. We are not going to cross-compile the Linux kernel at this point, but the compiler and the C library need to know what system calls are available, as well as the value of certain constants in the kernel's header files. These interfaces are illustrated in Figure 3.17, and the interdependencies among the last three components are shown in Figure 3.18. What is needed are kernel headers, which can be extracted without building the entire kernel yet.

So the high-level view of the process is as follows. Get the kernel headers, build the binutils for the PowerPC 440, build a static, minimal cross compiler, get Embedded GLIBC headers, recompile GCC, rebuild the full C library, and then build the full C compiler (and other languages) using the the "full" C library. Then we use our newly built cross-compiler to finish building binaries for the embedded system. Note that GCC 4.3 now requires two additional libraries: GMP and MPFR, which are libraries that implement arbitrary precision data types and multiprecision floating-point operations, respectively.

Downloading and Installing Crosstool-NG This process is long, laborious, and error-prone. A wonderful script called `crosstool.sh` created by Dan Kegel (see `http://www.kegel.com/crosstool/`) has automated this procedure for a variety of processors and software versions. For years, it served the authors of this text well. However, as

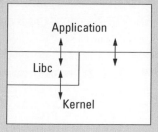

Figure 3.17. Interface relations.

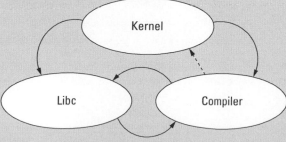

Figure 3.18. Dependencies between tools.

of this writing the latest vesion is 0.43 and that version is very out of date. GCC and GLIBC have evolved substantially since version is 0.43 of crosstool was released. Moreover, GCC's adoption of GMP and MPFR means that the script cannot compile the latest versions.

Inspired by Kegel's work, Yann Morin developed his own script to accomplish the same thing, automate the building of a cross-development tool chain. Kegel used a collection of data files with environment variables to manage the different processors and versions. Yann added a menuconfig-like interface. His system, called `crosstool-NG`, is invoked with the command

```
% ct-ng
```

and can be found at `http://ymorin.is-a-geek.org/projects/crosstool`. An interesting, but perfectly appropriate approach employed by Yann is to use a `./configure` script to check that the host system has the required versions of the system tools needed to configure and compile the tool chains. This project is very active at the time of publication and it is the tool we currently use. We specifically recommend the following setup. Starting with the `/opt` directory that we used to install the Xilinx tools in Chapter 1, we install the `ct-ng` command and software tools needed compile the GCC tool chains in the directory `/opt/crosstool-ng`. Next we use crosstool-NG to build different versions of the tool chain. For example, we might have one version that uses a software library to emulate the PowerPC's floating point operations and another version that assumes an floating-point unit has been synthesized. The directories might look as follows.

```
/opt/crosstool-ng/powerpc-440-linux-gnu
/opt/crosstool-ng/powerpc-440-fpu-linux-gnu
```

(Those directories contain the cross-compilers and cross-compiled libraries created by crosstool-ng.) Finally, we add a script `/opt/crosstool-ng/bin` to help create FHS compliant root filesystems and a `settings.sh` that, when sourced, will setup the user's PATH and MANPATH environment variables.

The first step is to download, configure, and install crosstool-ng. This may require downloading the latest version of some of the autotools because a few well-known Linux distributions have not yet upgraded to the required version. (If crosstool-NG passes the configure stage, then you can skip the steps related to autoconf, automake, and lzma given later.) We also install genext2fs at this time — it will be needed when we build a root filesystem image in section 3.B.5. This package can be found at `http://sourceforge.net/projects/genext2fs/files/`. The detailed step-by-step proceedure is given on the books Web page. The last step is to add two scripts that can be found on the book's Web site. The `settings.sh` script that we use also incorporates the path of a default cross-compiler and is illustrated in in Figure 3.19. The `mkfhs.sh` script creates a FHS compliant directory and populates it with some simple configuration files. This script can be copied into `/opt/crosstool-ng/bin` as well.

This is a one-time process and the tools installed thus far will let us create multiple tool chains and will support multiple projects. To build a simple PowerPC 440 cross-compiler suitable for the hard processor core on the ML-510's Virtex 5 chip, one would create a working directory and then run the `ct-ng` command. We will start with a PowerPC 405 configuration and then modify it for the 440. The general steps are

1. Create a default configuration with the command

```
ct-ng powerpc-405-linux-gnu
```

2. Edit the configuration with menuconfig

```
ct-ng menuconfig
```

```
#!/bin/sh
CROSSTOOL=/opt/crosstool-ng
export CROSSTOOL
# default toolchain (can be overriden on command line)
MT=${1:-powerpc-440-linux-gnu}
# generic rootfs scripts and ct-ng itself
# note we have to be first to get our version of autotools
PATH=${CROSSTOOL}/bin:${PATH}
# then add a specific tool chain
CROSSTOOL=${CROSSTOOL}/${MT}
PATH=${PATH}:${CROSSTOOL}/bin
export PATH
MANPATH=${MANPATH}:${CROSSTOOL}/share/man
```

Figure 3.19. The settings.sh to set up environment.

and then under the "Paths and misc options" menu change the prefix to be /opt/crosstool-ng/ ${CT_TARGET}. Also change "Emit assembly" and "Tune" options under "Target options" menu from 405 to 440. Change the "Tuple's vendor string" to 440 on the "Toolchain options" menu. And then, last, change the "C Library" from GLIBC to EGLIBC.

3. The last step is to build the cross-compiler tool chain with the commmand

```
ct-ng build
```

On an older desktop with a DSL modem, the process took 325 minutes to download and compile everything! A reasonably fast machine, with all of the software already downloaded, takes around 120 minutes. Detailed instructions with screenshots are on the book's Web site.

Now that we have a cross-development tool chain, we are ready to write an application. To do this, we need to cross-compile an operating system, create a root filesystem, and cross-compile the application. We describe these remaining steps next.

3.B.4. Cross-Compiling Linux

If you know how to compile a Linux kernel, cross-compiling Linux is simple. However, compiling the Linux kernel can be intimidating to people that are new at it because there are an enormous number of options and many of the choices assume a significant knowledge of commodity desktop hardware and its history. While many of the hardware cores described earlier (such as the 16550 UART) have Linux device drivers that support it directly, there are others that do not. Obviously, any custom hardware cores that you develop for a project won't have a Linux device driver for it!

The easiest way to get started and manage this complexity is to (1) rely on a Xilinx-maintained Linux repository that has Xilinx device drivers added to the kernel and (2) start with a known-to-work configuration and modify it. Fortunately, the 2.6 kernel series has a number of "default config" options that we can use.

Preparation   Before compiling Linux for the first time, there are a few things we need to do. The most important is to determine how to make Linux aware of our hardware design — including what hardware cores are included, how they are configured, and where they are located in the memory map. Also, as we begin to add more components to our Linux-based FPGA hardware, we describe some ways to organize these components.

In older versions of XPS and the Linux kernel, the common way of making device (hardware core) information from XPS available to the Linux kernel was by passing around an `xparameters.h` file. This file was generated by XPS as part of a Software Platform Settings option. By copying this file into the appropriate place in the Linux kernel's source, the critical information could be determined at compile-time. This information is only relevant for existing projects that use the "ppc" architecture in the Linux kernel.

In 2009, maintainers of the "ppc" architecture (which was mostly used for embedded systems) switched to using the architecture "powerpc." With this change, the way of communicating information about the hardware design in XPS to Linux changed as well. Instead of having an `xparameters.h` file at compile time, Linux uses a device tree file at boot. This approach allows the same compiled kernel to be used with multiple hardware designs — the designer simply configures the bootloader to pass a device tree to the kernel at boot. Alternatively, one can combine device tree information with the kernel at compile-time if one doesn't have a bootloader that supports device tree files.

With version 11.4 of the XPS, it is necessary to add device tree support to XPS. Linux (and closely related tools) uses a source code management system called 'git.' We won't go into the details of git but the commands are relatively simple.

After installing XPS, one needs to download the device tree generator from Xilinx following the instructions listed on the book's Web page.

Before downloading Linux, it is worthwhile to think a little bit about how to organize the growing number of components in a project. One way is illustrated in Figure 3.20. We create the hardware first in an `hw` directory, next we cross-compile Linux in a directory created by the `git` command, in the next section we'll create the software that goes on the root filesystem, and then we finally combine the hardware and software together into an ACE file.

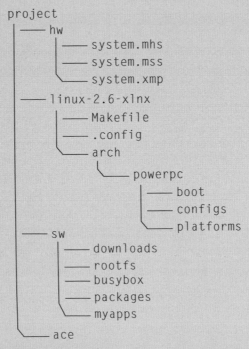

Figure 3.20. Typical directory structure of a project.

Building the Hardware Base System   We can quickly generate a hardware base system with the Base System Builder wizard from within XPS. Unlike the base systems generated previously, we must include a few additional components to support Linux. We will continue to use the Xilinx ML510 with the Virtex 5 FX130T FPGA. Create the base system described in Table 3.2.

In this base system we are preparing to support a Linux system running out of 512 MB of DDR2 memory, with Ethernet support, an RS232 UART interface running at 9600 baud, and and optional LCD display. We include 64 KB of BRAM for testing purposes. It is not necessary for future designs to include this much memory, although it is necessary to include some BRAM to boot up the initial system. The IIC EEPROM will be used in a future design in section 3.A.

BSB Wizard Page	BSB Settings
Board Selection	Xilinx Virtex 5 ML510 Evaluation Platform Revision C
System Configuration	Single-Processor System
Processor Configuration	400 MHz PowerPC processor with 100 MHz Bus
Peripherals	refer to Figure 3.21

Table 3.2 Base System Builder wizard options for a Linux base system.

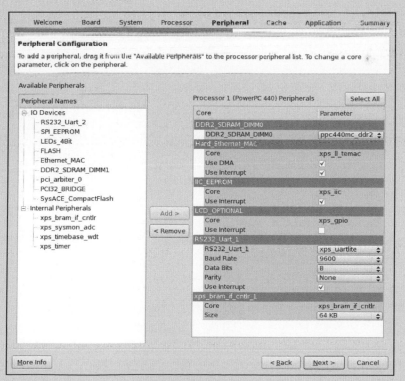

Figure 3.21. Peripherals needed for the "hello world" Linux system on the Xilinx ML510 development board.

Next we must *Export Hardware Design to SDK*, which will synthesize and generate the system's bitstream along with creating the SDK project and launching the SDK.

From within the SDK we must generate a new *Board Support Package* project. We can name this project `linux_bsp`. Set the *Board Support Package Type* to *device-tree*. If this option is not available, be sure to download and install the device tree generator as instructed in Section 3.B.4.

Under the Board Support Package Settings window, change the field "bootargs" for the console from

```
console=ttyS0 root=/dev/ram
```

to

```
console=ttyUL0 root=/dev/ram
```

which specifies that we will be using the UART lite.

Now a `xilinx.dts` file should exist in your SDK Project's directory under the following path:

```
% SDK/SDK_Workspace/linux_bsp/ppc440_0/libsrc/device-tree_v0_00_x
```

Remember this location as we will return to this file shortly when compiling Linux.

Compiling the Kernel To get the most recent update of the Xilinx-maintained kernel from the Xilinx GIT repository. Detailed directions are available online as well from the book's Web site.

Once the Linux kernel has been downloaded, we must configure it with a default Virtex 5 configuration. This is done to save us time navigating though the various menuconfig options. Change directories into the `linux-2.6-xlnx` directory and run the following make command

```
% cd linux-2.6-xlnx
% make ARCH=powerpc 44x/virtext5_defconfig
```

This copies the `arch/powerpc/virtex5_defconfig` configuration file in place of the `.config` file. Next we must copy our `xilinx.dts` file from our SDK directory to the DTS directory within the `arch/powerpc/boot/dts` directory. The DTS file specifies the components, their physical addresses, and other information pertaining to our hardware base system. When compiling Linux, the DTS file will provide the necessary information such that the system will be capable to correctly interface with our hardware.

With the DTS file in place and the default Linux configuration set for our PowerPC 440-based system, we can run *menuconfig* to set the "Default RAM Disk Size" to 16384. (This is found under the Block Device submenu under the Device Drivers menu.)

Now all that is left is to compile the Linux kernel.

```
% make ARCH=powerpc CROSS_COMPILE=powerpc-440-linux-gnu- simpleImage.xilinx
```

At this point we have a compiled Linux kernel that is configured to run with our specific hardware base system. Of course, a system with just a Linux kernel does not provide us with much. In fact, we will need to build the root filesystem before we will be able to boot the kernel on the Xilinx ML510 development board.

3.B.5. Building a Root Filesystem

Once we have the hardware, a cross-development tool chain, and a compiled kernel, the next step is to build a root filesystem. There are several reasonable options. If our FPGA board has a (removable) secondary storage device and the kernel has been configured for its interface, then we can use a partition on that filesystem to be our root filesystem. For example, many Xilinx FPGA boards have a CompactFlash drive that the SysACE (which programs the FPGA at power on) device uses to read stored bitstreams. Because the SysACE device only looks at the first partition, a Linux system can use another partition on the drive as a root filesystem. A second option is to use NFS — a Network File System protocol that allows a Linux system to mount a filesystem over a network (such as Ethernet). Of course, both of these options require that the proper kernel modules be compiled into the kernel. The NFS approach also requires that a DHCP and NFS servers be set up. Probably the easiest solution is to begin with is to build a RAMDISK image and include it with the kernel. Once this is mastered, switching to NFS for development work is generally faster. Having the root filesystem in flash makes the most sense when shipping the product.

One tells the Linux kernel where the root filesystem is by passing a command line argument to it. Table 3.3 shows the three options.

So a RAMDISK-based Linux system with the console on a UART Lite device (with a baud rate 115,200) would have the Linux command line

```
console=ttyUL0,115200 root=/dev/root
```

set during the kernel configuration, as we have already done.

Once we have a kernel, the next step is to create the root filesystem.

There are five basic steps to creating a root filesystem. The script we previously installed, mkfhs.sh (for "make filesystem hierarchy standard"), does three of them. You have to do one manually. Then genext2fs does the last step.

Specifically, the first thing is to create the sw directory in your project directory. In that directory run the script mkfhs. (If you sourced the settings.sh file in /opt/crosstool-ng, then this script will bin your path and the environment setup.)

The script is too long to include verbatim here but is available on the Web site. After checking to see that you don't already have a rootfs created and that environment is correct, the first thing that it does is populate a directory, rootfs, with every dynamically linked library (shared object) that is used by the cross-development tools. Then it goes through and removes all of general computing system libraries and files to save space on our embedded systems root filesystem. Then it strips the debugging information from the shared objects — we assume that no one will want to debug the C library on an embedded system! The last thing it does is unpack a bunch of typical configuration files – the bare minimum required to boot a system.

Next we need to add system binaries to our root filesystem. That is, add the usual Linux command line applications such as /bin/ls and the command line shell. It is possible to build the entire root filesystem from scratch; in fact, there are many suitable Web sites to explain all of the details. One in particular that we find very useful is http://www.linuxfromscratch.org.

Location	Kernel Parameters (examples)
RAMDISK	root=/dev/ram
NFS	root=/dev/nfs nfsroot=192.168.1.101:/export/rootfs/n01 ip=dhcp
CompactFlash	root=/dev/xsa1

Table 3.3 Command line arguments to specify the location of the root file system.

Alternatively, we quickly generate a root filesystem with the help of another package. BusyBox can be used to generate a majority of the binaries needed to boot and operate a Linux-based system on an FPGA, `http://www.busybox.net`. In you decide to use BusyBox (which is recommended for those relatively new to filesystems) we have detailed instructions on the book's Web site on how to download, configure, and cross-compile Busybox.

In addition to the system binaries, this is when you add any kernel modules you have developed and your embedded application to the root filesystem. For this demonstration, we can just use the familiar "Hello, World!" program. If the program is in a file named `hello_world.c`, then we can cross-compile the application with `powerpc-440-linux-gnu-gcc`.

```
% powerpc-440-linux-gnu-gcc -o hello_world_440 hello_world.c
```

Even though the build machine cross-compiled the application, it will not be able to execute the newly cross-compiled application *hello_world_440*.

```
% ./hello_world_440
  -bash: ./hello_world_440: cannot execute binary file
```

Instead, the executable must be copied into the root filesystem generated earlier. For convenience, copy the executable to the `rootfs/root` directory.

The last step is to create a filesystem image. This converts the subdirectory `rootfs` into a single file that is block-for-block identical to what would be found on the partition of a disk drive. One popular way of accomplishing this is to create an empty file and use the mount command with the loop back option. However, this requires root privileges and small mistakes as root can have catastrophic effects. A safer approach is to use the genext2fs command we installed earlier. Genext2fs was written by Xavier Bestel and allows a nonroot user to create an ext2 filesystem image from a directory and a device file. (The device file also allows the nonroot user the ability to change the owner of a file, set permissions, etc.)

After the root filesystem image is created, it is copied to the `arch/powerpc/boot` directory in Linux kernel. With the filesystem image in place, we can now combine Linux kernel previously created with our root filesystem using the command

```
% make ARCH=powerpc CROSS_COMPILE=powerpc-440-linux-gnu- simpleImage.initrd.xilinx
```

Note that "initrd" stands for "initial ramdisk." Ramdisk is another term for the filesystem image and some general purpose Linux distributions use an initial ramdisk to load required kernel modules before accessing the disk drives.

The command generates the file `simpleImage.initrd.xilinx.elf` in the `arch/powerpc/boot` directory. At this point, we have all of the software for our embedded system in a single file. The last step is to combine this ELF file with the hardware's `download.bit` bitstream file. This is accomplished with the `xmd` command and the `genace.tcl` script. One follows the same proceedure that was previously described in section 1.B.

3.B.6. Booting Linux on the ML510

Finally, we can copy the ACE file to the CompactFlash for our development board and boot the system. Linux will begin to boot with its output being displayed through the UART to a terminal program, such as minicom, as seen in Figure 3.22. The boot process should take less than a minute, depending on the configuration options selected. Once booted, the user can login with the default *root* login (which does not require a password).

```
(none) login: root

Greetings from Linux on the ML510 Development Board!

% pwd
/root
% cd /
% ls
bin         home        lost+found  proc        sys         var
dev         lib         mnt         root        tmp etc linuxrc
opt         sbin        usr
```

Figure 3.22.   Booting Linux on the Xilinx ML510.

Once logged in, the user can explore the Linux filesystem hierarchy and test the "Hello, World!" program that we added to our filesystem. Return to the /root subdirectory and the ls should show the hello_world_440 executable in the /root directory. You can test it with this command.

```
% ./hello_world_440
Hello World!
```

At this point, we have all of the rudimentary skills to develop application-specific Board Support Packages complete with cross development tools and custom root filesystems. Next we can begin to look at more sophisticated applications and begin to develop simple kernel modules to interact with our hardware.

Exercises

P3.1. Which of the following is more abstract?

- $o = a \cdot s' + b \cdot s$
- a 2MUX with a, b, and a select line

Why?

P3.2. Name specific examples that will make a design less cohesive.

P3.3. Decoupling may lead to duplicate hardware. From a system perspective, why is this a positive characteristic?

P3.4. If reusing software means that the developer doesn't have to write it, why do we say the reuse has a cost associated with it? Who pays that cost?

P3.5. What is the difference between an *instance* and an *implementation*? How is each denoted in UML?

P3.6. Consider a large combinational circuit that consists of five XORs, five ANDs, and five inverters. A proposed design divides this circuit into three modules: one module has all of the XORs gates, another has the AND gates, and a third has the inverters. Comment on the quality of this design.

P3.7. Suppose we have been asked to design a portable MP3 player. Draw a Use-Case diagram to identify the major functionalities of the system.

P3.8. How does a stand-alone C program that outputs "Hello, World!" differ from one running on a Linux-based system? Be sure to consider the compiler, the resulting executable, the operating mode of the processor, and run-time support provided.

P3.9. Does one need to create a root filesystem for a stand-alone C program? Is it required for a Linux-based system?

P3.10. How does a cross-compiler differ from a native compiler? Does one need both? Will a developer ever need more than one cross-compiler?

P3.11. Does the choice of the C library impact the choice of the operating system kernel? Does the C library impact the choice of a cross-compiler?

P3.12. What is the difference between a monitor and a bootloader? What does a monitor provide that is not found in a bootloader? What does a bootloader provide that is not found in a monitor?

P3.13. What is the address map? What makes the address map more dynamic in Platform FPGA design compared to a traditional microcontroller?

P3.14. What are the three components of a GNU machine triple? When can the triple appear with less than three components? Why do some appear to have more than three components?

P3.15. What are the typical steps involved in installing a standard GNU software package on a root filesystem?

P3.16. What is the difference between the directories `/bin` and `/usr/bin`?

P3.17. What is the output of the `genext2fs` command?

P3.18. What are the major differences between `menuconfig./configure` techniques for configuring software? Contrast what is done automatically for the developer and the number of options.

P3.19. Name three ways to mount a root filesystem. What are the advantages of each?

References

1394 Trade Association. *1394 TA specifications.* (2010 January). Also available at `http://www.1394ta.org/developers/Specifications.html`, last accessed June 01, 2010.

Alhir, S. Si. (1998). *UML in a nutshell.* Sebastopol, CA, USA: O'Reilly & Associates, Inc.

David, W. (2008). *OpenSPARC Internals.* Santa Clara, CA: Sun Microsytems, Inc.

Denk, W. *Das U-boot manual.* (2007). `http://www.denx.de/wiki/U-Boot`, last accessed May 2010.

eCos. *RedBoot user's guide.* (2004). `http://www.gnu.org/software/grub/`, last accessed May 2010.

Electronics Industries Association. (1969 August). *EIA standard RS-232-C Interface between data terminal equipment and data communication equipment employing serial data interchange.* Greenlawn, NY.

Futral, W. T. (2001). *InfiniBand architecture: development and deployment — A strategic guide to server I/O solutions.* Hillsboro, OR: Intel Press.

GNU Project. *GRand unified bootloader.* (2007). `http://www.gnu.org/software/grub/`, last accessed May 2010.

Grimsrud, K., & Smith, H. (2003). *Serial ATA storage architecture and applications: Designing high-performance, cost-effective I/O solutions.* Hillsboro, OR: Intel Press.

Hennessy, J. L., & Patterson, D. A. (2002). *Computer architecture: A quantitative approach.* San Francisco, California: Morgan Kaufmann Publishers, Inc.

Holden, B., Anderson, D., Trodden, J., & Daves, M. (2008). *HyperTransport 3.1 interconnect technology.* Colorado Springs, CO: Mindshare Press.

HyperTransport Consortium. *Hypertransport specifications.* (2010 January). Also available at `http://www.hypertransport.org/default.cfm?page=HyperTransportSpecifications`, last accessed June 01, 2010.

IBM. *IBM coreConnect*. (2009 December). Also available at `http://www-03.ibm.com/chips/products/coreconnect/`, last accessed June 01, 2010.

InterNational Committee for Information Technology Standards T13. *AT attachment storage interface*. (2010 January). Also available at `http://www.t13.org`, last accessed June 01, 2010.

PCIIG. *PCI express specifications*. (2010 January). Also available at `http://www.pcisig.com/specifications/pciexpress/`, last accessed June 01, 2010.

Poulin, J. S., Caruso, J. M., & Hancock, D. R. (1993). The business case for software reuse. *IBM System Journal*, **32**(4), 567–594.

USB Implementers Forum (USB-IF). *USB 2.0 specification*. (2010a January). Also available at `http://www.usb.org/developers/docs/`, last accessed June 01, 2010.

USB Implementers Forum (USB-IF). *USB 3.0 specification*. (2010b January). Also available at `http://www.usb.org/developers/docs/`, last accessed June 01, 2010.

Wulf, W. A., & McKee, S. A. (1995). Hitting the memory wall: Implications of the obvious. *Computer Architecture News*, **23**(1), 20–24.

Xilinx, Inc. (2008 December). *PLBV46 interface simplifications (SP026) v1.2*, last accessed June 01, 2010.

Xilinx, Inc. (2009a October). *Embedded processor block in virtex-5 FPGAs (UG200) v1.7*, last accessed June 01, 2010.

Xilinx, Inc. (2009b June). *Floating-point operator generator data sheet (DS335) v5.0*, last accessed June 01, 2010.

Xilinx, Inc. (2009c October). *MicroBlaze processor reference guide (UG081) v10.3*, last accessed June 01, 2010.

Xilinx, Inc. (2009d December). *Processor local bus (PLB) v4.6 data sheet (DS531) v1.04a*, last accessed June 01, 2010.

Xilinx, Inc. (2010 January). *PicoBlaze 8-bit embedded microcontroller user guide (UG129) v2.0*, last accessed June 01, 2010.

4

PARTITIONING

There is always a point where a certain effect is the greatest, while on either side of this point it gradually diminishes. [Law of Diminishing Returns]

Thomas R. Malthus
The Corn Laws

The decomposition of a software reference design into two components — the portion to be realized in hardware and the portion that will run as software on the processor — is the essence of the partitioning problem.

For systems based on Platform FPGAs, partitioning is a subset of the more general problem addressed by hardware/software codesign. In codesign, hardware refers to the broader definition of hardware (physical, nonsequential, and possibly analog components) and, as such, involves assigning components of the application to a wide variety of devices with a wildly varying set of characteristics. Some may also be analog devices, which are unlikely to be specified in software. With the advent of System-on-a-Chip (SoC) technology, many of these components are likely to be integrated on a single chip, but some may be discrete parts.

With the hardware/software codesign problem, the system designer must contend with issues such as interfacing continuous and discrete signals and electrical characteristics that are generally the purview of electrical engineers. Thus, hardware/software codesign is a very different (and much larger) problem than the partitioning of an application for a single, Platform FPGA. While some of the techniques described in this chapter may also be useful in hardware/software codesign, it is important that readers are aware that a complete treatment of codesign is beyond the scope of this chapter.

The learning objectives of this chapter are to:

- Understand the principles behind partitioning through the use of profiling and how to analyze performance while avoiding common profiling pitfalls.

- Build a formal mathematical formulation of the partitioning problem and provide a heuristic for maximizing the performance metric.
- Address issues associated with interfacing the hardware and software components of a decomposed software reference design.
- Focus on several pragmatic concerns not accounted for in the formal description of the problem.

Also note that this chapter only considers the static decomposition of a software reference design — *dynamically* decomposing a program is a research issue.

4.1. Overview of Partitioning Problem

To begin, let's consider an application as a set of instructions organized as a collection of control flow graphs that specify the order of execution. The idea of **partitioning** is the grouping of specific sets of instructions in an application and then mapping those groups to either hardware or software. The groups assigned to software are executed sequentially (on a processor), while those mapped to hardware are implemented as custom combinational or sequential circuits and are realized in the configurable logic resources of the FPGA. If the application is a fully functioning software reference design, the result of partitioning is known as **decomposition**.

A number of factors help guide partitioning decisions. These include the expected performance gains (due to moving software into hardware), the resources used in the hardware implementation, how often that portion of the application is used, and — perhaps most importantly — how much communication overhead the decomposition introduces. Other pragmatic issues surface as well, including how difficult it is to implement a particular set of operations in hardware. In an ideal world, the groupings, mapping, and translation of instruction groups to hardware would happen automatically. However, state-of-the-art techniques are largely manual, with a handful of analytical and empirical tools to assist us.

We define a **feature** as a connected cluster of instructions from the application's software reference design suitable for a hardware implementation. (We think of the hardware implementation as being an additional, application-specific architectural feature. Hence the name.) For the moment, we will leave *suitable* imprecisely defined to mean "the system designer anticipates that a hardware implementation will prove advantageous." For now we can think of a feature as a subroutine from the software reference design. However, to achieve good partitionings, we generally have to examine groupings that may be bigger than or smaller than the subroutines defined by the programmer. In practice, a feature can vary from a few instructions to a "kernel" of nested loops to a whole software module consisting of multiple subroutines. Because feature size affects performance directly and indirectly, the decision of whether to implement a feature in hardware depends on its improvement to the overall system performance and the resources it uses relative to the other candidate features. If determined to be worthwhile, then the feature's hardware implementation augments the hardware architecture.

To provide some context for what we are trying to do here, let's consider a motivating example. The graph illustrated in Figure 4.1 depicts the execution of a JPEG2000 encoder application (Adams, 2007). From the main subroutine, we can see that 11% of the time is spent reading the source image and 89% of the time is doing the encoding. Encoding is not a single subroutine but rather a score of subroutines. If communication were free and hardware resources limitless, one would simply move everything into hardware. Given that neither is true, we have to determine which subroutines (or parts of subroutines) will have the biggest impact on some performance metric.

Profiling

The graph illustrated in Figure 4.1 was created by profiling the software reference design. To **profile** an application means to collect run-time information on an application during execution. With profiling, the software reference design is executed with representative input and the time spent in various parts of the application is measured. Profiling can be accomplished in a number of ways. Although it has been around for decades, it remains an area of active research as to optimize sequential program execution at run time. Here, we assume the simplest technique that profiles an application by periodically interrupting the program and sampling the program counter. A histogram is used to count how often a program is interrupted at a particular address in the application. From histogram data, an approximate fraction of total execution time spent in various parts of the application can be computed.

The Unix tool `gprof` is a good example of a sample-based technique. The default granularity for `gprof` is subroutines but it can also be configured to approximate the fraction of time spent in various basic blocks. Its output includes a flat profile of the percentage of time spent in each subroutine and a dynamic call graph

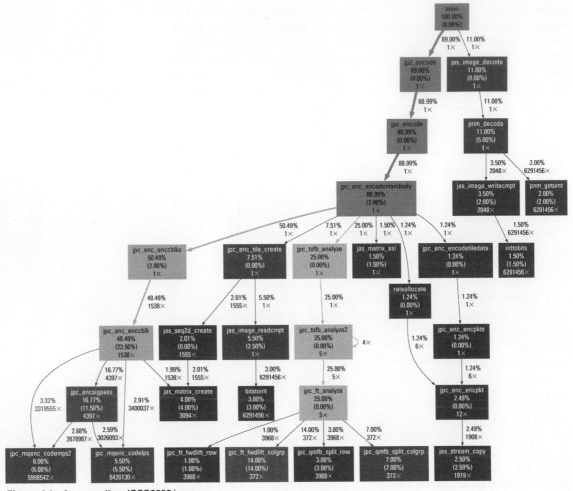

Figure 4.1. An encoding JPEG2000 image.

that indicates which subroutines were invoked. (We'll cover the practical details of using this tool in the gray pages of this chapter.)

Performance Analysis

Amdahl's law, which is well known in computer architecture, applies the law of diminishing returns to the usefulness of a single architecture feature. In short, it is written

$$\text{Speedup}_{\text{overall}} = \frac{1}{(1 - \text{Fraction}_{\text{enhanced}}) + \dfrac{\text{Fraction}_{\text{enhanced}}}{\text{Speedup}_{\text{enhanced}}}}$$

and it articulates the limits of the overall improvement to an application that a single enhancement can make on the execution time.

By generalizing this to include all potential enhancements and an arbitrary performance metric, we can characterize the limits of a collection of enhancements. If these enhancements are hardware features and we can anticipate the required resources potential gain for each feature, then we can develop a model that will guide the partitioning process.

Amdahl's law is not suitable because:

- It addresses a single enhancement and doesn't help us select a subset from a collection of potential features.
- It focuses solely on execution time — we generalize to any performance metric.
- It does not address resources required.
- It does not address communication costs.

In Practice

Of course, the model is only as good as the data used to drive it. Sample-based profiling requires representative input and is an approximation of time spent. Without implementing all potential features in hardware, it can be difficult to anticipate the resources required. Often, we can analytically determine how many clock cycles a hardware feature will take and we can calculate a metric, such as speedup. However, in general, it is impossible to know when an arbitrary process will complete. Rather than provide an automated solution, the analytical solution simply provides a framework to guide a system designer to a creative solution.

4.2. Analytical Solution to Partitioning

This section mathematically formulates the problem of grouping instructions into features and then mapping those features to either hardware or software. The final task — converting the features into FPGA implementations — is included throughout this book. Presently, the most common way of performing this translation is to manually design the core with an HDL using the software reference design as a specification. Many vendors have recently announced tools that are intended to automatically synthesize HDL cores from high-level languages; however, as of this writing their suitability for our purpose and effectiveness have not been established.

Many practical issues impact the actual performance of a system. Not all of these issues can be incorporated into an analytical model — so, at best, we can only expect a mathematical solution will produce an approximate answer to the partitioning problem. Moreover, many of the inputs to our model are estimates or approximations themselves, which further degrades the fidelity of our results. So why bother? By first solving the partitioning problem "on paper," so to speak, we get a partitioning that is close to optimal. From there, it is up to the designer to be artful in their use of guidelines and engineering expertise to refine the solution. In the end it is more efficient to use a combination of mathematical and ad hoc techniques to find an optimal solution than to simply rely on an engineer's intuition.

As mentioned earlier, performance is a complex metric. However, to simplify the presentation in this section, we focus on just one facet of performance, improving the rate of execution (or speedup) over an all-software solution. Specifically, the goal of this section is to find a partitioning of the system to maximize the speedup. Our choice of speedup is motivated by the fact that it has the most fully developed set of empirical tools and, despite the comments of Chapter 1, it is the primary goal of many designs.

The approach taken here requires that some measurements be taken from the software reference design (application) and that certain characteristics of the potential hardware implementation(s) are known. Some of this information can be derived from analysis of the source prior to execution, but often a good deal must come from executing test runs and profiling the application. The principles for performance metrics other than speedup are similar.

4.2.1. Basic Definitions

In Chapter 3, we modeled a *subroutine* of a software reference design as a Control Flow Graph (CFG). This notation will be used again to represent the flow between basic blocks of a subroutine. In addition, this chapter extends that notation to include a Call Graph (CG), which consists of a set of CFGs (one per subroutine)

$$\mathcal{C} = \{C_0, C_1, \ldots, C_{n-1}\}$$

where $C_i = (B_i, F_i)$ is the CFG of subroutine i. The application's (static) CG is then written

$$\mathcal{A} = (\mathcal{C}, \mathcal{L})$$

where $\mathcal{L} \subseteq \mathcal{C} \times \mathcal{C}$. Two subroutines are related $(C_i, C_j) \in \mathcal{L}$ if it can be determined at compile time that subroutine i has the potential to invoke subroutine j.

It is assumed that the basic blocks of every subroutine are disjoint: that is, every basic block in the application belongs to exactly one CFG. Moreover, it is assumed that a root node for the CG is implicit (i.e., one subroutine is distinguished at the start of execution). These are reasonable assumptions because, in practice, they are a natural artifact of how object files are created, stored, and executed.

On a practical note, not all executables can be expressed in this model. Certainly, system calls disrupt the control flow graph in unanticipated (and unrepresented) ways. Signal handling and interrupts are not represented. Some switch statements (i.e., case statements) are implemented as computed jumps. Thus, it may not be possible to determine all of the edges, F_i, in a given subroutine's Control Flow Graph, C_i, before execution. Finally, the object-oriented paradigm relies on run-time linkages to invoke virtual methods. By design, this paradigm prevents us from knowing all of the edges before execution. For now, we assume that the model is complete enough to express a software reference design and we will return to these practical issues in the next section.

A second practical note: we chose basic blocks because, to date, empirical evidence suggests that there is not enough parallelism at the basic block level to warrant attention at a finer grain level. If this changes in the future, it is relatively simple to recast the problem in terms of instructions (or smaller) computation units. However, gathering performance data on units less than a basic block may present a practical problem.

A common misconception is that a formal definition of partitioning only applies to the separation of an application into its hardware and software components. That is, the partition contains exactly two subsets. However, to make the problem more tractable, it is common to group operations into features first (a partition with a much larger number of subsets) and then map those features to either hardware or software. Assuming that the features are reasonably well clustered, then the decomposition of an application into hardware and software components can be driven by comparing the performance gains of features against one another.

First, let's formally define a partition. (We use this definition again in Chapter 5.) A partition $\mathcal{S} = \{S_0, S_1, \dots\}$ of some universal set U is a set of subsets of U such that

$$\bigcup_{S \in \mathcal{S}} S = U \tag{4.1}$$

$$\forall S, S' \in \mathcal{S} \mid S \cap S' = \emptyset \tag{4.2}$$

and

$$\forall S \in \mathcal{S} \cdot S \neq \emptyset \tag{4.3}$$

In prose, Equation 4.1 says that every element of U is a member of at least one subset $S \in \mathcal{S}$. Equations 4.2 and 4.3 say that the subsets $S \in \mathcal{S}$ are pair-wise disjoint and not empty. In other words, every element of our universe U ends up in exactly one of the subsets of $\mathcal{S}$ and none of the subsets are empty.

For example, consider the set of English vowels, $U = \{a, e, i, o, u, y\}$. One partition $\mathcal{X}_a$ of U is

$$\mathcal{X}_a = \{\{a, e, i, o, u\}, \{y\}\}$$

Figure 4.2 illustrates $\mathcal{X}_a$ graphically. Another example is

$$\mathcal{X}_b = \{\{a\}, \{e\}, \{i\}, \{o\}, \{u\}, \{y\}\}.$$

However,

$$\mathcal{X}_c = \{\{a\}, \{e\}, \{i\}, \{o\}\} \text{ and } \mathcal{X}_d = \{\{a, e, i\}, \{i, o, u, y\}, \{\}\}$$

are *not* partitions because $\mathcal{X}_c$ violates Equation 4.1 and $\mathcal{X}_d$ violates both Equation 4.2 and Equation 4.3.

We can also apply this to an application $\mathcal{A}$. If we assume our universe is the set of all of the basic blocks from every subroutine,

$$U = \bigcup_{C \in \mathcal{C}} V(C)$$

then the subroutines partition U. We will call this an application's natural partition where

$$\mathcal{S} = \left\{ \underbrace{\{b_0, b_1, \ldots, b_i\}}_{\text{subroutine } C_0}, \underbrace{\{b_i, b_{i+1}, \ldots, \}}_{\text{subroutine } C_1}, \cdots \underbrace{\{b_j, b_{j+1}, \ldots, \}}_{\text{subroutine } C_{n-1}} \right\}$$

Our task will be to reorganize the partition of basic blocks (and then map each subset to either hardware or software). We are

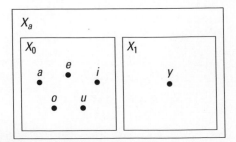

Figure 4.2. $\mathcal{X}_a$ illustrated graphically.

free to create new (nonempty) subsets, remove subsets, and move basic blocks around until we have a new partition. The result is a new $\mathcal{A}' = (\mathcal{C}', \mathcal{L}')$ inferred from the reorganized partition $\mathcal{X}'$. The second step is to map each subset of $\mathcal{X}$ to either hardware or software:

$$\mathcal{X}' = \left\{ \underbrace{\{b_j, b_{j+1}, \ldots,\}\{b_k, b_{k+1}, \ldots,\}}_{\text{software}} \cdots \underbrace{\{b_0, b_1, \ldots, b_i\}\{b_i, b_{i+1}, \ldots,\}}_{\text{hardware}} \cdots \right\}$$

4.2.2. Expected Performance Gain

As we know, performance is not just a single number. However, to explain how performance can be used to guide partitioning, we choose a single metric — execution rate. This is partially motivated by the fact that the performance gain is relatively easy to measure and because, of all the metrics commonly used, speedup is often the most important one.

In the software world, performance is modeled all the time. The idea of complexity (order) analysis — that $O(n)$ is better than $O(n^2)$ — is widely understood and serves as an implicit guide when programmers everywhere compare algorithms. Unfortunately, the hardware world lacks that kind of general guide. Moreover, performance gain for the application may reside in the accumulation of many small gains that would be lost in a direct application of complexity theory.

Therefore, we use profiling information to collect the total execution time as well as the fraction of that time spent in each subroutine. The product is the approximate amount of time needed to execute that portion of the application in software and we use it as the time we expect it will take in future execution runs. We use $s(i)$ to represent the expected execution time of one invocation of the subroutine (basic block) i. It is an approximation for a number of reasons, not the least is that it depends on the input data set for many applications. However, there are also sampling errors that may impact performance. (Again, we refer to Section 4.4 for practical ways of improving the accuracy of $s(i)$.)

Next we need to approximate the time that an equivalent implementation in hardware would take. In the case of a basic block, this is often much more precise. Without control flow (which, by construction, a basic block does not have), a synthesis tool can give a fairly accurate approximation of propagation time. Or, if the feature is pipelined, the number of stages is more precisely known. If the feature includes control flow but does not contain any loops, the longest path can be used as a conservative estimate. Features with a variable number of iterations through a loop present the biggest obstacle to finding an approximate hardware time. In this case, implementation and profiling with the hardware feature may be the only resort. Regardless, we assume that a suitable approximation $h(i)$ for the hardware execution time exists.

Finally, the interface between software and hardware requires time and that cost needs to be accounted for as well. We can approximate this cost by approximating the total amount of state that needs to be transferred (if instantaneous transfer is used) or the setup and latency cost (if continuous transfer is used). In either case, it is represented by $m(i)$ for feature $i \in \mathcal{H}$. (More on the specifics of implementing and calculating communication cost between hardware and software is presented in Section 4.3).

Execution rate is the speed at which a computing system completes an application, and in a Platform FPGA system we look to hardware for improvement in the execution rate. This gain, which compares a hardware and software solution against a purely software solutions, is typically measured as **speedup**. We use γ to represent performance gain — whether the metric is speedup or otherwise — and this will allow us to compare different features against one another to determine the best partitioning. (Consequently, any subset of basic blocks that does not yield a performance gain can be generally excluded from consideration. In other words, only subsets of basic blocks for which $\gamma > 1.0$ are considered candidate features.)

In general, we do not measure execution rate but rather the execution time, which is the inverse. So when considering whether a particular set of basic blocks should be mapped to hardware or software, we are interested in the gain in speedup, or

$$\gamma = \frac{\text{hardware speed}}{\text{software speed}} = \frac{\dfrac{1}{\text{hardware time}}}{\dfrac{1}{\text{software time}}} = \frac{\text{software time}}{\text{hardware time}}$$

More specifically, we are interested in the performance gains of individual features, so we define $\gamma(i)$, $i \in \mathcal{C}$

$$\gamma(i) = \frac{s(i)}{h(i) + m(i)} \tag{4.4}$$

where $h(i)$ and $s(i)$ are the execution times of a hardware and software implementation of feature i. The function $m(i)$ is the time it takes to synchronize affected, trapped state, that is, the time it takes to marshal data between processor and reconfigurable fabric.

Let's assume for the moment that we just use this single feature in our design. How much faster is the application? This depends on both the performance gain of the feature *and* how frequently it is used in software reference design. We can get this fraction of time spent in a particular feature, $f(i)$, from the profile information. The overall speedup of the application will be:

$$\Gamma = \left[\left(1 - f(i)\right) + \frac{f(i)}{\gamma(i)} \right]^{-1} \tag{4.5}$$

(Note the inversions again; we are moving between execution time and rates of execution to keep the sense of "performance gain.")

From this equation, we can observe that increasing the hardware speed of a single feature has less and less impact on the performance of the application as its frequency decreases. To increase the overall system performance of the application, we also want to look at another dimension: augmenting the system with multiple features that will increase the performance of individual components as well as increase the aggregate fraction of time spent in hardware. Recognizing this, we want to compute the speedup of multiple features in hardware. In other words, we want to evaluate the system gain of a set of features $\mathbb{D}$ where each member of the set contributes to the system performance based on the fraction of time spent in that feature. To estimate the performance of a partition, we can add features and rearrange the terms to get the overall expected performance gain,

$$\Gamma(\mathbb{D}) = \left[1 + \sum_{i \in \mathbb{D}} \left(\frac{f(i)}{\gamma(i)} - f(i) \right) \right]^{-1} \tag{4.6}$$

One might recognize Equation 4.6 as the the Law of Diminishing Returns first put forth by Thomas Malthus in the 16th century and Equation 4.5 is simply Amdahl's law that we stated at the beginning of the chapter.

4.2.3. Resource Considerations

Following Equation 4.6 *reductio ad absurdum*, one would attempt to continue to add features so that $\sum_i f_i$ approaches one. In other words, implement everything in hardware to maximize performance! Ignoring development costs (a very real constraint), there is a deeper issue of limited resources. On an FPGA, there are a finite number of resources available to implement hardware circuits. These resource are limited and most realistic applications will far exceed the resources available. Thus, given a large number of features, designers are often forced to limit the number of features implemented in hardware based on those that make the most impact.

One way of approximating resources is to count the number of logic cells required by each feature. So a chip will have a scalar value r_{FPGA} that represents the total number of logic cells available (after the required cores have been instantiated). Then $r(i)$ can be used to represent how many logic cells each feature i requires. A simple relationship,

$$\sum_{i \in \mathbb{D}} r(i) < r_{FPGA}$$

constrains how large $\mathbb{D}$ can grow.

However, we know that modern devices are heterogeneous. A typical Platform FPGA has multiple types of resources: logic cells of course, but also on-chip memory, DSP blocks, DCMs, MGTs, and so on. A vector,

$$\vec{r}_{FPGA} = \begin{pmatrix} r_0 \\ r_1 \\ \dots \\ r_{n-1} \end{pmatrix}$$

is a better representation.

For example, the Virtex 5 (XC5VFX130T) has (this is not a comprehensive list, just a subset of all of the available types of resources):

$$\vec{r}_{FPGA} = \begin{pmatrix} 11,240 \\ 298 \\ 320 \\ 20 \end{pmatrix} \text{ where } \begin{array}{l} r_0 \text{ is the number of CLBs} \\ r_1 \text{ is the number of BRAMs} \\ r_2 \text{ is the number of DSP Slices} \\ r_3 \text{ is the number of GTX Transceivers} \end{array}$$

We also promote the resource requirements function to a vector where $\vec{r}(i)$ represents the resources required by the feature i. Our new resource constraint equation is very similar:

$$\sum_{i \in \mathbb{D}} \vec{r}(i) < \vec{r}_{FPGA} \tag{4.7}$$

where $\mathbb{D}$ is the set of features included in the design.

Unfortunately, this model is somewhat idealistic as well. It does not take into account the fact that allocating some resources (such as BRAMs) may interfere with other resources (such as multipliers). Moreover, performance estimates are often based on the assumption that the resources allocated are in close proximity to one another. That is to say that routing resources are not an integral part of this model.

4.2.4. Analytical Approach

We now have the mathematical tools needed to describe the fundamental problem of partitioning. We first formally describe the problem in terms of the variables just defined and then describe an algorithm that finds an approximate solution.

Problem Statement

The basic idea is to find a partition of all of the basic blocks of the application and then separate those sets into hardware and software. Formally, we are looking for a partition $\mathcal{P}$ of all the basic blocks U of the application:

$$U = \bigcup_{C \in \mathcal{C}} V(C)$$

A subset, $\mathbb{C} \subseteq U$, where C is the vertices of the call graph $A = (C, L)$ of the software reference design, is called the candidate set and contains all of the potential architectural features, that is, the subsets of U that are expected to improve system performance if implemented in the reconfigurable fabric. Due to limited resources, we need to further refine this to a subset $\mathbb{D} \subseteq \mathbb{C}$ that maximizes our performance metric,

$$\Gamma(\mathbb{D}) \text{ is maximized, and}$$

$$\sum_{i \in \mathbb{D}} \vec{r}(i) < \vec{r}_{\text{FPGA}}$$

Algorithmically, one could approach this problem by finding all partitions of U, synthesizing and profiling each partition, and then quantitatively evaluating every Γ. However, a simple application might have anywhere from 10,000 basic blocks to upward of 10^7. This means that the number of partitions to evaluate is approximately 2^{10^7}! Clearly, a brute-force approach is intractable. However, when one considers the fact that most of the inputs are estimates, the idea of relying on a heuristic is not disagreeable.

Heuristic

The partitioning problem is essentially a matter of indirectly manipulating the parameters $f(i)$ and $\gamma(i)$ by rearranging the partitioning $\mathcal{X}$. Then we select elements of $\mathcal{X}$ that meet our resource constraints and maximize our overall system performance Γ.

A heuristic that can be applied informally is to start with the natural partition provided by the software reference design. That is, we use the subroutines of the original application.

We use our profiling tools to determine the fraction of time spent in each subroutine. Then, by listing the subroutines in decreasing fractions of time, we focus on those subroutines with a large f. We rely on our experience in developing hardware cores to estimate the γ — the performance expected from a hardware implementation — and we have a system gain estimate for each subroutine.

Next, we want to iteratively manipulate the partition $\mathcal{X} = \{X_0, X_1, \ldots\}$ by creating new basic block subsets, merging subsets, and moving basic blocks from subset to subset. The basic idea is to look for changes that will either increase the fraction f or increase γ of a potential feature.

Fraction of Execution Time

The way to increase the fraction of time spent in a subroutine is by making the feature larger. This can be accomplished by initially looking for relationships in the call graph or (after the partition has been manipulated) by looking for relationships in the control flow graph that bridge subsets.

For example, suppose an application spends 0.5% of its time $f(i) = 0.005$ in subroutine A and 0.025% in subroutines B and C. The fraction of time spent in A can be doubled by merging A, B, and C. (In software terms, functions B and C are inlined.) Of course, merging does not come without a price. First, it (usually) increases the number of resources used. This has the potential of negatively impacting system performance by displacing other features or simply becoming too large to fit on the chip. Second, as explained later, increasing the size of the subset may decrease speedup — again negatively impacting the overall system performance.

Performance Gain

To increase the performance gain of a feature, $\gamma(i)$, one needs to look at the control flow graph of the feature and evaluate whether a change will make it more sequential or more parallel. Often-times, algorithms that are inherently sequential (due to the flow or control dependences between the operations) perform better on a processor (because processors do not have the overhead of configuration transistors and because they have been carefully engineered for these circumstances). Simply adding basic blocks may have the undesirable effect of increasing the sequential behavior of the feature, lowering its γ. Conversely, making a feature *smaller* has the potential of increasing its performance gain.

In terms of the application, this means taking a subroutine X and breaking it into two subroutines $X - X'$ and X' where subroutine $X - X'$ invokes X'. If X' extracts the parts of X that can be improved in hardware and leaves the sequential parts in $X - X'$, then the γ of X' will be higher than the γ of the original X. Most likely, it will require fewer hardware resources as well.

To see how this works in practice, consider a subroutine that takes 93% of the execution time and provides a performance gain of $\gamma = 2$. As is, this would provide the application with a performance gain of $\Gamma = 1.869$. However, if part of the original subroutine can be extracted (and its parallelism increased), then it is entirely possible that the performance gain could go up by a factor of 10, $\gamma = 20$, while its fraction of execution time decreases to 83%. This partitioning gives us an overall system speedup of $\Gamma = 4.739$. Ahmdal's law teaches us to always try to increase the fraction of time spent in the "enhanced" portion of the code. However, as this example shows that, when limited, fungible resources are part of the equation, it is not always better.

It is also important to note that any change to the subset may affect performance by increasing or decreasing the marshaling costs. That is, merging two subsets could dramatically increase the performance gain because less data have to be explicitly communicated between the processor and the feature. However, the exact opposite effect is possible as well. Moving basic blocks from one subset to another may result in a larger affected state.

Thus, in summary, the general heuristic works by examining the CG and CFGs of the application and then making incremental changes to the subset of the partition. Changes are guided by:

- trying to increase the fraction of time spent in a subroutine while not dramatically increasing resources or decreasing performance; and
- trying to increase the performance without substantially reducing the fraction of time spent in a subroutine

4.3. Communication

When a single application is partitioned into hardware and software components, it is necessary for those components to communicate. It is now important to discuss a number of issues related to communicating state across partition boundaries and the mechanism for transferring control between hardware and software.

Consider an application consisting of several subroutines. One particular subroutine might involve many simple bit manipulations such as the error-correcting code shown in Listing 4.1. It is reasonable to expect that a hardware implementation of this subroutine will substantially outperform the best software

```
class HammingCheck {
  public static byte hammingCheck(byte d, byte check) {
    int my_check;
    byte err_loc;
    my_check=(((d>>7)^(d>>6)^(d>>4)^(d>>3)^(d>>1))&1) |
      ((((d>>7)^(d>>5)^(d>>4)^(d>>2)^(d>>1))&1)<<1) |
      ((((d>>5)^(d>>5)^(d>>4)^(d>>0))&1)<<2) |
      ((((d>>3)^(d>>2)^(d>>1)^(d>>0))&1)<<3);
    err_loc=(byte)(check^(byte)my_check);
    switch(err_loc) {
      case 0:
        //No error
        break ;
      case 1: case 2: case 4: case 8:
        //Error in parity bit (data OK)
        break ;
      case 3:
        return((byte)(d^0x80));
      case 5:
        return((byte)(d^0x40));
      case 6:
        return((byte)(d^0x20));
      case 7:
        return((byte)(d^0x20));
      case 9:
        return((byte)(d^0x8));
      case 10:
        return((byte)(d^0x4));
      case 11:
        return((byte)(d^0x2));
      case 12:
        return((byte)(d^0x1));
    }
    return(d);
  }
}
```

Listing 4.1. A simple error-correcting subroutine.

implementation and require relatively few FPGA resources. Hence, the subroutine hammingCheck is identified as a potential feature and its hardware implementation becomes a candidate for inclusion in the final system. However, we stress *candidate* because again, at this point, even though it may improve the system, we have a limited number of resources and it may not be part of the most beneficial decomposition.

Assuming that we have implemented this feature in hardware, the key question of this section is: how does this feature and processor interact? The set of rules that govern this interaction is called the interface; it is described in Chapter 2. Standard interfaces abound — such as how to pass arguments on the stack or the format of a network packet — and, most likely, readers have never considered defining their own low-level hardware interface. However, with Platform FPGAs, it is up to the designer to either choose a predetermined interface or design a new one for each feature. Often, this decision is crucial: the interface can make or break the performance advantages of using FPGAs.

Adding a feature to the system requires that the processor and the feature maintain a consistent view of the application's data. Collectively, all of the application's storage (variables, registers, and condition codes, as well as I/O device configuration and FPGA configuration) at any specific moment in time is called its state. This is similar to the concept of "the context of a process" in multitasking operating systems. However, when multitasking in the single processor world, the same hardware is time-multiplexed so that an application's state is generally accessible all of the time. Similar to the case where a collection of independent processors need additional mechanisms to keep state consistent, so do independent components in our Platform FPGA-based system. This issue is more complicated for features implemented in Platform FPGAs because (i) the embedded processors are not designed for use in multiprocessor systems, (ii) features have limited ability to directly access state inside the processor, and (iii) the nature of hardware design distributes state throughout the whole feature (instead of keeping it in centralized, named registers). Thus, interfacing is what allows a feature and the processor to communicate changes in state and is necessary to keep an application's data consistent. In developing the software reference design, a programmer is not likely to think about these issues. The specification uses a sequential model, which may or may not express any concurrency and a single memory hierarchy is assumed. However, if consistency is not explicitly maintained by the designer's interface between a feature and a processor, then the resulting system risks producing incorrect results. To reiterate the importance of the interface: every component of the system may work properly by itself but if the interface fails, the system as a whole may fail.

As a concrete example, consider the example in Listing 4.1 again. The interface could be as simple as transferring two bytes to the hardware core, signaling the core to start, and then waiting for the result. This highlights the two main issues of the interface:

- How is the feature invoked?
- How is the transfer of state handled?

The former is the subject of subsection 4.3.1 and the latter is the subject of subsection 4.3.2.

4.3.1. Invocation/Coordination

In a sequential computing model, the most common form of invocation is the subroutine call. The linkage usually takes the form of saving the return address and then branching to the address of the subroutine. Because the traditional (single) processor only maintains one thread of control, the caller is viewed as temporarily yielding control, which is returned when the subroutine finishes. Alternatively, the multithreaded programming model augments this invocation mechanism. Individual threads of control can invoke subroutines but also have the ability to fork a new thread of control and different threads communicate through a variety of well-known mechanisms (semaphores, monitors, etc.)

These two software concepts can be extended to hardware in terms of a coordination policy. Hardware is different from software in that there is no thread of control: generally, hardware is controlled by some state machine that is always "on". However, most state machines have some sort of "idle" state and transfer of control can be thought of as leaving and entering the idle state. So the idea of transferring control is conceptual — the real issue is coordination. In general, there are three common approaches to coordinating hardware and software components. The dominant approach is the coprocessor model; however, the multithreaded and network-on-chip models are gaining acceptance. Each is described next.

Coprocessor Model

The coprocessor model (also known as a go/done model or client/server model) is similar to traditional subroutine call linkage. In this model, the hardware sits in an idle state, waiting to provide a service to the processor. The processor signals for the feature to begin and then waits for a result from the feature. These signals are often called go and done (which give rise to one name for this model).

To implement the go/done model, every feature is designed to have an idle state. In the go/done model, every hardware component has an idle state where it is, in essence, off. That is, the designer treats the hardware like a sequential model and there is a time-ordered transfer of control. It has a go signal that can be implemented as a single I/O line or as part of a more complicated interface (such as a FIFO). The general protocol is for the processor to signal to the hardware to begin. The current (operating system) process blocks itself. When finished, the hardware component signals "done," which unblocks the invoking process.

The hardware goes back into the idle state, waiting for another go signal. A variation on this model has the hardware component interfaced to other hardware components or external I/O. In this case, the component may be concurrently gathering statistics or serving autonomous functions independent of the processor. (For example, the component may be counting external electrical pulses over a period of time to compute the frequency. The go/done interface is only used when the processor needs to know what current frequency is.)

An important consideration in the go/done model is how to handle the time while the feature is operating. There is a range of ways to handle it based on how long it is expected that the feature will take to complete and system overheads (such as a bus transaction or an operating system call).

Fixed Timing

For small features that take a fixed, short amount of time to complete, the most efficient mechanism is for the processor to simply do an appropriate number of "no-op" instructions and then retrieve the results (perhaps checking that done was actually signaled). Oftentimes, the hardware is so fast that by the time the state has been communicated, the hardware has results for the processor.

Spin-Lock

When the amount of time a feature runs is unknown, mechanisms such as a spin-lock can be used. The processor uses a loop to repeatedly check some condition, indicating hardware has completed its computation. This is also known as polling hardware for its results. However, checking the condition frequently (commonly implemented as a "done" signal) may have the unintended consequence of slowing the feature by taking bus cycles away, for example.

```
// Simple y = m*x + b Example
invoke_hw(m, x, b);
while(!done) {
  done = check_done_signal();
}
y = retrieve_results();
```

Blocking

The traditional way of handling this situation is to treat the feature like an I/O device. The "done" signal can be routed as an interrupt to the processor. Then the feature can transfer the inputs to

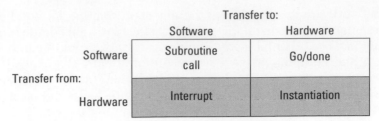

Figure 4.3. Transfer of control between hardware and software.

the feature and then tell the OS to deschedule it until the feature completes.

```
// Simple y = m*x + b Example
invoke_hw(m, x, b);
yield();
y = retrieve_results();
```

One limitation of the coprocessor model is that it provides an asymmetrical mechanism. Figure 4.3 shows the four combinations how software and hardware may transfer control. The shaded lower half of the figure shows transfer-of-control mechanisms that are not always supported.

Special Solution to Coprocessor Model

Transferring state from hardware to software is often impractical. One way of identifying candidate features is described here. By ordering the methods first, the algorithm focuses on the most likely locations for feature extraction.

A (static) call graph $G = (V, E)$ of the application is constructed. Recursion is detected by finding the strongly connected components of G and removed from consideration. For example, `print` and `list` in Figure 4.4 form a strongly connected component and are marked with an $*$. Next, the remaining vertices of G are assigned an ordinal by the rule,

$$
\mathrm{ord}(u) = \begin{cases} 0 & \text{if node } u \text{ is a leaf} \\ \max_{(u,v) \in E(G)} \left\{ \mathrm{ord}(v) \right\} + 1 & \text{otherwise} \end{cases}
$$

In other words, every node is labeled with the maximum integer distance from a leaf. For example, `sar` is 2 rather than 1 in Figure 4.4.

Multithreaded Model

Increasingly, the multithreaded model has emerged as an important programming model. In systems with multiple processors, an

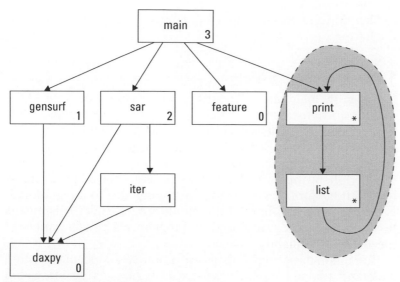

Figure 4.4. Example call graph.

operating system can manage multiple threads of control (operating within the same process context) to improve performance. In other situations, such as graphical user interfaces, the model has proved to be a productive programming one.

In the threading model, the true parallel nature of the hardware is acknowledged and the designer recognizes that the processor and hardware components are both running continuously. Coordination of these components is handled by the communication primitives that have been developed for concurrent processes: semaphores, messages, and a shared resource such as RAM. In this case, every hardware component is considered to be running as a parallel thread. A set of system-wide semaphores are used to keep hardware and software threads consistent. Instead of sitting in an "off" state, hardware conceptually sits in a blocked state waiting on a semaphore or for a message.

Network-on-Chip Model

Again, borrowing from the multiprocessor world, researchers are beginning to investigate a completely distributed state and rely on a message-passing to explicitly transfer state (subsection 4.3.2). In this model, coordination is implicit with the transfer of state. However, because this assumes that software reference design has been constructed with explicit message-passing commands, we do not consider it in an embedded systems textbook. Nonetheless, we have seen that high-performance computing techniques continue to filter down to the embedded systems, so it may be important in the future.

4.3.2. Transfer of State

Historically, maintaining consistent state across discrete FPGA devices and processors has been especially challenging. This is because each FPGA device had its own independent memory hierarchy, state is widely distributed in the design, and there is a wide variety of interconnects between FPGAs and processors (from slow I/O peripheral buses to high-speed crossbar switches). This is true with Platform FPGAs as well. (Although due to integration, it is possible to share more of the memory hierarchy.) This means less state has to be transferred and there are a greater variety of mechanisms available.

To explain these mechanisms, we define two concepts: affected state and trapped state. These concepts help us decide which parts of the application state need to be explicitly communicated between processor and feature.

Affected State

One of the first questions to ask is "What data need to be kept consistent?" For example, software compilers generate a large number of temporary variables that are used to hold intermediate values, and these values are often kept in registers (although they can be "spilled" into main memory). However, after serving their purpose as intermediate values, these data no longer contribute meaningfully to the state — only the final computation matters. Hence, some state is independent by construction and there is no need to transfer it. Other portions of state are simply not relevant to every computation; that is, not every portion of the application refers to every data value used. If we exclude these parts of the application state, what remains is the affected state. More precisely, we can define *affected state* as application data, during a transfer of control, that has the potential of being either read or written by either the feature or the processor. In other words, the affected state is the state we need to keep consistent.

For example, consider the (contrived) subroutine `translate` (a) in Figure 4.5 and a corresponding hardware implementation (b). This feature affects four `int` values in the application's state. This includes the values stored in `x`, `m`, and `b`, as well as the return value of the function. For most processors, the compiler would have generated a temporary variable (call it `t1`) to hold the product $m \times x$ to be stored in the general-purpose register of the processor. In hardware, this value would be expressed on a bus

```
int trans(int x, int m, int b) {
    int y ;
    y = m*x + b ;
    return y ;
}
```

(a)   (b)

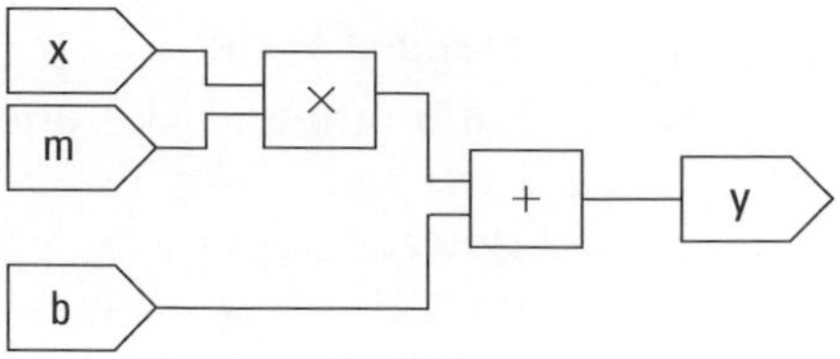

Figure 4.5. Translate.

```
extern int m, b ;
int trans_and_inc ( int &x ) {
    int y;
    y = m*x + b ;
    x++ ;
    return y ;
}
```

(a)   (b)

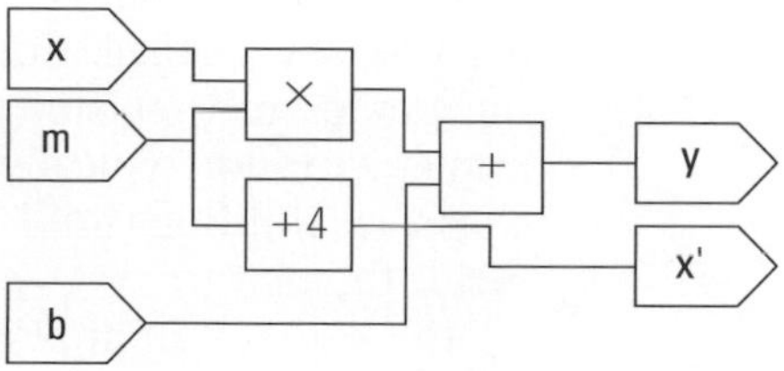

Figure 4.6. Translate and increment.

of signals between the multiplier and the adder (with the possibility of a register inserted). These values (corresponding to $t1$ and y in the software) are part of the state but are not necessary to maintain consistent state. Consequently the interface does not have to accommodate them.

The subroutine translate might leave the false impression that the arguments and return values are the only parts of state affected by the subroutine. In functional programming languages this is true — it is called referential transparency — and it allows functions to be evaluated in local context with no global state to worry about. However, the dominant programming paradigm for embedded systems is imperative. In the imperative paradigm, identifying state can be much more difficult. (Indeed, for a wide range of reasons — from high-performance computing to computer security — identifying affected state is a major challenge.) Consider the subroutine[1] of Figure 4.6 and the corresponding visual representation of state. Ignoring the presence of a cache, this example shows that part of the state resides in main memory. Because C handles arrays by reference and permits pointers to be used to freely access all of a process's state, there are numerous ways for a single subroutine to affect the global state.

---

[1] We are using a C++ shortcut here; alternatively int *x and (*x) could be substituted for int &x and x in the body.

Trapped State

Unfortunately, the affected state can be quite large (especially if arrays, complex data structures, or pointers are involved). However, when part of the memory hierarchy is shared, not all data have to be explicitly transferred. For example, if all of the affected state currently resides in primary storage (RAM) and both processor and features have access to this part of the memory hierarchy, then there is no reason to explicitly transfer data.

However, modern processor designs are built with significant memory caches that can capture part of the state without updating primary storage. Compilers naturally take full advantage of the processor's register set to hold state. Likewise, a hardware implementation typically embeds data throughout the design; registers and flip-flops hold intermediate values between clock cycles inside the design.

The state of an application is stored in several parts of a system. In the processor core alone, application data might reside in a general-purpose register, a cache, or (if an MMU is present and enabled) in a translation look-aside buffer. Outside the processor, application state might exist in on-chip Block RAM, in a subsystem buffer, or in external RAM. In order for a feature to implement part of the application, it will need to interact with part of the state. Sometimes this is accomplished easily because the state is in a common location (such as a common memory hierarchy). However, if that state is in a register or in the processor's cache, it is usually not accessible to units outside the processor. We call this **trapped** state and it has to be explicitly transmitted across the interface. (The symmetric situation exists as well. Features are not required to be stateless — some of the application's data may reside in a feature's internal registers and that too would have to be communicated across the interface.) Figure 4.7 illustrates two places where state can be trapped.

This gives us a general idea of state, but before we begin to formalize the transfer-of-state problem, it is worth interjecting a brief discussion of pointers and the challenges faced in determining affected state. Often pointers are a source of frustration. They often confound automated translation techniques because the translator cannot tell if two pointers might refer to some location. Pointers also have the reputation of frustrating humans trying to analyze a piece of code (either to debug or to extend it). In both cases, it is a matter of intent — what did the original programmer wish to accomplish? In our situation, we can partially sidestep the intent problem without declaring pointers off limits because the sole purpose of a software reference design is to express intent. Pointers tend to obfuscate when used to improve performance in a clever way. But that effort is misspent in a reference design.

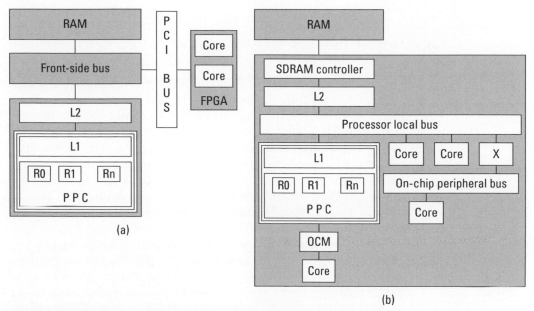

Figure 4.7. Two examples of where state can be trapped (a) traditional and (b) Platform FPGA.

So, while theoretically it is impossible to determine exactly what the affected state is, we will assume that in practice it is feasible because we have access to the original intent. Applying these manual techniques to "dusty deck" codes (older software that is still in use) may require more effort.

Transfer-of-State Problem

Affected state that is trapped needs to be explicitly communicated. This is done by a process called marshaling. ***Marshaling*** groups elements of the application's affected state into logical records that are explicitly transferred. If we assume that the processor has control initially, there are up to four different kinds of records that may be used. The first two kinds are used to transfer an initial set of elements to the feature (Type-I, for initial) and a final set of elements back to the processor (Type-F, for final). These transfers take place when the feature is loaded and unloaded. Because all of the systems described are statically configured, these records are typically transferred at the very beginning of the application and when the application exits. The other two kinds of marshal records are used repeatedly — one when the feature is invoked (Type-CI, for copy-in), and the other when it completes (Type-CO, for copy-out).

For example, consider an MPEG encoder core. A matrix of quantization parameters is typically specified once for an entire

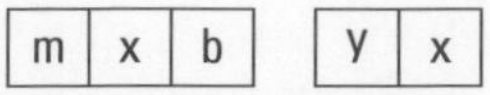

| m | x | b | | y | x |

Figure 4.8. Type-CI and Type-CO marshal records.

application and those values do not change. Rather than transfer that state every time the encoder is invoked, those values can be transferred once at the beginning. For an example, consider the subroutine described previously in Figure 4.6(a). The Type-CI and Type-CO marshal records are shown in Figure 4.8. As an example of a Type-F record, suppose a core implemented in the configurable logic accumulates a value in a global variable (it may be statistics about its operation or samples it was given). A typical use of a Type-F record would be to read that variable after the last invocation.

A translation process may be incorporated with marshaling. For example, floating-point numbers in the processor may be converted to fixed point when transferred to a feature. More complicated transformations, such as converting a linked list into a linear array of values are often appropriate. The grouping of elements is logical because the assembly does not strictly mean that elements are copied to a continuous location of memory. For small transfers, this is often simple and efficient. However, for many high-performance and multimedia applications, this is impractical. We describe a variety of techniques later.

The simplest way to transfer state is to copy the entire record, stop the processor until the feature completes, and then transfer the entire record back. This is considered a *push*, as the caller transmits data before transferring control. Alternatively, if a transfer record already exists and is in a known location in memory, then the caller can transfer control and the callee *pulls* data. One practical caveat: if the processor writes data to RAM and there is a cache, it has to also flush the cache.

For example, suppose the core is designed to find one element in an array. Depending on the algorithm, but under most circumstances, only a fraction of the array will be needed to complete the operation. Transferring the whole array when invoked is an expensive and unneeded operation.

Although simple, an instantaneous transfer of state is not always efficient or feasible. For example, oftentimes the size of the potential affected state is large but the actual data used are small. In this case, the data access pattern is random and the logical transfer record is large, but actual data are small. In this case, it is better for the interface to service the feature continuously. This occurs when both the logical record and the actual data record are large. If a large portion of data is expected to be used, then it makes sense to develop a regular access pattern that transmits data continuously. In such a scheme, the feature begins working on part of the transfer record while data are in transit.

In general, the choice of instantaneous versus continuous transfer is determined by the goal of the feature. Instantaneous is used when the feature is designed to reduce the latency of the task and the transfer record is small. Alternatively, continuous is used when the feature is designed to increase the throughput and latency reduction is not possible or not enough to meet performance targets.

Continuous transfer usually involves setting up DMA transfers and the incorporation of FIFO buffers. As such, it can increase the area and sometimes increase the latency. However, for many multimedia, signal processing, and scientific applications, throughput is the most important criterion.

One more fact is important with respect to transferring state. Often the best implementation in hardware uses a different format of data or possibly a different data structure. Translation is often an integral part of marshaling data.

Now that we understand the fundamental concepts of communication — coordination and state — we can take a formal approach to solving the partitioning problem. Armed with a detailed understanding of how to interface hardware and software components, we are now ready to tackle the partitioning problem directly.

4.4. Practical Issues

As mentioned at the beginning of Section 4.2, a mathematical model is suitable for getting us "in the ballpark." However, there are numerous practical caveats and pitfalls that can confound the analytical approach. In general, experience and a few rule-of-thumb-type guidelines aid the system designer. This section highlights a number of these issues not addressed by the formal analytical solution.

4.4.1. Profiling Issues

A not-so-subtle assumption with analytical formulation is that it uses profile information to approximate the fraction of time that an application spends in a subroutine or basic block. However, a number of situations could generate misleading results.

Data-Dependent Execution

Some applications are stable in the sense that the fraction of time spent in each subroutine does not change substantially when the input is changed. However, for a number of applications — especially when the size of the input data set is changing or when the application responds to external events — a single data set will not be representative. In these cases, it is likely that the importance of various subroutines (based on their fraction of execution time) will change with respect to one another.

To detect this situation automatically is difficult. It requires the system designer to understand the fundamental operation of the application. For example, with applications that use many dense-matrix algorithms, the designer can analyze the complexity of various routines (i.e., "big-Oh" notation). For example, it may be the case, for small data sets, that two subroutines may take a similar fraction of time, but as the n increases, a $O(n^3)$ subroutine will take an increasingly larger fraction of time relative to a $O(n^2)$ subroutine. Harder to analyze, but no less important, are applications that depend on external events. An example of this might be a handheld device: for some users, the phone will be the dominant mode, whereas for others Internet access might be the dominant mode.

There are a number of ways to address these types of situations. One is to manually analyze the algorithms and understand the basic operation of the application. Another important tool is to collect profiling information based on a number of different (size and usage) data sets. A third is to try to separate the application, perhaps along module boundaries, and profile each module independently.

Correlated Behavior

Another subtle issue arises in some applications that explicitly use timed events. That is, many profilers implicitly assume that the application will make steady, asynchronous progress toward a solution. However, some applications are written to specifically incorporate time (such as event-loop and real-time systems). Because many profiling systems use an interval timer to sample the program counter, they rely on the application to be statistically uncorrelated to the timer. However, if the application is also executing operations in a regular, periodic fashion, then this is not true and profiling results may mislead the designer. For example, if events in the application are regular (occuring every 10 ms) and the samples are regular (the program counter sampled every 10 ms), then the profiler will not provide a statistically accurate description of the system. It could either over- or underrepresent a periodic event at the expense of other components in the software reference design.

Phased Behavior

It is well known that applications exhibit *phased* behavior: over the entirety of their execution, control moves between clusters of related operations. In other words, the execution of an application exhibits locality. Indeed, this characteristic, quantified by Denning & Kahn (1975), is commonly exploited in the design of computing systems. However, because of this, it may be difficult for the designer to identify important features from the summary numbers reported by a profiler.

As an example, consider an application that has three routines, *A*, *B*, and *C*. Assume each account for 33% of the execution time, each has the possibility of being sped up by 50%, and the routines are ordered such that *A* runs to completion before *B* begins and *B* completes before *C* begins. If there is only room for one routine in the FPGA resources, then the maximum speedup for the system is 12%. However, if the phased behavior is recognized by the system design, then there are options. One is to look at commonalities among the three cores and try to find commonality. A second approach is to time-multiplex the hardware. No automated approaches exist to do this, but the last chapter of this book discusses how to perform run-time reconfiguration to "share" certain hardware resources.

I/O Effects

Unforunately, most profilers do not account for I/O. This means that if the application spends a lot of time waiting for external events (such as network activity, reading/writing to disk, or waiting for user input), then improving the performance of computationally intense parts of the application may not yield the expected overall system performance gain.

For example, suppose an application consists of three subroutines — A, B, and C — and it takes 75 seconds to run. The profiler may report that 80% of the execution time is spent in subroutine C; however, that does *not* mean subroutine C takes 60 seconds! Why? Another useful utility `time` will help explain. Running the same application results in an output shown in Figure 4.9.

```
% time ./simple
real    1m14.991s
user    0m21.959s
sys     0m0.523s
```
Figure 4.9. Time utility output.

Note that the "wall clock" ("real") time is 75 seconds but the user time is only 22 seconds and the system time is 0.5 second. The user and system times represent how much time is actually spent inside of the application (in "user space"), and how much time is spent in the operating system ("kernel space"), respectively. Where's the missing ≈ 52 seconds? This is time that the application spent waiting on some external event. If the application is waiting for a user keystroke, that time is not counted against the user or system time — even though it does contribute to the overall wall clock time. Thus, a better estimate of the time spent in subroutine would be 80% of 22 seconds (17 seconds). Also any performance improvements gained from subroutine C will only improve 17 seconds of the total 75 seconds — not the original estimate of 60 seconds! So, while most profilers exclude I/O time and focus on just the computation time, the analytical formulation presented earlier implicitly assumes that we are trying to improve the computation rate of a compute-bound process.

Number of Calls

`gprof` and other profilers will also keep track of how often subroutines are called. It is important to note how often a subroutine is called because it can cause the profile information to vary wildly between runs. For example, it is common for applications to handle "special cases" differently. For example,

```
if( delta_t>MAX_SWING ) {
  dim_display() ;
} else {
  compute_accel() ;
  update_display() ;
}
```

This code might be used to hide results when certain conditions suggest that data are not useful. Depending on data used to profile the software reference design, it may turn out that `compute_accel()` is only called once and has a small fraction of execution time. However, in practice, one would expect that the else-part of this code is the dominant path. The number of calls is a major clue to the engineer that perhaps the data set is not describing the behavior of the system accurately.

4.4.2. Data Structures

Software reference designs naturally reflect a bias toward software implementations. This is understandable, as programming is (appropriately) taught within the context of an ordinary sequential computing model. With such a model, the difference between

```
while( i!=NULL ) {
  proc(i) ;
  i = i->next ;
}
```

and

```
while( i<n ) {
  proc(x[i]) ;
  i = i + 1 ;
}
```

is negligible: both take $O(n)$ steps. However, with hardware, the latter is much more desirable. At minimum, a large prefetch window is possible. However, as we will learn in the next chapter, if we can determine that, in the various iterations of the loop, each `proc(x[i])` is independent, then we could potentially improve the design through pipelining or regular parallelism so that the execution time approaches $O(1)$ steps.

What this example shows is another flaw in the analytical model of the previous section. The software reference design serves extremely well as a specification. However, if we simply profile the execution of the reference design, we do not capture the implementation benefits. Because these benefits come from algorithmic changes, it seems unlikely that automated techniques will reveal these advantages. Consequently, it falls on the system designer to understand both the software algorithm implemented in the reference design and how it might be reimplemented in hardware.

There are several places where common software structures can be rearranged to yield better hardware designs. For example, look for data structures that use pointers (linked lists, trees, etc.) and determine if they can be implemented "flat" structures (such as vectors and arrays). Flat structures produce regular memory accesses that can subsequently be prefetched or pipelined. Another place to look for performance gains is in bit widths. Software programmers generally assume a fixed, standard bit width even though they only use a few bits of information. This is because processors are slow at extracting arbitrary bits from a standard word. However, hardware excels at managing arbitrary bit widths. Finally, software has fixed data paths and (relatively) low bandwidth between components. When possible, consider leveraging the large bandwidth provided by the FPGA's programmable interconnect.

4.4.3. Manipulate Feature Size

The simplest change involves breaking up subroutines into smaller subroutines. This has the effect of isolating computationally intensive parts of the subroutine (making the hardware feature smaller and easier to implement).

Alternatively, there are times when aggregating subroutines is needed. For example, if two subroutines each account for 25% of the application's time, then they are strong candidates for hardware. However, if implemented individually, each invocation incurs the cost of interfacing. If these three subroutines are related (for example, one is also invoked after the other), then combining them into a single subroutine saves the cost of invocation. This happens frequently because ordinarily software programmers will break up subroutines to make the code more maintainable and the run-time cost of a subroutine is completely negligible. However, the cost of interfacing is not always negligible and it is worth the designer's effort to investigate the situation.

Other changes are more substantial. Often the hardware implementation has a significant performance gain because it is better and has application-specific data formats. To make software more efficient in these nonstandard data formats, programmers will frequently employ data structures that do not map well to hardware. Thus, it is worth looking at the intent and considering replacing data structures to better exploit the hardware.

Chapter in Review

The aim of this chapter was to give readers a better understanding of the principles behind partitioning a software design into a more efficient and, if possible, faster hardware implementation. One technique introduced was profiling. We will actually see this tool more in Section 4.A; however, it provides a good indicator to what part(s) of the application may map well to a hardware implementation. Analytically, we built a mathematical formula of the partitioning problem and built a heuristic to maximize the performance.

Practical Expansion: Partitioning

In these gray pages, there are two main learning objectives. First, we describe a few of the practical issues of using profiling and the tool `gprof`. Then we build on the basic Linux system described in the previous chapter to add kernel modules.

4.A. Profiling with Gprof

As mentioned in the white pages, one can get a statistical approximation of how much time a program spends in a subroutine relative to the overall execution time. In this section, we will look at how to use one specific profiler, demonstrate its use for a few applications, and then describe some more advanced tools for analyzing the behavior of a software reference design. Specifically, we will look at an application profiler and a system-wide profiler.

One of the easiest tools to use is the profiler that comes with GCC, called `gprof` (Graham *et al.*, 1982). Gprof gathers three pieces of information during program execution: the number of times a subroutine is called, the execution time spent in each subroutine, and the dynamic call graph. (The static call graph described earlier in the white pages includes an edge between two vertices if there is the potential for one subroutine to call another. The dynamic call graph is a subset of vertices — excluding edges if a particular subroutine call was not invoked during execution.)

The first step in anlyzing a software reference design is to run it without `gprof` using the `time` command. (Time is a stand-alone application but is also built into some command line shells.) For example,

```
time sort -r /usr/share/dict/words > /dev/null
```

returns

```
real  0m0.657s
user  0m0.624s
sys   0m0.012s
```

What this means is that it took the sort command 0.657 second (wall clock time) to reverse the list of words in a dictionary file. Of the wall clock time, 95% of the time (0.624 second) was spent in the user's program and 5% was spent outside the user space (either blocked on I/O or executing kernel code). In contrast, if we run

```
time cat /usr/share/dict/words > /dev/null
```

the output is

```
real  0m0.009s
user  0m0.000s
sys   0m0.008s
```

In this case, most of the time was spent outside of the user application. What is the difference here? In the case of sort, the application has a fair amount of work to do — sorting a bunch of strings, which is substantial compared to the I/O (moving data from the disk to the application). Computationally, the `cat` command is extremely simple — most of its time is literally spent alternating between the read system call and the write system call.

There are two immediate implications of this information. If the software reference design spends a substantial port of time in the operating systems (waiting on I/O, doing an intense number of system calls), then profiling the application as it is may be worthless. That is, looking for hardware accelerators in an application that is I/O-bound (instead of compute-bound) won't yield any speedup. Second, the issue may be that, in the rapid prototyping stage, I/O was used to substitute for an instrument in the embedded system. For example, radar data may be the primary input to the system but the radar is not connected to the development machine running the software reference design. To simulate the radar, real data can be collected and stored in a data file. In this case, the software reference design is artificially I/O-bound and it may be possible to tweak the program to better represent its behavior in the embedded system. (A third possibility is that the application will be I/O-bound in the final embedded system. If this is the case, then this analysis tells the designer to shift emphasis from the computation to how the data is delivered. See Chapter 6 for more about managing bandwidth.)

Assuming the basic timing analysis suggests that the application is compute-bound, the next step is to profile the application. This is straightforward with GNU Compiler Collection — it will work even on the cross-development tools. (Of course, to get useful results can be a little more challenging; see next section.) There are three basic steps:

1. Add -pg to the CFLAGS environment variable. With a typical GNU package, this can be accomplished when the package is configured

   ```
   ./configure CFLAGS=-pg
   ```

   Two potential snags to watch out for: (1) watch for the final link command — if the Makefile calls the ld command directly, you must include the -pg option to that command as well! Add LDFLAGS=-pg to the configure command will do this. (2) If there was a default CFLAGS in the configure (-g or -O2 are common) this *overrides* the existing options rather than augments them.

2. Run your application with inputs that are representative of what the embedded system will experience. The only catch here is that you have to have write priviledges in the current working directory at the end of the application. As long as the application exits normally, the there will be a gmon.out file with your profiling data. Note that gmon.out is overwritten every time you run the application.

3. The program gprof command interprets this data and produces two reports on the standard output. The command looks like this:

   ```
   gprof -b myprog > myprog.stats
   ```

   The -b option suppresses a long summary of the different comments. You may want this initially but after using gprof for a while, this option gives you just the data. (This command, as written, assumes that myprog and gmon.out files are in the current working directory.)

The output consists of two tables. The first table is a "flat profile." The second shows the dynamic call graph. To understand what a developer can learn from these tables, let's look at a several examples.

The first example comes from an implementation of an encryption/decryption standard called AES. This was compiled with the -pg option. Then the application was used to encrypt a 217MB CompactDisk image. The output from the profiler was gmon.out, and the gprof command, shown below, was used to convert the data file into a human-readable file.

```
gprof -b aespipe > data.stats
```

The first table from data.stats is shown in Figure 4.10. This is either very good news or disappointing news! It is clear from the flat profile that 8.12 seconds of the 8.65 seconds that the program ran was spent inside of one routine, aes_encrypt! If the AES routine was part of a bigger project, then this clearly identifies the routine to implement in hardware. However, if you were trying to partition AES on its own, then the single routine is all you have and none of the mathematical analysis of the white pages will help you.

```
Each sample counts as 0.01 seconds.
  %   cumulative   self              self    total
 time   seconds   seconds    calls  ms/call ms/call  name
93.87      8.12      8.12 14145726     0.00    0.00  aes_encrypt
 5.55      8.60      0.48    27629     0.02    0.31  doEncryptCBC
 0.29      8.62      0.03        1    25.00   25.00  aes_set_key
 0.29      8.65      0.03                            aes_decrypt
 0.00      8.65      0.00    55282     0.00    0.00  rd_wr_retry
 0.00      8.65      0.00       93     0.00    0.00  sha512_write
 0.00      8.65      0.00        1     0.00    0.00  getPass
 0.00      8.65      0.00        1     0.00    0.00  get_FD_pass
 0.00      8.65      0.00        1     0.00    0.00  sGetPass
 0.00      8.65      0.00        1     0.00    0.00  sha512_final
 0.00      8.65      0.00        1     0.00    0.00  sha512_hash_buffer
 0.00      8.65      0.00        1     0.00    0.00  sha512_init
 0.00      8.65      0.00        1     0.00    0.00  sha512_transform
```

Figure 4.10. AES flat profile.

Let's look at another example. The Finite-Difference Time-Domain (FDTD) application simulates the propagation of an electromagnetic wave through some medium. For example, it might be used to simulate signals on the traces of a printed circuit board. After running it with profiling turned on we get the flat profile shown in Figure 4.11. One thing to notice is that six routines get called 20 times. This suggests that these routines are part of the main loop that iterates 20 times — in this case, it was the number of time steps simulated. It also tells you that these routines are the major procedures in the application. In this case, we probably also want to look at the dynamic call graph. Besides simply knowing where the time is spent, we would also like to know the invocation patterns to make sure we don't spend all of our time marshaling data back and forth. We use a tool called Gprof2dot and the GraphViz package to convert the second table into a dynamic call graph. The resulting graph is shown in Figure 4.12.

The last example, shown in Figure 4.13, presents an interesting case. It converts a PNM format image into a PGX format image. (PGX is a nonstandard format used by the JPEG2000 Verification Model software.) It is interesting to note that 53% of the time is spent decoding the PNM image. If we were to speed up the JPEG2000 encoding side, the best possible speedup would be ×2.1 and that's assuming no communication cost and nearly instantaneous execution time in hardware!

We wrap up this discussion with a few final hints about profiling and related software tools. There is an art to profiling that involves figuring out what to profile. For example, sometimes it makes sense to profile a large subsystem in isolation. Most embedded systems would us JPEG2000 as a library, but in order to profile it, we needed a simple application that reads, converts, and writes a single image. Had the entire system been profiled together, a single subroutine may not have been so easy to spot. This was appropriate because we envisioned it being used with a camera that was continuously producing data. In another application, it might not be appropriate.

Another issue that one might see with gprof is the presence of a mysterious function called mcount. This is the function that gets linked in when the application is compiled with -pg and it is used to keep track of how many times a subroutine is called. It is very short, so if it takes a large fraction of the run time, that suggests that the application is doing lots of subroutine calls and that the subroutines are short.

```
Each sample counts as 0.01 seconds.
  %   cumulative   self              self     total
 time   seconds   seconds    calls  ms/call  ms/call  name
16.97      2.42      2.42       20   121.02   121.02  amp1
16.83      4.82      2.40       20   120.02   120.02  amp2
16.41      7.16      2.34       20   117.02   117.02  far2
11.01      8.73      1.57       20    78.51    78.51  amp3
10.94     10.29      1.56       20    78.01    78.01  far3
10.66     11.81      1.52       20    76.01    76.01  far1
 4.70     12.48      0.67     2920     0.23     0.23  h_inner13
 2.88     12.89      0.41     1920     0.21     0.21  e_inner13
 2.52     13.25      0.36        1   360.05   360.05  grid
 1.82     13.51      0.26        1   260.04   260.04  tstep
 1.19     13.68      0.17      778     0.22     0.22  asuby
 1.05     13.83      0.15        2    75.01    75.01  initial
 0.98     13.97      0.14        2    70.01   250.04  SYMMLQ
 0.56     14.05      0.08     2334     0.03     0.03  DAXPY
 0.42     14.11      0.06     1560     0.04     0.04  DDOT
 0.42     14.17      0.06        1    60.01    60.01  free_space
 0.35     14.22      0.05     1556     0.03     0.03  DCOPY
 0.14     14.24      0.02       10     2.00     2.00  Je1
 0.07     14.25      0.01       10     1.00     1.00  Jm1
 0.07     14.26      0.01        1    10.00    10.00  rectangle
 0.00     14.26      0.00       40     0.00     0.00  wave
 0.00     14.26      0.00        1     0.00     0.00  tst_params
```

Figure 4.11. Finite-Difference Time-Domain simulation.

Using compiler optimizations requires some special attention as well. On one hand, if the application requires a certain level of optimization (-O3, for example) then one should profile it with optimizations turned on. However, the -O3 will automatically inline small subroutines — a change that will obviously distort the results.

Although we have focused on gprof, an Open Source profiler, there are others. Both AMD and Intel offer profilers that are especially well suited to their processors. AMD's product is called CodeAnalyst and can be found at this Web site: http://developer.amd.com/cpu/CodeAnalyst. Intel's product is called VTune and can be found at http://software/intelcom/en-us/intel-vtune. Two other Open Source profilers are called oprofile and Tprof. Part of the oprofile package is integrated with the Linux kernel and offers the ability to dynamically start and stop profiling. Another nice feature of oprofile is that the application does not have to be recompiled. Tprof is part of the Performance Inspector collection of tools (http://perfinsp.sourceforge.net/) and has the ability to run in two modes. Its time-based sampling is similar to gprof (periodically the process is interrupted). The second mode is event-based sampling. In this mode, the sampling interval is determined by how often a specific event occurs. Events that are counted could be cache misses, branches taken, or instructions retired. Increasingly, commercial processors are including counter registers throughout the processor and it seems likely that more event-based profiling will become an important tool to understanding the behavior of applications.

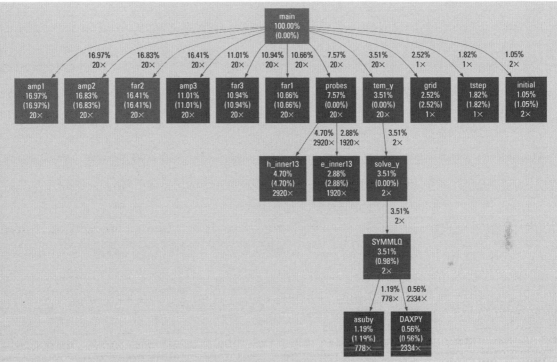

Figure 4.12. Finite-Difference Time-Domain dynamic call graph.

Finally, it is worth mentioning two very useful tools. A Python script called `Gprof2dot` (`http://code.google.com/p/jrfonseca/wiki/Gprof2Dot`) converts the output of `gprof` into a graph language called Dot. GraphViz (`www.graphviz.org`) reads a graph expressed in Dots and renders it visually in various graphics formats. These tools were used to create the figures in this section.

4.B. Linux Kernel

4.B.1. Kernel Modules

There are two ways of extending the functionality of the the Linux kernel. The first is known as the "in-tree" approach. In this case, you get a copy of the kernel, pick a directory in the source code, and begin adding code to existing files or creating new files. The second approach is called "out-of-tree" and as the name suggests, you create new files in a new directory outside of the Linux source directory. The major difference is that with out-of-tree development, one must learn the interface and develop the extension as a kernel module.

It is slightly counterintuitive but the latter approach is usually preferable for first-time Linux developers. Regardless of which approach is used, one still has to learn the kernel build system and usually it will be necessary to learn the kernel module interface anyway. So the disadvantages are few but the big advantage is that an out-of-tree kernel module can be compiled and recompiled on its own. This makes testing easier. When the internal state of the Linux kernel is not corrupted

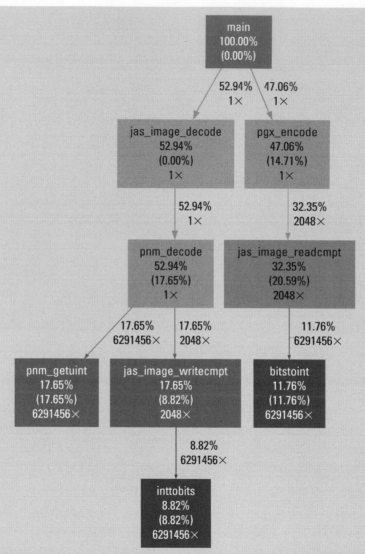

Figure 4.13. JPEG2000 example.

by a bug, then it is possible to remove the kernel module, fix the source and recompile it, and then insert it again into the running system. Not having to reboot the system every time to fix small mistakes is much faster.

To use the kernel build system in the out-of-tree approach, one has to create a simple Makefile. The generic Makefile illustrated in Listing 4.2 is similar to the Makefile described in Corbet *et al.* (2005): the variable LINUX points to where the Linux source tree is and DRIVER_NAME is the name of the source file. In addition, we include the ARCH and CROSS_COMPILE to cross-compile the kernel.

```
LINUX =../../ linux −2.6−xlnx
DRIVER_NAME=hello.c

# If KERNELRELEASE is defined, we've been invoked from the
# kernel build system and can use its language.
ifneq ($(KERNELRELEASE),)
        obj−m := $(DRIVER_NAME).o

# Otherwise we were called directly from the command
# line; invoke the kernel build system.
else
    KERNELDIR ?= $(LINUX)
    PWD := $(shell pwd)

default:
    $(MAKE) −C $(KERNELDIR) M=$(PWD) ARCH=powerpc \
        CROSS_COMPILE=powerpc−440−linux−gnu− modules
endif

clean:
    rm −rf *.ko *.o *~ modules.order Module.symvers \
        *.mod.c .tmp_versions
```

Listing 4.2. A simple kernel Makefile.

The minimum components of a kernel module are shown in Listing 4.3. After including the header files, there are four preprocessor macros used to put identifying information into the kernel module's object file. Next comes two subroutine — one is invoked when the kernel module is inserted, the other when the kernel module is removed. The kernel module does not use a reserved function name (like main in C) to identify the init and exit functions of a kernel module. Instead, the developer chooses them and tells the kernel with the last two preprocessor macros.

```
#include <linux/module.h>
#include <linux/kernel.h>  /* Needed for KERN_ALERT */
#include <linux/errno.h>

MODULE_AUTHOR("Ron Sass <rsass@uncc.edu>");
MODULE_DESCRIPTION("Simple Demo");
MODULE_SUPPORTED_DEVICE("(none)");
MODULE_LICENSE("GPL");
```

```
/*************************************************************************/
/* insert/remove kernel module routines
*/
/*************************************************************************/

/**
 * Called when module is inserted into kernel
 */
static int __init mycore_init (void) {

    printk(KERN_ALERT "Hello, World!\n") ;

    return 0;
}

/**
 * called when module is removed from kernel
 */
static void __exit mycore_cleanup(void) {

    printk(KERN_DEBUG "Goodbye, cruel world!\n") ;

}

module_init(mycore_init);
module_exit(mycore_cleanup);
```
Listing 4.3. A simple "Hello, World!" kernel module.

At this point, if one wanted to test their kernel module, one would create a base hardware system and compile a Linux kernel for this hardware. Then one uses the Makefile here to compile the `hello.c` into a "kernel object" called `kernel.ko`. This file is copied to the root filesystem before a filesystem image is created. Then you follow the steps described in the last chapter to create an ACE file or download the bitstream and kernel.

Once the system is booted, the developer can log in on the console and use the commands `insmod ./hello.ko` to insert a kernel module. The command `rmmod hello` removes it. (Note: In the former, the argument is a file name; the argument to `rmmod` is the name of a module.)

4.B.2. Address Spaces

One of the advantages of using a full-featured operating system is that each process gets its own address space. One of the disadvantages is that every process gets its own address space. In exchange for isolating software faults, it means we have to manage multiple address spaces in our kernel modules. This includes (at minimum) the physical address space (with respect to the system bus), the kernel's address space, and the current running process's (virtual) address space. The minimum one needs to know about moving data from one space to another in a device driver are these six routines:

1. `request_mem_region(PHY_ADDR,LEN,"mycore")` / `release_mem_region(PHY_ADDR,LEN,"mycore")`
2. `ioremap_nocache(PHY_ADDR,LEN)` / `iounmap_nocache(PHY_ADDR,LEN)`
3. `copy_from_user(KER_ADDR,USER_ADDR,LEN)`
4. `copy_from_user(USER_ADDR,KER_ADDR,LEN)`

The first pair of routines (1) mainly serve to catch misconfiguration errors. They simply alert the kernel that there is a range of physical addresses that this kernel module "knows" about and is claiming exclusive access to. (The third parameter gives a short text description of what is at that memory location.) If the two versions of a kernel module are accidentally installed twice, the second one will get an error when it requests the same memory range — this mechanism gives us a chance to catch that common mistake.

Requesting a range of physical addresses does not mean that the kernel has access to them. Even the kernel's memory references are translated by the MMU into physical addresses. Consequently, we have to set up the MMU to translate our physical addresses into the kernel's address space. Fortunately, the kernel provides `ioremap_nocache` to do this for us. (Whenever we are using physical addresses to communicate with an FPGA core, we want to be absolutely sure that the cache is turned off in that address range.)

The last two operations also protect a kernel module developer from a large degree of complexity. When the kernel is interacting with an application, the application will pass data through the system call interface. If that data is an addresss, then the kernel module will see it as a virtual address in the *user process's* address space. Any local variables in the kernel module will be in the *kernel's* address space. These two routines make it easy and efficient to move data between these two, distinct address spaces.

If the hardware device that you are trying to access was interfaced to the bus using the Add/Import Peripheral wizard, then the control registers are all going to be adjacent and start from some base physical address. Often the easiest way to keep track of the locations is to use a structure, such as the one below, and a pointer.

```
typedef struct {
    volatile unsigned int reg0 ;
    volatile unsigned int reg1 ;
    volatile unsigned int reg2 ;
    volatile unsigned int reg3 ;
    volatile unsigned int reg4 ;
    volatile unsigned int reg5 ;
    volatile unsigned int reg6 ;
} mask_regs_t ;

mask_regs_t *regs ;
```

Then, assuming `base` and `len` are set to the base physical address and length of the memory region, the `regs` pointer can be set with

```
regs = (mask_regs_t *)ioremap_nocache(base,len) ;
```

To write a 0x00000001 to the third registers, one would write

```
regs->reg2 = 0x000000001 ;
```

Dynamic memory allocation in the kernel is common. However, special care has to be taken because there are significant performance issues when the kernel is responsible depending on how the memory is allocated. For small, infrequent allocations — such as module insertion — then `kmalloc` is very similar to the C library's `mmaloc`. For frequent or large memory allocations, there are several mechanisms, and that is beyond the scope of this section.

4.B.3. Application View

Once the application has been partitioned and some parts now exist as hardware implementations, there needs to be some mechanism for the software and hardware to communicate. Since the application has no way of directly addressing the hardware's physical address and it cannot update its own MMU page tables, it must rely on an operating system. This functionality is typically provided in the device driver associated with the core. Thus, it is up to the designer to either provide a device driver or find a generic one that is suitable.

Before returning to creating a kernel module to implement a device driver, let's review how the application can interact with a (char) device driver.

There are two main ways that an application communicates with a kernel (and indirectly with the hardware core). The first is a streaming interface that is appropriate for transferring large data sets or continuous transfers. The second approach is appropriate for small transfers when latency is critical or the transfer is occasional. (There is a third approach that we won't cover here, which is appropriate if the hardware core resembles a block-oriented storage device — such as a disk drive.)

The first approach uses the `open`, `read`, `write`, and `close` system calls. `Open` takes an arbitrary file name and creates a file descriptor. In our case, we are going to pass a special file called a *device node*. This file is created with the `mknod` (or with `genextfs2`) and, by convention, goes in the `/dev` directory. For example,

```
ls -l /dev/ttyS0
crw-rw-rw- 1 root uucp 4, 64 Oct 19 11:14 /dev/ttyS0
```

shows a very common device file. The first character on the line indicates that this is a char device driver (versus a normal file which has -). Notice the numbers "4, 64" where the size of the file is normally found. These are the major and minor device numbers (respectively). This means that the device node was created with the following command.

```
mknod /dev/ttyS0 c 4 64
```

The first argument is the name of the file, the second is whether it is a char or block device (we will only discuss char devices), and the last two arguments are the device numbers. The device numbers are used to connect a device file (and the application) to a specific device driver.

In a general-purpose system, there is a protocol for dynamically allocating device numbers that involves a separate process. This is necessary because the total number of hardware devices is growing and there are a fixed number of device numbers. However, in an embedded system we can safely assume we know all of the hardware that will be in our system, so we can make a static assignment and use an operating system call that may disappear in the future (see below).

The way an application accesses the kernel device driver is through this device node, for example,

```
char buff[100] ;

fd = open("/dev/mycore",O_RDWR) ;
// fill buff with some data to be sent to the core
write(fd,buff,100) ;
close(fd)
```

The operating system will do a number of security checks and then transfer control to a subroutine in the device driver associated with /dev/mycore (typically called mycore_write). The difference between PIO and DMA is how mycore_write processes buff. If the device driver programmatically writes each 32-bit of word of buff to the hardware core, that is PIO. The alternative is for the device driver to copy entire contents of buff to a block of kernel memory and then set up a DMA engine (a dedicated piece of hardware used to transfer the data across the bus, independent of the processor). The former is fast and simple, while the latter frees up the processor for other tasks. However, DMA requires a significant amount of setup and an interrupt to indicate when the transfer is complete. Thus, for small amounts of data (hundreds of words), PIO makes sense. For long transfers (large buffers of thousands of words), DMA can improve overall system performance.

A third I/O option available on Unix-based systems, such as Linux, is called memory-mapped I/O. This mode works by having the device driver manipulate the application's MMU page tables so the application can read or write to its virtual memory, and the MMU maps those addresses to the core's physical address, effectively bypassing a system call to the operating system. An example of this mode is shown below.

```
char *buff ;

fd = open("/dev/mycore",O_RDWR) ;
buff = (char *)mmap(0, 4096, PROT_READ|PROT_WRITE, MAP_SHARED, fd, 0) ;
// fill buff with something for the core
munmap(buff,4096) ;
close(fd)
```

Note that the application doesn't allocate any space for the buffer; the mmap system call simply creates a new virtual page for the application. The arguments to mmap are straightforward. (The reason we use 4096 is simply to emphasize that mmap will map the whole page.) An application can request 100 bytes (as before) but it will get a whole page.

4.B.4. Char Device Driver

In general, the main way of interacting with the Linux kernel is through a registration process that involves passing a data structure with function pointers. In the case of a "char" device driver, this is called the file operations structure and it loosely corresponds to the system calls that the application makes. A simple file operations structure might have four to seven entries. The first entry identifies the device driver; the rest correspond to functions in the device driver. The idea is that whenever a kernel needs some device driver-specific functionality, it calls one of these functions and the device driver is expected to complete the operation.

Our goal here is not to detail all of the function arguments and required functionality. That would simply be repeating what is widely available in books and on the World Wide Web. Rather, our goal here is to highlight the functions that are relevant to an embedded systems developer. To provide a framework, consider the following file_operations structure.

```
static struct file_operations fops = {
  .owner   = THIS_MODULE,
  .open    = mycore_open,
  .read    = mycore_read,
  .write   = mycore_write,
  .mmap    = mycore_mmap,
  .llseek  = mycore_llseek,
  .release = mycore_release,
};
```

The first field, .owner, is boilerplate — you always include it to identify the device driver to the kernel. The open and release fields roughly correspond to the application's open and close system calls. The major difference occurs when more than one process might be simultaneously using the device driver. The open file operation is called every time a new process invokes the open system call. The release file operation is called when the *last* process invokes the close system call.

The read and write file operations are analogs of the system calls by the same names; llseek corresponds to the lseek system call. In short, read is a request from the kernel to move data from the developer's hardware core to an application's memory buffer and write is a request to move data from an application's memory buffer to the hardware device. Both file operations provide a virtual memory address and a length. However, these is the application's virtual memory address (not a kernel address), so one has to use the copy_from_user and copy_to_user functions to access the application's buffer. When the hardware is accessed with a char device driver, the kernel treats it like a large sequential file and keeps track of the current position in that stream as an offset from the beginning of the stream. The application can explicitly change the current offset with the lseek system call and the llseek file operation is called to let the device driver know. In all of these file operations, the device driver can reject the request, causing the application to receive an error. If the device driver does not need to do anything — for example, when the application invokes lseek — then that file operation field can be omitted from the stream.

The last file operation, mmap, is often very useful if the application and hardware communicate in small (one- or two-word) transactions. Instead of the device driver acting as an intermediary for every transaction, the application issues the mmap system call to request that the hardware's physical address region be mapped to the application's virtual address space. Fortunately, Linux does most of the work — it sets up the MMU page table, checks permissions, and allocates a free range of pages in the application's virtual memory space. However, the Linux kernal does not know the physical address of mycore, so it calls mycore_mmap to complete the job. It passes the usual file pointer argument and a vm_area_struct that has the application's virtual memory page table information. Our job is simply to finish setting up the MMU with the core's base address. This is accomplished with a call to remap_page_range as shown below.

```
static int mycore_mmap ( struct file *filp, struct vm_area_struct *vma ){
  int rc = 0 ;
  ⋮
```

```
    vma->vm_flags |= VM_RESERVED;
    if( remap_page_range(vma->vm_start, base,
            vma->vm_end-vma->vm_start,
            vma->vm_page_prot) ) {
      rc = -EAGAIN;
    }
    return rc ;
}
```

Most of the information (start of virtual address page, length, page protections) has already been taken care of; we simply provide a base physical address.

One important detail to note: here we use `remap_page_range`, which updates an *application's* page table. With PIO or DMA, the device driver often uses the `ioremap` subroutine call. The difference is that `ioremap` updates the *kernel's* page tables, that is, it maps the kernel's virtual address space to the core's physical address space. Use `ioremap` when you want to make the core visible to the device driver; use `remap_page_range` when you want to make the core visible to an application.

4.B.5. Summary

A complete example of a simple adder core, a memory-mapped device driver, and an application that uses the core is on the book's world wide Web site. Using this adder core example (with two inputs and one output), a simple test was set up that invoked the core 6 million times and the rate (transactions/second) measured. The results are shown in Figure 4.14. For simple cores with a handful of arguments, it is clear that the MM is the most efficient mode of transfer. However, for more complex cores — especially ones that include the continuous transfer for state — more complex device drivers are needed. For an example of this, see the RCADE device driver, also on the book's world wide Web site.

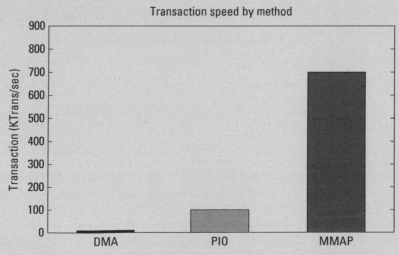

Figure 4.14. Performance of various transfer methods.

Exercises

P4.1. Suppose an application was profiled and it was discovered that the do_work function accounted for 50% of the execution time. If do_work was implemented in hardware, what is the largest speedup one could expect for the codesigned system?

P4.2. Consider the following snippet of a C program.

```
int  x ;
int  y ;
int div ( int m , int n ) {
    if( n==2 )       y = m>>1 ;
    else if( n==4 ) y = m>>2 ;
    else             y = m/n ;
    return y ;
}
void sub1 ( int a, int b ) {
    int t1, t2 ;
    t1 = a-b ;
    t2 = a+b ;
    x = div(t1,t2) ;
}
```

(a) If div was determined to be a feature implemented in hardware, what is the affected, trapped state when it is invoked?

(b) Suppose div was merged into sub1 and together, implemented in as a hardware feature. Does it change the affected, trapped state?

P4.3. Consider subroutine fmul below.

```
float fmul ( float a, float b ) {
  return a*b ;
}
```

Give a practical explanation of why fmul might not be considered a candidate feature.

P4.4. Suppose subroutine crc below is a candidate feature,

```
int crc ( int input ) {
  return .... ;
}
```

Name three quantitative reasons it may not end up mapped to hardware.

P4.5. A software reference design has one main loop that takes 600 seconds to complete. The loop was executed on the target processor and each subroutine was timed. Table 4.1 shows this data. Each subroutine is called once and $s(i)$ is

i	$s(i)$	$h(i)$	$m(i)$	$r(i)$
work	250	25	5	$[\,500 \quad 20 \quad 20\,]^T$
gen	100	50	1	$[\,200 \quad 0 \quad 10\,]^T$
smooth	100	35	10	$[\,100 \quad 0 \quad 0\,]^T$
pack	100	10	5	$[\,100 \quad 10 \quad 0\,]^T$

Table 4.1 Profile information for item 5.

the time it takes in seconds. The time to perform the same functions in hardware and the marshaling costs (in seconds) are also shown in the table. The resource usage of CLBs, BRAMs, and DSP blocks are also shown.

(a) Calculate the expected performance gains, $\gamma(i)$, and rank the subroutines.

(b) If all four subroutines are implemented in hardware, what is the overall expected improvement in speed, Γ?

(c) Ignoring routing concerns and assuming the target device has

$$r_{\text{FPGA}} = \begin{pmatrix} 600 \\ 20 \\ 20 \end{pmatrix}$$

resources, what hardware/software partition yields the maximum speedup?

P4.6. Suppose a scientific application was profiled using gprof. The data for the top fifteen functions was collected and is shown in Table 4.2. (You may want to consult with gprof manual page to answer these questions.)

(a) Approximately how long did this application run?

(b) What is the difference between "self ms/call" and "total ms/call"? If these numbers are identical, what does this suggest?

(c) Suppose amp1 can be implemented in deeply pipelined core that takes 220 clock cycles to complete. It takes an additional 40 clock cycles to copy the data in and another 40 clock cycles to copy the data out. If the frequency is 100 MHz, what is the expected gain if amp1 is implemented in hardware?

(d) Suppose the same hardware function can be used to accelerate amp1, amp2, and amp3? What would be the overall expected system gain?

% Time	Cumulative seconds	Self seconds	Calls	Self ms/call	Total ms/call	Name
16.97	2.42	2.42	20	121.02	121.02	amp1
16.83	4.82	2.40	20	120.02	120.02	amp2
16.41	7.16	2.34	20	117.02	117.02	far2
11.01	8.73	1.57	20	78.51	78.51	amp3
10.94	10.29	1.56	20	78.01	78.01	far3
10.66	11.81	1.52	20	76.01	76.01	far1
4.70	12.48	0.67	2920	0.23	0.23	h_inner13
2.88	12.89	0.41	1920	0.21	0.21	e_inner13
2.52	13.25	0.36	1	360.05	360.05	grid
1.82	13.51	0.26	1	260.04	260.04	tstep
1.19	13.68	0.17	778	0.22	0.22	asuby
1.05	13.83	0.15	2	75.01	75.01	initial
0.98	13.97	0.14	2	70.01	250.04	SYMMLQ
0.56	14.05	0.08	2334	0.03	0.03	DAXPY
0.42	14.11	0.06	1560	0.04	0.04	DDOT

Table 4.2 Flat profile information for item 6.

P4.7. Subroutine C consumes 85% of the execution time of an application. However, an engineer refactored the application so that a portion of subroutine C is moved into a new subroutine, C1 and then subroutine C calls C1. In this arrangement, subroutine, C1 accounts for only 80% of the execution time. C1 is mapped to hardware. How does a change like this impact expected system gain?

P4.8. Suppose the performance metric of interest was to cost savings so long as a minimum execution rate was met.

(a) How does this effect the formal framework presented in Section 5.2?

(b) Does this change γ and Γ?

(c) Describe a heuristic to replace the one described in this chapter to maximize this performance metric (i.e. minimize cost).

P4.9. Suppose the performance metric of interest was energy consumption.

(a) What information is relevant at profiling to build a performance gain metric?

(b) Rework the performance gain metrics γ and Γ with this information.

(c) Will the heuristic described in this chapter be appropriate?

References

Adams, M. D. (2007). *The jasper project home page.* http://www.ece.uvic.ca/mdadams/jasper/ last accessed May 2010.

Corbet, J., Rubini, A., & Kroah-Hartman, G. (2005). *Linux device drivers, (3rd ed.).* Sebastopol, CA, USA: O'Reilly & Associates, Inc. URL: lwn.net/Kernel/LDD3.

Denning, P. J., & Kahn, K. C. (1975). "A study of program locality and lifetime functions." In *SOSP '75: Proceedings of the Fifth ACM Symposium on Operating Systems Principles* (pp. 207–216). New York, NY, USA: ACM Press.

Graham, S. L., Kessler, P. B., & McKusick, M. K. (1982). "Gprof: A call graph execution profiler." In *SIGPLAN '82: Proceedings of the 1982 SIGPLAN Symposium on Compiler Construction* (pp. 120–126). New York, NY, USA: ACM.

5

SPATIAL DESIGN

Things which are equal to the same thing are equal to each other.

Euclid
Elements of Geometry

So far we have concentrated on designs with a single thread of control. That is, each instruction or operation is executed fully prior to execution of the next. This chapter introduces the idea of spatial parallelism, which allows for multiple threads of control. In computer science terms, this is more generally known as concurrency. Here, we use ***spatial*** to highlight the fact that in our hardware designs, the multiple threads of control are actually able to operate simultaneously. This is in contrast to early forms of concurrency/parallelism where multiple threads were time-multiplexed on a single control unit.

Although concurrency has been around since the very first electronic computing machines, the general perception is that managing, debugging, and maintaining multiple threads of control are significantly more difficult than doing the same with one thread. Thus, for years many researchers in the computing community have strived to avoid implementing explicitly parallel designs. However, in exchange for the extra effort, explicitly parallel solutions offer a much larger design space. This larger design space provides us more opportunities to meet the complex performance metrics of a modern embedded system.

For the unfamiliar, parallelism is an easy concept to grasp. Consider performing a search over a large array of numbers. Computer scientists may perform one of a variety of search algorithms. The simplest is a linear search, comparing against each element in the array until the number is found or the end of the array has been reached. Unfortunately, the simplest solution often does not yield the best performance, and in this case a linear search results in $O(n)$. A more sophisticated approach is binary search, which relies on the fundamental concept of divide and conquer. In a recursive implementation, a binary search splits the search space in half repeatedly until the number is found. On a sequential processor this would require sequential comparisons, resulting in a worst-case $O(log(n))$ performance.

In hardware, a sequential implementation using a single comparator would also result in a worst-case performance of $O(n)$. By implementing a similar binary search algorithm in hardware the performance increases to $O(log(n))$; however, by increasing the number of comparisons occurring at the same time (in parallel) this performance can be improved significantly by replicating the logic to implement a fully parallel search implementation. This solution, however, comes at the expense of additional resources. Consider a situation where there are sufficient resources to perform all of the necessary comparisons in parallel. In this case we would be able to improve the performance to $O(1)$. While it may not always be feasible to implement such a large number of parallel operations, it is worth considering how to design systems that are able to take advantage of the exceptional parallelism available within FPGAs.

The learning objectives of this chapter are quite straightforward:

- We begin with an overview of the principles and terms related to parallelism to give readers a firm understanding of the concepts that will be applied throughout this chapter.
- Next, we use these principles to help in our understanding of what to look for when trying to find parallelism in software reference design that is typically expressed as a sequential algorithm.
- Finally, we look at practical methods to implement parallel designs on Platform FPGAs and common issues that can arise if design considerations are not met.

5.1. Principles of Parallelism

The terms *parallel* and *serial* originated with electrical terms used to describe the organization of electrical components. For example, two resistors, R_1 and R_2, can be arranged in series or in parallel, as illustrated in Figures 5.1(a) and 5.1(b), respectively. However, this analogy is slightly broken. Electrically, both components in these figures are actively contributing to the system behavior, which is the output characteristics of the circuit, at all times. However, when describing serial computation, the operations are often broken up into individual instructions, which may share the same physical components of the system. It is more accurate to describe serial computation as *sequential* computation. Therefore, the terms **sequential** and **parallel** can be defined as follows.

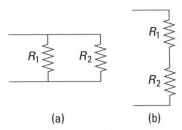

Figure 5.1. Resistors in (a) series and (b) parallel.

Sequential: The same components are resused over time and the results are the accumulation of a sequence of operations.

Parallel: Multiple instances of components are used and some operations may happen concurrently.

The advantage of sequential implementations is that they efficiently use hardware resources, whereas parallelism can be used to reduce the time to completion of a task at the expense of additional hardware.

To make these concepts clear, consider a task T that is composed of a set of eight subtasks $\{t_0, t_1, \ldots, t_7\}$. Executed sequentially on a single component, task T takes 8Δ to complete. This is illustrated in Figure 5.2(a). In contrast, these eight subtasks could be assigned to four units as illustrated in Figure 5.2(b), and it only takes 3Δ to complete task T. The obvious savings here is that the application completes $2.1\times$ faster; however, it requires $4\times$ the resources.

A good question to ask at this point is: "Why not move t_7 to unit 3 in time slot $[1\Delta, 2\Delta]$? Then the application will complete $4\times$ faster. The answer lies in how the application is specified. If the programmer who developed the application assumed that it would be executed sequentially (t_i is executed before t_{i+1} — perhaps because t_{i+1} uses a result from t_i) then any parallel execution of T's subtasks must respect this ordering. This highlights the fundamental issue of this chapter: Given a sequential software reference design, our job is to discover the parallelism and implement a suitable execution structure.

5.1.1. Granularity

Consider our task T and its set of subtasks $\{t_i\}$ again. If T is our software reference design, we must first determine how to define the substasks. *Fine-grain* parallelism means that subtasks t_i are relatively small, maybe the equivalent of a few instructions. *Coarse-grain* parallelism refers to relatively large subtasks, such as an entire subroutine.

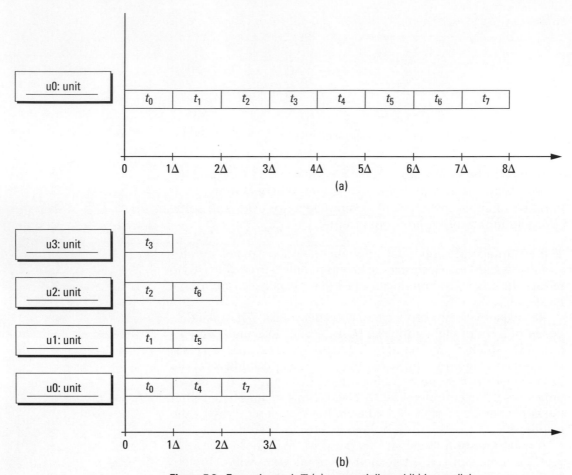

Figure 5.2. Executing task T (a) sequentially and (b) in parallel.

Often, natural boundaries based on the expression of the reference design are used. For example, we think of an application consisting of modules and packages, packages consisting of classes, classes consisting of subroutines or methods, and so on down to processor instructions consisting of n-bit arithmetic operations. These have common names based on how large the subtasks are.

Bit-Level Parallelism

If given that task T operates on scalar values (integers, real numbers, etc.) then the scalar will have some bit pattern to represent specific values. For example, if we are using large integers (± 2 billion) we would typically use a 32-bit two's complement notation. Bit-Level Parallelism (BLP) would break the 32-bit scalar into fields (subsets of the 32-bit representation) and then organize parallel subtasks (t_i) to work on each field in parallel. The advantage of bit-level parallelism is that it tends to be independent of the application and one parallel implementation can be used over and over in many general-purpose applications. The disadvantage is that the number of bits in the scalar tends to limit the parallelism.

Instruction-Level Parallelism

The ubiquitous general-purpose processor today has an instruction set architecture (ISA) that defines the various operations available to it. In Instruction-Level Parallelism (ILP) the task T is typically a program and the subtasks t_i are the individual instructions. A large body of work in computer architecture describes different techniques for detecting this parallelism. The advantage is that the programmer can develop sequential codes and then the software/hardware system performance increases. This term is reserved for parallelism found in single threads of control.

Iteration-Level Parallelism

Also known as Data-Level Parallelism (DLP), this granularity is similar to bit-level parallelism except that instead of scalars, data are vectors, and instead of a single operation, the subtask might be a sequence of instructions. For example, the mathematical operation called *inner product* takes two vectors and produces a scalar,

$$c = \sum_i a_i \times b_i$$

If written as a loop in a sequential language,

```
c = 0 ;
for( i=0 ; i<N ; i++ ) {
  c = c + a[i]*b[i] ;
}
```

each iteration of the loop happens one after the other. However, as shown later in this chapter, some of the iterations can overlap and be executed in parallel.

Thread-Level Parallelism

In iteration-level parallelism, each subtask is performing the same sequence of instructions on separate datum in a vector. Thread-Level Parallelism (TLP) is different from the previous three in that the programmer specifies the parallelism explicitly in multiple threads. Each subtask is a thread and does not necessarily perform the same sequence of instructions.

All of these terms are generally applied to parallelism found in software implementations. We adopt the terms here because many embedded systems designs start as software reference designs. Also, they effectively convey a sense of "size" or complexity that mirrors a hardware implementation. Granularity is not a precise measure but serves to help teams communicate the intent. A hardware engineer may strive to achieve a level of parallelism based on the software reference design's expressible granularity.

5.1.2. Degree of Parallelism

Another important concept, which is not always treated precisely but is useful nonetheless, is a software design's ***degree of parallelism***. Degree of parallelism (DOP) can refer to (i) the number of concurrent operations at a single moment of time, (ii) a time-varying function over the entire application, or (iii) the average of a time-varying function over some task *T*, which may not include the entire application. For example, the degree of parallelism for some software reference design is illustrated in Figure 5.3. From the diagram, we can see that the maximum DOP is 6, which occurs between time 5–7. We can also see that the average DOP is 3. We calculate the average DOP as the area under the curve. What this tells us is that, if these are iteration-level subtasks, having more than six cores would be overkill. Likewise, if hardware resources were scarce, having four cores instead of six would achieve 78% of the performance with 66% of the resources.

5.1.3. Spatial Organizations

In the early 1970s, Flynn proposed a taxonomy for describing computer organizations (Flynn, 1972). Although his taxonomy was intended for large multichip computers, it turns out that the same taxonomy can be useful for characterizing parallel computation

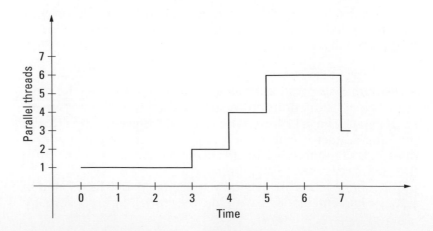

Figure 5.3. Illustration of the degree of parallelism of a fictitious application.

in a Platform FPGA-embedded system. His taxonomy is organized around two axes that describe a stream(s) of instructions in one dimension and a stream(s) of data in the other. Each dimension has two cases, single stream or multiple streams, which leads to four cases:

Single Instruction Stream, Single Data Stream (SISD) architecture is the single-core, von Neumann-style processor model used in many general-purpose computers. In this situation, a single processing unit consumes a single sequence (stream) of data.

Single Instruction Stream, Multiple Data Stream (SIMD) architecture uses simple Processing Elements (PEs) that are all performing instructions from a single instruction stream in lock step. Early implementations of this type of architecture used a single instruction decode circuit that broadcasted control signals to all of the PEs. Associated with each PE is an independent local memory hierarchy. Because each PE's memory is separate, there are, in effect, multiple data streams being processed simultaneously.

Multiple Instruction Stream, Single Data Stream (MISD) architecture breaks up a single stream of instructions (or sub-instructions) into independent operations that can be executed in parallel. These independent streams of instructions are then assigned to separate processors, and a single stream of data is passed from processor to processor via explicit communication channels.

Multiple Instruction Stream, Multiple Data Stream (MIMD) is what most computer engineers think of first when they think of a parallel computer: multiple independent processors, each with its own memory and own instructions. The processors might communicate via some shared memory hierarchy or via explicit communication channels. Either way, it is characterized by multiple independent processors and multiple independent blocks of memory.

This two-dimensional taxonomy is designed to classify hardware organization. It has served its purpose well, even though at times various cases have appeared to be no longer commercially viable, only to then see the model arise again in a different context. For example, there was a great deal of excitement about SIMD machines built with custom IC chips a couple of decades ago. However, the industry faltered when it became clear that MIMD architectures built from Commodity Off-The-Shelf (COTS) components were much less expensive. But, since then,

SIMD concepts have returned — finding their way into special instruction sets added to commodity processors and Graphical Processor Units (GPU) used in video cards.

Other names for specific cases in Flynn's taxonomy have surfaced. For example, the original conception of MISD was that the processors operate in lock step, that is, data flow along (fixed) communication channels in a prescribed, synchronized pattern. Others have dubbed this organization as "systolic architectures" because it resembled blood flowing along arteries and veins.

Likewise, the principle of SIMD is echoed in the vector processing unit of *vector processors* and in pipelined processors. Although not implemented with processing elements and distributed memory, vector processing units are built to exploit the fact that the same operation is applied to many data elements. This includes direct memory access to main memory and operations that are deeply pipelined. Likewise, it bears a striking resemblance to *stream processors* that have emerged more recently. We expand on this later on in the chapter.

A separate issue from the hardware organization is the programming model. For example, even though MIMD hardware *could* execute separate instruction streams, it is possible to replicate the same instruction stream on every processor. This programming model, called Single Program Multiple Data (SPMD) (or data parallel programming), has proven to be a productive way to write *scalable* parallel applications.

Arguably the biggest difference between vector and stream processors is in the programming model — how involved the programmer is in setting up and controlling the parallelism. In vector processors, the programmer just had to follow a handful of guidelines and the compiler was primarily responsible for detecting and implementing vector instructions. In stream processors, the programmer is usually tasked with explicitly identifying "kernels" in their code — each which has its own local memory and will ultimately be mapped to an independent processor.

In the case of building custom-embedded systems designs for Platform FPGAs, we have the luxury and responsibility of building any parallel architecture we want. Likewise, we are also responsible for coming up with the application to use our architecture! To help readers understand the options and trade-offs, we next describe some of the common organizations in a little more detail.

Vector and Pipelined Parallelism

Suppose we have a task T that can be divided into a set of subtasks $\{t_0, t_1, \ldots, t_{n-1}\}$ where T might be some operation such as

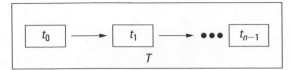

Figure 5.4. Pipeline parallelism.

"multiply two integers $x \times y$." In cases where the subtasks are sequential steps, each subtask can be assigned to a processing station, or hardware core. Visually, this is shown in Figure 5.4.

If we assume that each task buffers its outputs with a register and clocks this core at $\Delta = \max_{t \in T}\{\mathrm{delay}(t)\}$, then it takes n cycles to produce a result. However, the design T can accept a new input every cycle. If we assume a sequence of inputs I that produces a sequence of outputs $O = \langle o_0, o_1, \ldots o_{m-1} \rangle$, then T's throughput is the number of outputs per cycle or

$$\frac{m}{m+n}$$

Assuming that the number of subtasks is fixed and is much less than the desired number of results $n \ll m$, then we observe that the throughput approaches 1 as $m \to \infty$.

Often in embedded systems, we focus specifically on the throughput of hardware cores because the number of inputs is large. Although we do not always think of embedded systems as devices that move large vectors of inputs around, it is very common for these systems to continuously sample their environment. This produces a sequence of inputs I — even though data may not be stored permanently in primary or secondary storage.

How is this related to parallelism? By starting a new operation before the last one completes, we are effectively overlapping consecutive operations of T. The time-space diagram shown in Figure 5.5 illustrates this for four subtasks. Time advances along the x axis in Δ increments, and the activity of the subtasks is shown along the y axis. If $n \ll m$, then it can be seen that — for a larger percentage of the processing time — the degree of parallelism is the number of subtasks. If the time to complete task T sequentially is $\approx 4 \times \Delta$, then we can see that the degree of parallelism translates into a speedup of the core.

Control Parallelism

Control parallelism, also known as irregular or ad hoc parallelism, examines a single stream of instructions to find operations that can be performed across multiple components in parallel. In some cases, this can be done automatically. More often, though, it is

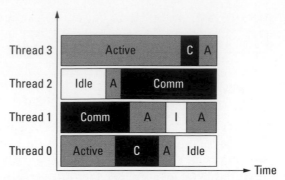

Figure 5.5. Time-space diagram of pipeline parallelism with threads in the idle (I), active (A), or communication I/O bound (C/Comm) state.

up to the designer to block and identify the operations that may be performed simultaneously. Control parallelism is more clearly visible in implementations from hardware description languages. VHDL, for example, can specify explicit components to operate on each individual task. We will discuss HDL implementations more in the last section of this chapter.

Pipelining can be used at any granularity from bits up to complex modules. Likewise, control parallelism can be utilized at different granularity as well. To understand this form, we look at sequences of simple algebraic assignments, that is, imperative programs with no conditional branches or loops. Consider the following program that computes the roots of the polynomial $ax^2 + bx + c$ with the quadratic formula:

$$x_0, x_1 = \frac{-b \pm \sqrt{b^2 - 4ac}}{2a}$$

The sequential code would be:

```
(s1)    descr = sqrt(b*b-4*a*c)
(s2)    denom = 2*a
(s3)    x0 = (-b+descr)/denom
(s4)    x1 = (-b-descr)/denom
```

The inputs to our entity will be a, b, and c. The outputs will be $x0$ and $x1$.

The basic idea behind control parallelism is that some of the statements used to compute $x0$ and $x1$ can be performed simultaneously while still producing the correct answer. So, for example, the instructions $s1$ and $s2$ can operate in parallel, but $s3$ and $s4$ cannot begin until the first two finish. However, once $s1$ and $s2$ have finished, then the last two can be performed separately. Assuming that each phase of these two steps takes one clock cycle, this formulation of the solution would have a constant degree of parallelism of two. Graphically, this is illustrated in Figure 5.6.

However, this becomes more interesting if we change the granularity from statements to operations. The basic operations that we'll use are $\times$, $\pm$, $+$, $-$, and $\sqrt{}$. For the moment, let's just consider the last few operations to compute $x0$ and $x1$ assuming that we have signals that have computed the discriminate and denominator. Figure 5.7 illustrates this control parallelism.

Two important considerations arise from these examples. The first is timing. Earlier we assumed that every statement took one clock cycle. However, chances are that the square root operation

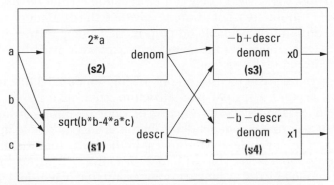

Figure 5.6. Control parallelism found in a quadratic formula (statement granularity).

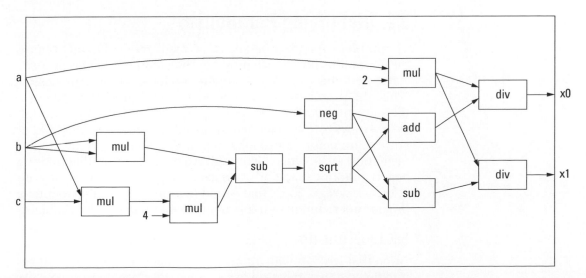

Figure 5.7. Control parallelism found in a quadratic formula (operation granularity).

will take more time than calculating the denominator. Also, we just waved our hands to arrange these blocks because, as humans, we can see the relationships in this very simple equation. We will need a much more formal way of working with these relationships — especially when we introduce loops and conditionals. The issue of timing in pipelines is so important that the entire next chapter addresses it. We formalize the relationships between operations later in this chapter.

Data (Regular) Parallelism

Unfortunately, *ad hoc* parallelism is not very scalable. If a design is created today that has been divided up by operations or by statements, the division may not be sensible for the next generation of components. For example, dividing the operations based on the amount of available on-chip memory may restrict future designs when more on-chip memory resources are available, causing a redesign of the system.

Data parallelism focuses on organizing the parallel threads of control around a regular data structure, such as an array, and loops. In this model, if each iteration of a loop operating on an element of an array is independent, then it is easy to divide up the array among the available resources. In the event that the design is migrated to a larger chip with more resources, the array can be again divided by the available resources. All that is required is the circuitry to implement the additional components. This is much more feasible than redesigning the entire system.

5.2. Identifying Parallelism

The problem of finding parallelism is made more difficult because the ubiquitous imperative programming model is so closely associated with the sequential compute model of a typical processor. While the problem itself may have a large degree of parallelism, the software reference design may not. We can visualize our job as first moving from a sequential specification to a higher level of abstraction before translating to a more concrete parallel implementation as shown in Figure 5.8. We must mention that nothing here is guaranteed to find the maximum parallelism in a software reference design. To do that would require us to be able to infer a programmer's intent — a seemingly impossible to solve problem.

5.2.1. Ordering

Given that we start with a software reference design, our problem is to detect the parallelism in a sequentially specified application. We do this by looking for general patterns in the application that

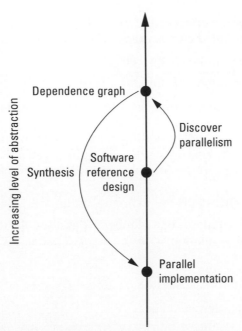

Figure 5.8. Implementing parallelism from a software reference design.

map to certain forms of parallelism. Once the parallelism has been discovered, then we can decide how to make use of it by building the appropriate execution structures in the FPGA fabric. We will start with the assumption of simple sequential tasks with no conditionals or loops. After formalizing the concepts we will extend our notion to conditionals and then loops.

Let's return to the idea of a single task T composed of a sequence of n subtasks $\langle t_0, t_1, \ldots, t_{n-1} \rangle$. We say that the subtasks are ***totally ordered*** because we start t_i before t_{i+1}, for all i. However, we know from the discussion of irregular parallelism that this order is unnecessarily restrictive. We define a ***legal order*** of subtasks to be an ordering that produces the same results as total order T for all inputs. For example,

```
a = (x+y) ;   // (s1)
b = (x-y) ;   // (s2)
r = a/b ;     // (s3)
```

We could divide this into

$$T = \langle s1, s2, s3 \rangle$$

and (without proof) it should be clear that T and T'

$$T' = \langle s2, s1, s3 \rangle$$

are both legal orderings.

Indeed, we observed earlier in the discussion of irregular parallelism that $s1$ and $s2$ can be executed in parallel. This idea of

- "$s1$ has to complete before $s3$" and
- "$s2$ has to complete before $s3$"

but no specific order is imposed for $s1$ and $s2$ is the idea of a ***partial order***.

Therefore, the goal of finding parallelism is to take a software reference design, usually written in an imperative language, and find a legal partial order from the expressed total order.

5.2.2. Dependence

If we treat the elements of a sequence T as a set of substasks, then we can define a relation R on the set T. That is, $R \subseteq T \times T$. If, in English, we say:

$$(t_i, t_j) \in R \quad \text{if subtask } t_i \text{ comes before subtask } t_j$$

then the total order would be

$$(t_i, t_j) \in R_{\text{tot}} \quad \text{if } i < j$$

R_{tot} as written is transitive, but one could reduce the number of ordered pairs and keep the total order by expressing it this way:

$$(\forall 0 \le i < n \,|\, (t_i, t_{i+1}))$$

What we really want is a precedence relation that is not defined by the sequence of operations in the sequential software reference design. What we want is something like:

$$(t_i, t_j) \in R \quad \text{if subtask } t_i \text{ must complete before subtask } t_j$$

But what does "must complete before" mean? This means that task t_i produces some result that must be communicated to task t_j before t_j completes.[1] For imperative programs, these results are communicated through state, which is essentially the register file and primary memory (RAM) of a processor. Further refining our precedence relation, we might say

$$(t_i, t_j) \in R \quad \text{if subtask } t_i \text{ writes a value that subtask } t_j \text{ reads}$$

Thus, the last thing we need to do to formalize this problem is to define two sets for every subtask:

$$OUT(t_i) \quad \text{the values that } t_i \text{ writes}$$
$$IN(t_i) \quad \text{the values that } t_i \text{ reads}$$

---

[1] We say "completes" because there may be sub-subtasks in t_j and this allows for some overlap of t_i and t_j.

Now we can formally write the (true) dependence relation,

$$(t_i, t_j) \in D \quad \text{if } OUT(t_i) \cap IN(t_j) \neq \varnothing \qquad (5.1)$$

In English, one would say that "t_j depends on t_i."

Note that the definition talks about *values* that a subtask does not need, for example, the memory location of a variable. Imperative programs complicate the matter because they frequently reuse memory locations. An equivalent program to an earlier example would be:

```
a = (x+y) ;   // (s4)
r = a ;       // (s5)
d = (x-y) ;   // (s6)
r = r/b ;     // (s7)
```

In this case, s7 rewrites the location of variable r. So the value that s5 writes only exists for a period of time. This means that there is another ordering restriction in this code — we cannot execute s7 before s5 for two reasons. First, if s5 was executed after s7, the value of r that s7 wrote would be lost. Second, we have a (true) dependence in that the s7 IN set includes r, which is in the s5 OUT set. This example introduces the concepts of output and antidependence. As written, s4 is output dependent on s6, as s5 has to complete before a can be overwritten. Similarly, the value stored in a has to be used by s5 before s6 executes, which is called antidependence. However, the key here is "overwrite." Human and automatic processes can be used to avoid this problem with relatively little overhead. It comes down to renaming variables rather than reusing the variables. In fact, there is a very common compiler intermediate representation called Static Single Assignment (SSA). By construction, SSA avoids output and antidependence (Cytron *et al.*, 1991). Thus, we will ignore these "false" dependencies and focus on the "true" dependencies defined in Equation 5.1.

Often a hardware designer does not have to formally construct the IN and OUT sets and graphically render the dependence relation. However, academically it is useful to make the concepts concrete. Returning to the first program example ($s1 - s3$), we can tabularize the IN and OUT sets as shown in Table 5.1.

subtask t	$IN(t)$	$OUT(t)$
s1 a = (x+y);	$\{x, y\}$	$\{a\}$
s2 b = (x-y);	$\{x, y\}$	$\{b\}$
s3 r = a/b;	$\{a, b\}$	$\{r\}$

Table 5.1 Subtasks and their IN and OUT sets.

The OUT sets in these cases are all single variables because of our choice of granularity. Of course coarser units or certain operations (divide, for example) will produce larger sets.

Although a brute-force exhaustive enumeration approach is not necessary, or advisable, in practice we can do this for three subtasks. That is, we need to compute $OUT(t_i) \cap IN(t_j)$ for all i and j (except when $i = j$, as it does not make sense that a subtask would be dependent on itself).

	s1 OUT	s2 OUT	s3 OUT
s1 IN		$OUT(s2) \cap IN(s1) = \{\}$	$OUT(s2) \cap IN(s1) = \{\}$
s2 IN	$OUT(s1) \cap IN(s2) = \{\}$		$OUT(t_3) \cap IN(t_2) = \{\}$
s3 IN	$OUT(s1) \cap IN(s3) = \{a\}$	$OUT(s2) \cap IN(s3) = \{b\}$	

Thus, dependence relation is all of the nonempty sets:

$$D = \{(s1, s3), (s2, s3)\}$$

which is illustrated graphically in Figure 5.9.

Note how the dependence relation, when expressed graphically, relates to hardware design. Computation is organized spatially, and dependence relation indicates the communication. One can easily envision creating a VHDL entity, where the inputs and outputs are ports. A structural VHDL architecture would instantiate each of the computations.

To extend this model to include conditional statements is relatively straightforward. In an *if/then/else* statement, as shown in Figure 5.10, the conditional is simply an expression that produces a value. The first subtask in the *then* group of statements depends on this value, and the first subtask in the *else* group of statements depends on this value. Graphically, we can use the hardware notion of a multiplexer with the condition determining which values to propagate. A `while-loop`, shown in Figure 5.11, shows a single implementation of the loop body as a block diagram. The `while-loop` is similar to an *if/then/else* statement in that they both evaluate a condition and execute code based on the condition; a while loop presents a greater chance for parallelism due to loop unrolling. The *if/then/else* simply multiplexes

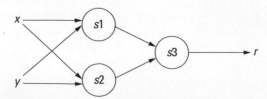

Figure 5.9. Graphic illustration of dependence relation D.

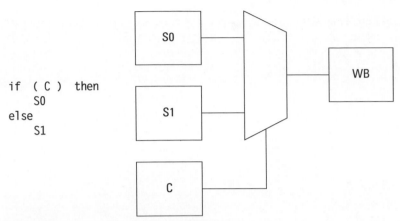

```
if ( C ) then
    S0
else
    S1
```

Figure 5.10. Graphic illustration of *if/then/else*.

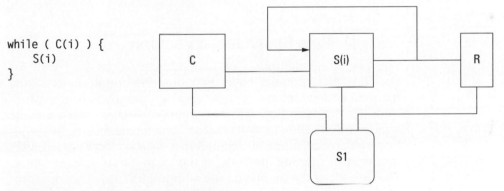

```
while ( C(i) ) {
    S(i)
}
```

Figure 5.11. Graphic illustration of a while loop.

between the two parallel code bodies and unless the if-then-else is executed many times, there may not be any significant gains in performance to implement the design in hardware. The while loop, however, can actually be represented, as it is done in Figure 5.12. Now truly parallel implementations can be executed for a significant performance gain.

Induction variables are integer or floating-point variables that are incremented or decremented the same amount for each for-loop iteration. Recognizing their relationship to the loop control variables often lifts data dependence restrictions and permits further optimization, as shown in the following example:

```
for ( i=0; i<n; i++ ) {
    a[j] = b[k];
    j++;
    k+=2;
}
```

```
while ( C(i) ) {              while ( C(i+0) or C(i+1) or C(i+2) ) {
    S(i)                          if ( C(i+0) )
    i = i+1                           S(i+0)
}                                     if ( C(i+1) )
                                          S(i+1)
                                          if ( C(i+2) )
                                              S(i+2)
                                  i = i+2
                              }
```

Figure 5.12. Unrolling of a while loop.

becomes

```
_Kii2 = k;
_Kii1 = j;
for ( i = 0; i<n; i++ ) {
    a[_Kii1+i] = b[_Kii2+i*2];
}
```

5.2.3. Uniform Dependence Vectors

The preceding definition of dependence works very well for stretches of sequential code with possible conditionals. While the dependence relation can be used on iterative control structures (like for-loops) as well, that notation usually doesn't provide enough information to really exploit spatial parallelism. For loops, designers usually look at dependence vectors. Dependence vectors start by treating the body of the loop(s) as a point in space. The bounds of the loop(s) define a boundary in space. For example, the singly nested loop here has points on a line, as illustrated in Figure 5.13.

```
for( i=0 ; i<10 ; i++ ) {
  a[i] = c * x[i] ;
}
```

Note that the first point is on the origin and that each of the 10 iterations are marked with a point $i = 1$, $i = 2$, and so on up to $i = 9$. Each point on the line represents the computation

```
a[i] = c * x[i] ;
```

If there were multiple statements in the body of the loop, there would still be just a single point.

A more complicated example that uses the (unconventional for C) bounds starting at 1 is shown here:

```
for( j=1 ; j<=3 ; j++ ) {
  for( i=i ; i<=(j+5) ; i++ ) {
    c = x[j][j] ;                 // (s1)
```

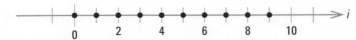

Figure 5.13. Singly nested loop iteration space.

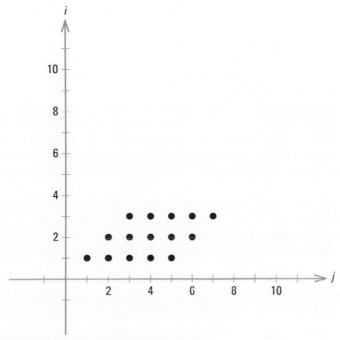

Figure 5.14. Doubly nested loop iteration space.

```
    a[i][j] = x[i][j]/c ;        // (s2)
  }
}
```

In this example, there are three rows and five columns. However, because the inner *i* loop bounds are not constants (but rather
expressions of the outer loop *j*), the iteration space is not rectan-
gular, as Figure 5.14 illustrates. Also notice that inside the body
that *s2* depends on statement *s1*. This dependence is repeated in
every iteration but it doesn't impact the execution of independent
iterations. This is called *loop-independent dependence* and it is
perfectly reasonable to use early formulation. However, consider
this third example:

```
sum = 0 ;
for( i=0 ; i<10 ; i++ ) {
   sum = sum + x[i] ; // (s3)
}
```

Figure 5.15. Loop-carried dependence.

In this case, statement $s3$ depends on itself — but in a different iteration. This is where the iteration space becomes helpful. We highlight this relation between iterations with arrows, as shown in Figure 5.15. Note that the axis has been removed to make the dependence arrows clear. To distinguish this kind of dependence from the earlier description, we call this *loop-carried dependence*.

Now, iteration $i = 1$ depends on iteration $i = 0$, iteration $i = 2$ depends on iteration $i = 1$, iteration $i = 3$ depends on iteration $i = 2$, and so on. These are regular, uniform relations that can be generalized to iteration $i + 1$ depending on iteration i. In terms of arrows, we illustrate this as $i \rightarrow (i + 1)$ for all i within the bounds of space.

Singly nested loops are a rich source of parallelism. However, nested loops quickly magnify the degree of parallelism limited only by the loop-carried dependencies. If the dependencies are uniform, this can be summarized with distance vectors. Consider the following example:

```
for i=1 to 3
  for j=1 to 5
    x[i] = x[i] + samp[i][j]
```

This results in an iteration space with the dependence vectors illustrated in Figure 5.16. Notice opportunities for parallelism. If one looks at just the rows, there is the opportunity for pipeline parallelism. If one looks at the columns, there is unsynchronized parallelism. The 12 (uniform) dependence vectors in this iteration space can be summarized with the distance vector $(1, 0)$. This is because iteration $(j + 1, i)$ depends on iteration (j, i) — taking the point-by-point difference leaves $(1, 0)$. If one had a triply nested loop, then the distance vectors would have three components (one for each dimension).

5.3. Spatial Parallelism with Platform FPGAs

As Chapter 2 has shown, unlike traditional microprocessors, Platform FPGAs offer a wide variety of configurable logic resources that can be used in parallel, extending the computing power beyond the conventional ALU-centric processor. Platform FPGAs also offer the ability to reduce the number of integrated circuits needed in a design by migrating the logic onto the FPGA. This

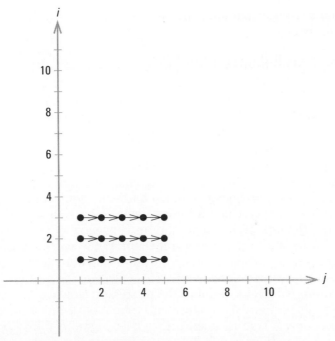

Figure 5.16. Doubly nested loop iteration space with dependencies.

section looks more closely at how to exploit the inherent parallelism within the FPGA. We also highlight some of the issues commonly associated with parallel implementations. Chapters 6 and 7 focus on addressing these issues more specifically.

Let's begin with an example of the differences between a microprocessor (single-core sequential processor) and a Platform FPGA with respect to parallel processing. A microprocessor may use interrupts and timers to switch between sampling input data (maybe a temperature sensor or antenna signal). In many cases the interval between samples is sufficiently large to allow a single controller to switch between each input, acquire a sample, and perform the necessary computations based on the inputs. However, as timing constraints become more tightly coupled, a single controller may not suffice. An FPGA provides the capability to not only interface with a multitude of external devices, but to perform intermediate (an often more complex) computations. These computations are able to occur in parallel and at a higher sampling frequency than conventional microprocessors (we use frequency here in terms of how often input data are sampled, not necessarily operating frequency of the computation or hardware component). Then, when you consider that input signals may be computed on by different function units (i.e., different filter cores for DSP), the processing capabilities of an FPGA begin to overshadow a microprocessor.

5.3.1. Parallelism within FPGA Hardware Cores

Chapter 3 presented two types of design approaches, bottom-up and top-down. We will start with a top-down approach to exploiting parallelism within FPGA components (hardware cores). We specifically chose to start here because we can rely on the principles presented at the beginning of this chapter when identifying and implementing portions of the design in hardware.

Using hardware description languages, we can begin to learn how to express parallelism within a hardware core. For less familiar readers, here is where we must begin to break away from the strictly sequential coding mentality and instead focus on the HDL's inherent nature to describe parallelism. When writing C code, a designer may start with an algorithm or pseudocode to express the flow of the operation at hand. This results in the sequencing of operations, which are fed into the processor. Now we must think less of this sequential structure and more about "What needs to happen now?" That is, how can we best organize the system to perform operations in parallel?

In the top-down approach we identify which operations can occur in parallel without specifying how the operations will be implemented. This is not to say that implementation does not matter; however, it does provide us with a clearer picture of what operations should be occurring. For example, if we are giving an if-then-else statement, we could design a circuit that will operate on both the `if` code and the `else` code in parallel and use the `if` condition to select between the two outputs. This was discussed earlier in the chapter, but as an initial example of parallelizing code with a hardware core, it may be one of the easier concepts to grasp and implement.

Consider another example: if we are given a `for-loop`, which iterates $10\times$ over a portion of code, without knowing the specifics of how to implement each portion of code, we could consider beginning the design by instantiating 10 cores to operate in parallel. This works when the computation in each iteration does not contain loop-carried dependencies, but it does provide us with a general example of how to conceptualize and implement `for-loop` in parallel.

Unfortunately, it may not always be as easy as instantiating multiple copies of the same hardware core. As with loop-carried dependencies, the loops (and resulting hardware cores) now must pass the dependencies between iterations. Consider a previous example again:

```
sum = 0 ;
for( i=0 ; i<10 ; i++ ) {
    sum = sum + x[i] ; // (s3)
}
```

The previous approach of instantiating 10 hardware cores to perform the internal loop computation will not work due to the dependency on sum, which is the result of past iteration additions. Instead we need to parallelize the operation in a hierarchical fashion. By overlapping parallel add operations, we can perform half of the addition operations in parallel, then with three more additions we have our result. So, compared to sequential implementation, which requires 10 sequential add operations, this parallel implementation requires four operations for a speedup of $2.5\times$.

While these examples are simplistic in nature, they are useful when analyzing an application to determine its feasibility to be implemented in hardware. They also showcase that while we might not be able to always parallelize every line of code, we can assemble the computation to perform as many parallel operations as possible to achieve speedups in the design. If the designer can think about the computation in terms of what can be occurring concurrently, then the ability to exploit parallelism within the system becomes less difficult.

Turning our attention to VHDL, parallelism is expressed in a variety of manners. First, within the architecture section of a design, code not within a process block operates concurrently. This may include portions of the circuit that consist of combinational logic. The comparator shown in Listing 5.1 performs three operations in parallel to identify equality, greater than, and less than relations between two inputs, A and B.

This first example is used to emphasize that each operation is occurring at the same time, continuously. That is, if the inputs A or B change, the outputs will change based on the new inputs. Moreover, unlike with sequential code, there is no need to "invoke" hardware as some subroutine or function call outputs. The fact that hardware is *always running* may be a tricky concept to grasp

```
architecture BEH of comparator is
begin
  -- Less Than Comparison
  LT <= '1' when (A < B) else '0';
  -- Greater Than Comparison
  GT <= '1' when (A > B) else '0';
  -- Equality Comparison
  EQ <= '1' when (A = B) else '0';
end BEH;
```

Listing 5.1. An 8-bit magnitude comparator in VHDL.

```
architecture BEH of comparator is
begin
 -- Infer D Flip-Flops
 register_proc : process (clk) is
 begin
  -- If there is a change in the clk signal and
  -- that change is from a 0 to a 1 (a rising edge)
  if ((clk'event) and (clk = '1')) then
   -- Register the value of my_signal_next
   my_signal <= my_signal_next;
  end if;
 end process register_proc;
end BEH;
```

Listing 5.2. Inferring D flip-flop with process block in VHDL.

at first, but it will finally afford your designs the freedom to perform operations in a more tightly integrated manner.

Now, with the addition of a `process` block, we can more explicitly state the portions of code that are running in parallel. On first inspection, a process block does not immediately indicate that there is any additional functionality associated with it. However, with behavioral HDL, we are able to take advantage of a richer set of operations in a more convenient way. For starters, Listing 5.2 illustrates how to use process blocks to infer the use of D flip-flops to improve the design's timing requirements. Synchronous and asynchronous resets can be included through the addition of another `if` statement. It is recommended (specifically for Xilinx FPGA parts) to try to avoid using asynchronous reset signals because it requires the reset signal to be routed across the chip, which may impact the ability of the design to meet the timing requirements. Listing 5.3 is an example of a synchronous process, and Listing 5.4 is an example of an asynchronous process. These simple concepts are important when beginning to design more complex systems.

Obviously, not every line of code can run in parallel, and in those cases we must be able to support sequential operations. This can be accomplished through simple handshaking signals indicating when one operation has completed prior to enabling the next operation to continue. This is similar to the go/done model presented in Chapter 4. Instead of using it to transfer control from software to hardware, we can use it within a component (or between components). We have already seen some of these handshaking signals in Section 3.A when discussing the single-precision floating-point adder design. The handshaking can vary in form, but we will consider it from a producer/consumer perspective. A producer is generating data and a consumer

```
architecture BEH of comparator is
begin
 -- Infer D Flip-Flops
 register_proc : process (clk) is
 begin
  -- If there is a change in the clk signal and
  -- that change is from a 0 to a 1 (a rising edge)
  if ((clk'event) and (clk = '1')) then
   if (rst = '1') then
    -- Reset (Synchronous)
    my_signal <= '0';
   else
    -- Register the value of my_signal_next
    my_signal <= my_signal_next;
   end if;
  end if;
 end process register_proc;
end BEH;
```

Listing 5.3. Inferring D flip-flop with synchronous reset process block in VHDL.

```
architecture BEH of comparator is
begin
 -- Infer D Flip-Flops
 register_proc : process (clk, rst) is
 begin
  if (rst = '1') then
   -- Reset (Asynchronous)
   my_signal <= '0';
  elsif ((clk'event) and (clk = '1')) then
   -- Register the value of my_signal_next
   my_signal <= my_signal_next;
  end if;
 end process register_proc;
end BEH;
```

Listing 5.4. Inferring D flip-flop with asynchronous reset process block in VHDL.

is operating on newly generated data (in the adder example, the producers are the input FIFOs and the consumer is the single-precision floating-point adder). The consumer must indicate when it is ready for data, rfd, and the producer indicates when new data are valid, nd.

This of course is not the only way to implement handshaking. As stated earlier, a variation of the go/done model allows us to begin to sequence each portion of code and use a signal (perhaps done) to indicate when computation of that portion of code is complete. In strictly sequential operations, the done signal would activate the next portion of code to run. A good designer may realize that within each operation, there may be parts that can run concurrently with the preceding operation. This leads to the best-case scenario where all of the computation can be done in parallel, which is ideal but very unlikely. If this event should arise, the done signal would not need to be set.

Alternatively, sequential operations can be controlled through the use of finite state machines (FSM) where each state may represent one or more instructions (or operation). In this case the sequential execution is encoded in the state transitions. State machines are a very useful tool within hardware designs to help with coordinating concurrent operations and, when necessary, providing sequencing capabilities. As mentioned in Chapter 4, a state machine may consist of idle, running, and done states. Note: we do not imply that FSMs *only* consist of these states, or always have to include these states; instead we use these states to indicate three common conditions occurring within state machines. In the idle state, the FSM waits for some condition (perhaps the go signal) before proceeding to the running state where the operation is executed. Upon completion, the FSM may signal any number of dependent components via a done signal, as was discussed earlier. Clearly, the running state can be expanded to include more than just one state; however, the state machine is never *in* more than one state at a time.

For example, a very simple multiply-accumulate (MAcc) operation must perform a multiplication followed by an addition, which accumulates the results of all of the multiplies. Before the addition can proceed, it must wait for the multiplication's result. Once the go signal is received the FSM will perform the multiplication in the mult_a_b state; then, in the next clock cycle, the FSM will perform the addition in the accumulate state. We should note that this is not the preferred implementation of a MAcc; it is used here for demonstration purposes.

This leads into the next point of discussions, which is pipelining. Often designs contain some computation sequence. In the previous example the MAcc sequence consisted of a multiply followed by an addition. We used a state machine to sequence the multiplication and addition; however, by doing so we have fixed the computation rate to two clock cycles per input (assuming that each operation can be computed in one clock cycle). If our design contains 500 pairs of inputs (A × B) to multiply and accumulate, the final result would be valid after 1000 clock cycles.

To improve the performance we introduce pipelining into the design. We modify the design by turning each state in the state machine into individual process blocks (or separate components). We denote this by using squares to represent process blocks/components in our diagram instead of circles, which represent states in the finite state machine.

Let's focus just on the multiply and accumulate portion of the system. We still assume that both the multiplier and the adder require the same amount of time to perform their computation (Chapter 6 addresses pipeline balancing). The multiplier receives its input and produces an output to the adder. The latency of the first result is two clock cycles; however, unlike the sequential implementation, which requires 1000 clock cycles for 500 inputs, the pipeline approach is able to perform the same operations in 502 clock cycles, for $\approx 2\times$ speedup.

Implementing pipelined systems may be easier than sequential finite state machines because a designer may chain together components to build complex operations. This chain can be (if it is not already) pipelined with the addition of registers between each component. Granted, there may be more work if each component internally is not pipelined, but if a designer is aware of the construction of the system from a top-down design view, pipelining the internal components is as simple as was discussed previously.

5.3.2. Parallelism within FPGA Designs

Now that we have covered how to convey parallelism within hardware core designs, we will look at how to construct systems to achieve parallelism within FPGA designs. Because each application is different, we can only speak in general terms of how to build parallel FPGA designs. We will revisit previous design rules and consider how they apply to the system as a whole rather than within a component or hardware core.

From the beginning of this book we have emphasized that, strictly speaking, there is no difference between an FPGA and a Platform FPGA. We use the term *platform* to indicate that the FPGA is used as more than simple glue logic or to interface general-purpose processors with many I/O signals. FPGAs are to be used as a foundation to build designs upon. In this case, the FPGA *is* performing the computations rather than propagating the signals to/from the processor.

Strictly speaking, the most straightforward method to performing parallel computation is to add multiple instances of the same hardware core. At the top level, each instance is a hardware core that would connect to a system bus or the same interconnect as the original hardware core. This is explicit parallelism; there are physically more components running in parallel. However, in order to take advantage of this new compute power, we must divide the data set among the compute cores.

At a finer granularity we can replicate multiple instances of a component within a single hardware core. The difference in this approach is that the parallelism is typically being achieved over a more tightly controlled section of the system. This can simplify the interaction between a processor and the hardware core, but it can also create a bottleneck in the event that each component needs to gain access to the system bus or other shared resources through a common interface.

Top-down design allows us conceptualize the overall flow of the program. It provides us with a way to group together parallelizable code (such as was done for the for-loop example). We use process blocks to explicitly state the operations occurring in parallel. From an abstract view we can consider process blocks to be instances of subcomponents performing the specific computation. Therefore, we can mix and match designs that contain process blocks and component instantiations because they are fundamentally doing the same thing. That is, they are adding additional circuitry to provide parallel computation.

Now we can more concretely explain why modular design and designing for reuse are so important. Rather than replicate the same code within one large design (say, with multiple process blocks), we can instantiate a component multiple times. If a component is used frequently within the design, it stands to reason that future designs can reuse that component.

Chapter in Review

This chapter covered material relating to spatial design and parallelism within Platform FPGA systems. Understanding and expressing parallelism in various levels of granularity help set the stage for how to identify potential parallelism within a design. Then we discussed some of the common practices of implementing spatial parallelism in a Platform FPGA. Describing parallelism within an FPGA design can be an easy task when first building a design. A hardware core that needs to perform a certain amount of computation can quickly be replicated and the explicit parallelism identified.

Section 5.A discusses some techniques associated with VHDL to more efficiently generate parallelism within a hardware core. We also describe various debugging tools, such as ModelSim and Xilinx ChipScope, to help designers track down and fix bugs quickly. These topics are covered within this chapter because up until now we have not focused heavily on moving functionality into custom compute cores within the FPGA. Because our focus has begun to shift toward utilizing the FPGA's resources for more than just a processor-memory model, a discussion of these debugging tools is warranted.

Practical Expansion: Spatial Design

5.A. Useful VHDL Topics for Spatial Design

Chapter 2 covered some basic HDL notation to help act as a catalyst to learning a particular HDL. This was done because when learning a new language, it helps to have a few references. Of course, we wholeheartedly agree that there are far better textbooks available for learning a particular HDL.

What we aim to do in this section is to cover some of the slightly more advanced topics that can be useful with spatial parallel designs and with Platform FPGAs. We should also mention that it is not our goal within this section to specifically talk about how to achieve faster clock rates, reduce resource utilization, or increase throughput. Those concepts have already been discussed (and will continue to be discussed) throughout the remainder of this book. What we are interested in is how to describe large designs with less code and to increase the ability to reuse code. Our focus will remain with VHDL, although many of these concepts will carry over to Verilog or other HDLs.

5.A.1. Constants and Generics

On first inspection, constants and generics may not appear to be directly related to parallel design principles. We will see that their use directly applies to scalable designs by providing a convenient mechanism to change various parameters from top-level components. For example, rather than setting a fixed data width when declaring a signal, we can use a constant or generic to set the width:

```
-- Using a fixed data width can lead to scalability problems
signal incorrect_way : std_logic_vector(31 downto 0);
-- We strongly suggest using constants/generics whenever possible
signal suggested_way : std_logic_vector(C DATA_WIDTH-1 downto 0);
```

Constants are used to avoid "magic numbers" in the design. So, how do constants differ from generics? Constants can be defined within a component or within a package that is used by one or more components. Generics, however, are passed through the system's hierarchy. We want to avoid using constants within a single VHDL file whenever possible and try either to use generics or to use packages.

During the entity declaration the generics and ports are specified. The generics are assigned a default value, which is used when its instantiating component does not specify a value.

```
entity adderFSM is
  generic (
    -- Data Width of the Operands and Result
    C_DATA_WIDTH : natural := 32);
  port (
    a, b       : in  std_logic_vector(C_DATA_WIDTH-1 downto 0);
    a_we, b_we : in  std_logic;
    clk, rst   : in  std_logic;
    result     : out std_logic_vector(C_DATA_WIDTH-1 downto 0);
    valid      : out std_logic);
end adderFSM;
```

In some cases it may be necessary to share a constant/generic between more than one component. Rather than passing the generic between every component, using a *package* can be more convenient. An example of using a package define some constants is

```
package test_package is
  constant C_DATA_WIDTH : natural := 32;
  constant C_ADDR_WIDTH : natural := 32;
  constant C_BYTE_WIDTH : natural := 8;
  constant C_NUM_BYTES  : natural := C_DATA_WIDTH/C_BYTE_WIDTH;
end test_package;
```

To use this package, all a hardware core must do is include it along with the other libraries and packages needed at the beginning of the file.

```
-- Traditional Library and Packages used in a hardware core
library ieee;
use ieee.std_logic_1164.all;
use ieee.std_logic_arith.all;
use ieee.std_logic_unsigned.all;

-- Included our new package with Data, Address, and Byte constants
use work.test_package.all;

-- Entity that inputs data and outputs bytes
entity byteSerializer is port (
  data_in           : in  std_logic_vector(C_DATA_WIDTH-1 downto 0);
  data_in_we        : in  std_logic;
  byte_data_out     : out std_logic_vector(C_BYTE_WIDTH-1 downto 0);
  byte_data_out_we  : out std_logic);
end byteSerializer;
```

How to best use constants, generics, and packages depends greatly on the design and the designer, but our suggestion is to plan for reusability and scalability. Soon we will see how generics can aid in the rapid generation of multiple instances of a component.

5.A.2. User-Defined Types

Similar to packages, user-defined types give us additional flexibility, such as grouping related signals together. The three examples we wish to quickly address are types and subtypes, record types, and user array types.

Types and Subtypes It has already been shown that types are used for declaring finite state machine state names rather than having to use some form of binary encoding:

```
type FSM_TYPE is (wait_a_b, wait_a, wait_b, add_a_b);
```

Subtypes are a convenient way to further simplify declaring signals of a specific type. We would use the new subtype in place of the type it is declaring. For example, we declare DATA_TYPE so that we can use it everywhere we would normally use std_logic_vector(DATA_WIDTH-1 downto 0).

```
subtype DATA_TYPE is std_logic_vector(DATA_WIDTH-1 downto 0);

-- Without Subtype
signal old_way : std_logic_vector(DATA_WIDTH-1 downto 0);

-- With Subtype
signal new_way : DATA_TYPE;
```

Now, not only can we type less when declaring a signal, but if we decide later on we need all of the DATA_TYPE signals to be of a different type, we can change the subtype declaration instead of searching through all of the code and replacing each individual signal declaration.

Record Types Along with types, we can create our own record types. These are similar to *structs* in C. Records can be used to group like signals together, making it easier to pass the group of signals between components or to create registers containing multiple data elements. For example, instead of port-mapping every signal, we could create a single record:

```
type locallink_if is
  record
    data      : DATA_TYPE; -- from our subtype example
    sof_n     : std_logic;
    eof_n     : std_logic;
    src_rdy_n : std_logic;
    dst_rdy_n : std_logic;
  end record;

-- Declare a signal of the record type
signal locallink_tx_port : locallink_if;

-- Accessing each element in the record
tx_data  <= locallink_tx_port.data;
tx_sof_n <= locallink_tx_port.sof_n;
```

User Array Types Arrays in VHDL are a useful tool, especially when designing large systems with multiple instances of each component, such as those found in spatial parallelism designs. For example,

```
-- Declare an array of locallink_if records
type RECORD_ARRAY_TYPE is array (15 downto 0) of locallink_if;
-- Declare a signal of the array type to be used in the system
signal my_record_array : RECORD_ARRAY_TYPE;
```

We have now created an array of 16 elements, where each element is of the previously defined record type, locallink_if. Using array indexing, we can retrieve a specific element within the array just as we would do with arrays in C/C++.

Let's say, instead of 16 elements, we want to be able to define a variable sized array that can be set during the declaration of the signal rather than having to create two different arrays. In order to support these different array sizes we must change how we specify the array range. Previously, we had set the array range to

```
array (15 downto 0)
```

We would now replace this with our variable range described by

```
array (integer range <>)
```

The entire declaration now looks like

```
type RECORD_ARRAY_TYPE is array (integer range <>) of locallink_if;
-- Declare a 16 element array
signal my_record_array_16 : RECORD_ARRAY_TYPE(15 downto 0);
-- Declare a 32 element array
signal my_record_array_32 : RECORD_ARRAY_TYPE(32 downto 0);
```

As you can see with only a single `array type` we can create arrays with a different number of elements, but where each element has the same type.

5.A.3. Generate Statements

The generate statement is one of the most useful tools when creating generic, scalable, and reusable designs. Specifically, we use generate statements to explicitly state parallelism of a design when needing multiple instances of the same component. For this example we will use a doubly nested generate statement to create a two-dimensional array of multiply-accumulate units, which can be used for matrix-matrix multiplication. This actually will generate C_NUM_ROWS×C_NUM_COLS of the `macc_component`.

```
-----------------------------------------------------------------------
-- Multiply Accumulate Units Generate Statement
-----------------------------------------------------------------------
MACC_ARRAY_COL_GEN: for i in 0 to (C_NUM_COLS-1) generate
  MACC_ARRAY_ROW_GEN: for j in 0 to (C_NUM_ROWS-1) generate
    macc_i_j : macc_component
      port map(
        clk         => clk,
        rst         => rst,
        -- Row and Column Data Input to MAcc Array
        col_data    => col_data_array(i),
        row_data    => row_data_array(j),
        row_valid   => row_valid(i),
        col_valid   => col_valid(i),
        -- Results Output
        macc_result => macc_data_array(i)(j)),
  end generate MACC_ARRAY_ROW_GEN;
end generate MACC_ARRAY_COL_GEN;
```

Unlike a C/C++ instance of a `for-loop` where the code is evaluated sequentially during the program's execution, these `for` generate statements will result in multiple copies of the component being instantiated in the design. As a result, we are left with a more scalable design. In fact, if we include generics to be able to specify the number of rows and columns of multiply-accumulate units to generate, we can scale the design to maximize the amount of resources to yield the greatest performance.

   If you notice, both the clock (`clk`) and the reset (`rst`) signals are common among all of the component instances. The row and column arrays are indexed by the `for` generate to connect the corresponding multiply-accumulate component. The result is fed out of each multiply-accumulate component to a two-dimensional array (`macc_data_array`).

We can also use generate statements to include or exclude code or components. Using the `if` generate statement, we can check a Boolean condition to determine whether a segment of code or an instance of a component will be included in the design, for example, if we had an entity with a generic to allow a user to include or exclude debugging circuitry (we will cover how the Xilinx ChipScope debugging tools work in the next section). Based on the value of the generic we may want to include or exclude debugging.

```vhdl
entity byteSerializer is
  generic (
    USE_CHIPSCOPE    : boolean := false);
  port (
    data_in          : in  std_logic_vector(C_DATA_WIDTH-1 downto 0);
    data_in_we       : in  std_logic;
    ...
```

Within the architecture section of the `byteSerializer` code we can generate the ChipScope components based on the `USE_CHIPSCOPE` generic boolean value.

```vhdl
CHIPSCOPE_GEN : if (USE_CHIPSCOPE) generate
  icon_i: chipscope_icon
  port map (
    CONTROL0 => ila_control
  );

  ila_i: chipscope_ila
  port map (
    CONTROL => ila_control,
    CLK     => Bus2IP_Clk,
    TRIG0   => slv_reg0,
    TRIG1   => slv_reg1,
    TRIG2   => slv_reg2,
    TRIG3   => slv_reg3(0 to 7)
  );
end generate CHIPSCOPE_GEN;
```

5.A.4. Design Constraints

We saw in Section 3.A how design constraints such as voltage standards and location can be applied using the UCF file. Using attributes in VHDL, we can also apply similar design constraints. We can also specify things such as FSM encoding and how the embedded multipliers and Block RAMs are inferred. This is especially useful in large designs that use generate statements. VHDL attributes are applied to signals, instances, or architectures. The biggest advantage of using VHDL attributes over applying all of the design constraints in a UCF file is that the designer specifying the attributes uses the names in the HDL design, whereas names in a UCF file refer to signal and instance names after the design has been through the map stage.

The constraint from Section 5.A:

```
Net fpga_0_clk_1_sys_clk_pin LOC = L29 | IOSTANDARD=LVCMOS25;
```

could be applied at the top level HDL file of the project with the following code.

```
architecture struct of top is
  -- Defining the LOC attribute to be of type string
  attribute LOC        : string;
  -- IOSTANDARD attribute is also of type string
  attribute IOSTANDARD : string;

  -- Applying the newly defined attributes to the clock pin
  attribute LOC of fpga_0_clk_1_sys_clk_pin : label is "L29";
  attribute IOSTANDARD of fpga_0_clk_1_sys_clk_pin : label is "LVCMOS25"
```

One useful constraint is to tell XST how to handle multipliers in a design. This code will tell XST to use pipelined block multipliers in every instance of the `macc_component`.

```
attribute MULT_STYLE : string;
attribute MULT_STYLE of macc_component : component is "pipe_block";
```

For more details of the allowable attributes, refer to Xilinx, Inc. (2009c).

5.B. Debugging Platform FPGA Designs

Although debugging hardware is not directly related to spatial designs, oftentimes spatial designs grow large and additional debugging techniques can be helpful. Using Platform FPGAs with a processor and Linux, we have the ability to use the Gnu Debugger (gdb). Unfortunately, gdb does not help us much when we have a bug in the hardware design. Over the course of this section we discuss some of the methods and tools that can be used to assist the designer in tracking down and fixing problems in their hardware designs.

5.B.1. Simulation

Typically, a designer will simulate the whole design (or portions of the design) before ever generating a bitstream and running it on the FPGA. The simulation can offer a tremendous amount of information regarding the functionality of the system. For complex designs, this is useful in testing all aspects of the design comprehensively and limiting the number of synthesis runs.

Through simulation we can test our design with a variety of input stimuli and analyze the generated output waveform for each signal in the design. Initially, viewing the simulation waveform may be a bit overwhelming, especially when testing designs with hundreds and thousands of signals, but once the designer becomes familiar with the simulation tool (such as ModelSim), a design can be put through its paces all before ever being run on the FPGA.

A fallacy that often exists in writing HDL is "if it works in simulation, it will work on the FPGA." While this may be true for some designs, it is because the designer is making sure the system will behave in simulation just as if it were running on the FPGA. Without planning ahead, a quickly rushed design may look fully functional in simulation, but behave differently on the FPGA.

We recommend that you follow a modular hierarchical design approach when designing complex systems. By this we mean building a design from the bottom-up, testing each component thoroughly before moving on to the next level of the design. Once all the components are connected together, a final system-level simulation would involve checking for unexpected outputs due to component inter-dependencies.

```
============================================================================
*                            HDL Analysis                                  *
============================================================================
Analyzing Entity <adderFSM> in library <work> (Architecture <beh>).
WARNING:Xst:819 - ''adderFSM.vhd'' line 44: One or more signals
are missing in the process sensitivity list. To enable synthesis of
FPGA/CPLD hardware, XST will assume that all necessary signals are
present in the sensitivity list. Please note that the result of the
synthesis may differ from the initial design specification.
The missing signals are:
   <b_reg>, <a_we>
```

Figure 5.17. HDL Analysis section of the XST synthesis report.

If a simulation test bench exists for the system (or device) that is malfunctioning, we can set up the same input stimuli and observe how the device is functioning. Two possibilities now exist: (1) the simulation matches the malfunctioning hardware or (2) the simulation runs correctly.

In the event that the simulation is also incorrect, the designer must identify where within the design the error exists. Because the simulation waveform can present all of the signals in the design, the designer has the ability to verify the functionality during each clock cycle. This can be accomplished through use of the `assert` command in VHDL. Assert checks a condition and, if it fails, displays a message to the simulation tool for the user to view.

Unfortunately, the more likely case is that the simulation will not match the hardware's functionality and instead show the correct, expected output. This stems from the fact that the simulation tool and the synthesis (and corresponding netlist builder, MAP, and PAR) tools can analyze the design differently.

To help curb these simulation versus synthesis problems we cover some of the more common, yet difficult to track down, differences between the two. Our goal here is not to cover every type of problem, but to help readers become more familiar with some of the issues at hand.

Sensitivity List   In VHDL a process must have all of the signals that will cause the process to be re-evaluated in its sensitivity list. This includes any inputs into a process which may be used for signal assignment or even condition checking. In simulation, a missing signal in the sensitivity list means even though a particular signal's value changes elsewhere in the design, it will not necessarily force this process to be re-evaluated. The designer may actually want the simulation's functionality; however, the synthesis tools (especially the Xilinx Synthesis Tool) will identify the missing signal(s) from the sensitivity list and assume the signal *should* exist in the sensitivity list. From the HDL Analysis section of the XST synthesis report, this is reported as a warning, as seen in Figure 5.17.

One approach to combating this problem is to combine the synthesis and simulation processes together. By this we mean while designing each component in the system, the designer should not only simulate the design, but also run the synthesis tool and look for warnings such as the one previously listed. This can be done using the Xilinx Synthesis Tool (xst) for synthesis, either through ISE directly or from the command line scripts presented in Section 2.B.1. While this may seem like a lot of work, if the end goal of the design is to run in hardware, it only makes sense to build the design incrementally and test both its functionality in simulation and check for warnings and errors in the synthesis report.

Latches versus Flip-Flops   Chapter 2 discussed some of the differences between latches and flip-flops in a design. We encouraged readers to use flip-flops in the design in place of latches. A latch can be unintentionally inferred as a result of excluding the assignment of a signal in all branches of the process, as in the `else` case of an `if-then-else` statement. For example:

```
if (a_we = '1') then
  result <= a_reg + b_reg;
  valid  <= '1';
  fsm_ns <= done;
else
  -- Missing result assignment HERE
  valid  <= '0';
  fsm_ns <= wait_for_a;
end if;
```

Here, because the `result` signal is only assigned a value in the `if` case, an 8-bit latch will be generated. In simulation the effect may not be noticeable because if a signal is not assigned a value, the simulation tool usually assumes the signal stays the same. In hardware this is not the same as generating a latch. If you recall, a latch is sensitive to the positive (or negative) level of the clock, which means at anytime during that level if the value changes, the latch is updated. This leads to tighter timing constraints and could result in setup/hold violations. This problem can arise if the designer is less familiar with the synthesis tools and more familiar with simulation. Bad habits formed when learning HDLs where only simulation is used can be a cause for headaches later on in the design process. An example of the warning generated during synthesis where a latch is generated is shown in Figure 5.18

The obvious approach to avoid inferring latches is to include the signal assignment in all of the branches. For programmability and readability this is the recommended approach. Another option is to assign default values to the signal outside of the branches, which will force the signal to a default value. For example, we will assign the missing signals outside of the `if-statement`.

```
-- Default assignment
result   <= (others => '0');
valid    <= '0';
-- If true, overwirte result/valid signals
if (a_we = '1') then
  result <= a_reg + b_reg;
  valid  <= '1';
  fsm_ns <= done;
else
  fsm_ns <= wait_for_a;
end if;
```

Inferring Components   Similar to differences between latches and flip-flops, the simulation and synthesis tools can differ greatly in how the two infer a component. This case does not arise very often, as a designer will typically instantiate a component to be used in the design. The most common reason for there to be problems is if there are differences between the simulation version of a component and its generated netlist.

An example is when inferring a Block RAM. The designer may think the HDL will cause the synthesis to infer a BRAM and code the interface accordingly. When simulating the design, the inferred BRAM will behave as the designer intends. However, when the design is synthesized it is possible that the way the BRAM is trying to be inferred does not actually result in the use of a BRAM. In fact, it could be inferred as straight flip-flops, a distributed RAM, a shift register, or a variety

```
*                          HDL Synthesis                          *
===================================================================
WARNING:Xst:737 - Found 8-bit latch for signal <result>. Latches may
be generated from incomplete case or if statements. We do not
recommend the use of latches in FPGA/CPLD designs, as they may lead to
timing problems.
INFO:Xst:2371 - HDL ADVISOR - Logic functions respectively driving the
data and gate enable inputs of this latch share common terms. This
situation will potentially lead to setup/hold violations and, as a
result, to simulation problems. This situation may come from an
incomplete case statement (all selector values are not covered). You
should carefully review if it was in your intentions to describe such
a latch.
```
Figure 5.18. HDL Synthesis section of the XST synthesis report showing generated latches warning.

```
*                          HDL Synthesis                          *
===================================================================
INFO:Xst:738  HDL ADVISOR - 32736 flip-flops were inferred for signal
<memory_example>. You may be trying to describe a RAM in a way that is
incompatible with block and distributed RAM resources available on
Xilinx devices, or with a specific template that is not
supported. Please review the Xilinx resources documentation and the
XST user manual for coding guidelines. Taking advantage of RAM
resources will lead to improved device usage and reduced synthesis
time.
```
Figure 5.19. HDL Synthesis section of the XST synthesis report showing inferred warning.

of other components. This does not necessarily produce an error in the synthesis flow, but it can produce a message in the synthesis report, as seen in Figure 5.19.

Ultimately, many of these simulation versus synthesis problems exist due to differences between simulation and synthesis tools. As designers, we can help reduce the problems by being aware of the end goal, that is, will the design be purely a simulation model or will the design run on a particular FPGA device? Because we are interested in running our design at some point on the FPGA, it stands to reason that we should invest extra time and energy to make sure we are writing good quality, synthesizable HDL and checking throughout the design process for any indications that errors or problems will arise when the design is run on the FPGA.

5.B.2. Software Addressable Registers

When using a Platform FPGA with a processor, it may be possible to do some debugging through software addressable registers. By this we mean reading and writing to a hardware core's address space to set and retrieve status information. Far more limiting than a simulation model, it does provide a simple way to check the status of the system quickly. What this amounts to is the ability to record the signals in registers and read them back in software, like how we use the print statement when debugging software.

A quick example is a system with eight FIFOs. A status register can be added to report on each FIFO's full status. If the system unexpectedly halts, this status register can be read to determine which of the FIFOs is full, which would provide insight to the designer as to where the problem or bug exists in the system.

```
fifo_full_status_reg <= fifo_0_full & fifo_1_full & fifo_2_full &
                        fifo_3_full & fifo_4_full & fifo_5_full &
                        fifo_6_full & fifo_7_full;
```

We have seen software-addressable registers already in Section 3.A where a template hardware core was created with three software addressable registers (`slv_regX`). Adding a few additional software registers can make for easier debugging. In the case of state machines, it may be useful for the processor to check the current state:

```
with fsm_cs select
  fsm_status_reg <= "00" when wait_a_b,
                    "01" when wait_a,
                    "10" when wait_b,
                    "11" when add_a_b,
                    "ZZ" when others;
```

Of course, there are some drawbacks to using software registers as a form of debugging. First off, software accessible registers can only capture the signal's value at certain time points. Continuous recording and reporting back to the processor is not feasible. Second, each status signal (or any probing signals) must be added to the addressable register space. In a 32-bit address space, this may not be of much consequence, but it is something to be aware of. Third, each status register may consume additional resources, adding to the total size of the component. Fourth, checking the status registers repeatedly can introduce unwanted noise in the system, occupying the processor and bus, and preventing the actual system to run at full capacity. In fact, it is possible that the software status register's introduction to the system could impact the system to either hide the problem or introduce new problems. Finally, if the designer identifies additional signals to be included in a status register, the whole design needs to be resynthesized, which can be a costly price to pay for debugging the system.

5.B.3. Xilinx ChipScope

What we really want when debugging a hardware system is a way to view the values of each signal as if we had an oscilloscope hooked to the FPGA circuit. By this we do not mean connecting an oscilloscope to the actual pins of the FPGA; what we mean is to actually probe the FPGA fabric, where the design is implemented. A variety of tools exist for this purpose, one of which is included within the Xilinx ISE tool chain.

The ChipScope Pro Analyzer Tool from Xilinx is a set of software tools designed to help analyze and debug an FPGA system. The basic principle is to use on-chip memory (BRAM) to store samples of signals over the course of a predetermined amount of time using one or more trigger conditions to start the sampling process. Then, through a host PC, we can connect to the on-chip memory and retrieve stored data to be displayed back to the designer as waveforms similar to a simulation waveform.

The goal of this section is to cover how ChipScope can be included into an existing hardware core. We also cover how to interface with the ChipScope controller through the host PC using a JTAG programmer. The result is a powerful tool that enables a designer to probe and identify malfunctions in the circuit design in far less time than with other software register-based solutions.

Integrated Controller and Analyzer At the heart of the ChipScope system is an integrated controller (ICON) component that is instantiated within the hardware core under test (Xilinx, Inc., 2009a). Think of the ICON as the on-chip equivalent of an oscilloscope's interface to the probes.

The integrated logic analyzer (ILA) is the equivalent of a traditional logic analyzer's probes (Xilinx, Inc., 2009b). The ILA buffers multiple signals for playback later by the ChipScope application running on the host PC. We discuss how to generate both the ICON and ILA components through the use of the Xilinx CoreGen tool. This process is similar to how we generated FIFOs and the single precision floating-point unit.

The integrated controller is used to connect one or more integrated logic analyzers to the host PC application through the JTAG programmer. Up to 15 ILAs can be connected through a single ICON component. Currently, only one ICON is supported per hardware design. To generate the ICON component, launch CoreGen and create a new project for the specific FPGA device; in our case it is the Xilinx Virtex 5 FX130T.

Generating the ICON Expand the Debug & Verification directory to expose the ChipScope Pro directory. Expanding the ChipScope Pro directory, we will see a list of ChipScope components that can be generated, including the ICON and ILA. Select the ICON to run the ICON generator wizard, shown in Figure 5.20. Only a few sets of options exist for the user to configure the ICON. For now we will set the number of control ports to one (1) to support a single ILA connection. The ICON can support up to 15 separate connections. The reason for so many connections is due to the limitation the ILAs have on the number of signals they can sample. We will discuss this limitation more shortly. For now generate the `chipscope_icon` component.

The integrated logic analyzer is used to sample a specific number of signals over a predetermined sample period. Each ILA is limited to at most 16 ports, with each port width being limited to 256 bits. The sample depth is adjustable based on the FPGA device (which is based on the BRAM construction). With the Virtex 5 the sample data depth starts at 1024 and increases by a factor of two from there. That being said, the ILA quickly consume BRAM resources. If you have a design that is BRAM limited, adding ChipScope can be a tricky adventure. To generate the ILA, launch CoreGen, if it is not already open.

Generating the ILA As with the ICON, we will expand both the Debug & Verification directory and the ChipScope Pro directory. Select the ILA to run the ILA generator wizard, shown in Figure 5.21. Leaving the component name as `chipscope_ila`, let's add four trigger ports; three of them will be 32 bits wide and the last will be 8 bits wide. Leave the sample depth at 1024 to minimize the number of BRAMs used. Figure 5.22 shows the configuration for trigger port 1, which we said would have a trigger port width of 32 bits. Set the remaining three trigger ports and generate the `chipscope_ila` component.

Modify Hardware Core for ICON and ILA After the ICON and ILA have been generated we should see that two netlists have been generated (along with a collection of other files, some of which are discussed shortly). Referring back to Section 2.A, we must insert the `chipscope_icon.ngc` and `chipscope_ila.ngc` netlists into the hardware core's project directory. To recap the steps, perform the following:

1. Create the `netlist` directory (if one does not already exist)
2. Copy the ICON and ILA netlists to the directory
3. Create the Black Box Description (BBD) file in the data directory
4. Modify the Microprocessor Description (MPD) file

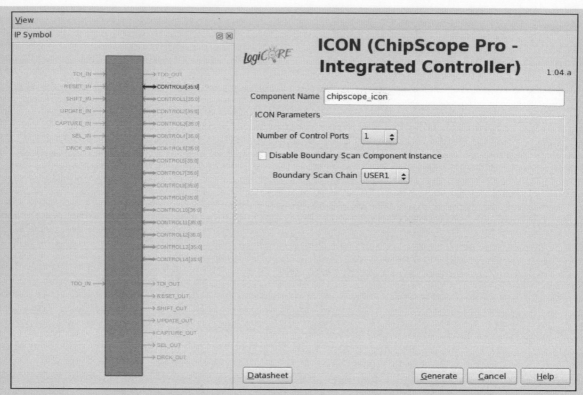

Figure 5.20. Xilinx CoreGen ICON generator wizard setup page.

Once that is complete we must declare and instantiate the components within the hardware core. For example, if using a Xilinx-generated template hardware core, we must modify the user_logic.vhd file. First, insert the declarations for the ICON, ILA control signal, and ILA component. The ILA control signal is an internal signal that will connect the control port of the ICON to the control port of the ILA.

```
-- ChipScope ICON Declaration
component chipscope_icon
PORT (
  CONTROL0 : inout std_logic_vector(35 downto 0));
end component;

-- ChipScope ILA Control Signal
signal ila_control : std_logic_vector(35 downto 0);

-- ChipScope ILA Declaration
component chipscope_ila
PORT (
  CONTROL : INOUT STD_LOGIC_VECTOR(35 DOWNTO 0);
  CLK     : IN STD_LOGIC;
  TRIG0   : IN STD_LOGIC_VECTOR(31 DOWNTO 0);
```

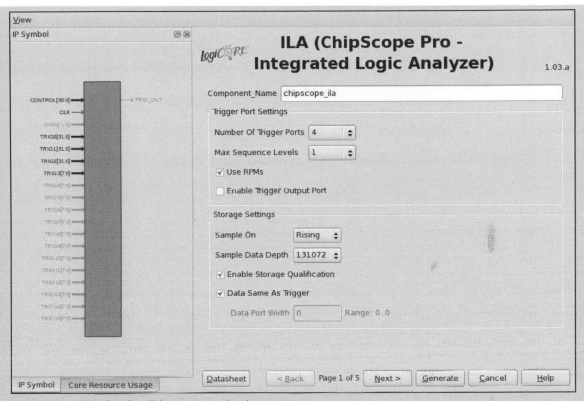

Figure 5.21. Xilinx CoreGen ILA generator wizard setup page.

```
TRIG1   : IN STD_LOGIC_VECTOR(31 DOWNTO 0);
TRIG2   : IN STD_LOGIC_VECTOR(31 DOWNTO 0);
TRIG3   : IN STD_LOGIC_VECTOR(7 DOWNTO 0));
end component;
```

Upon declaring the components and signals we can instantiate both the ICON and the ILA. During instantiation of the ILA, we must set the trigger ports to sample each specific signal. In this sample we will be sampling slave registers 0, 1, 2, and 3 (although we will only sample 8 bits of slave register 3). Then all that is left is to synthesize the design and run the ChipScope Pro Analyzer, which we will cover next.

```
icon_i: chipscope_icon
port map (
  CONTROL0 => ila_control
);

 ila_i: chipscope_ila
port map (
 CONTROL => ila_control,
 CLK     => Bus2IP_Clk,
 TRIG0   => slv_reg0,
```

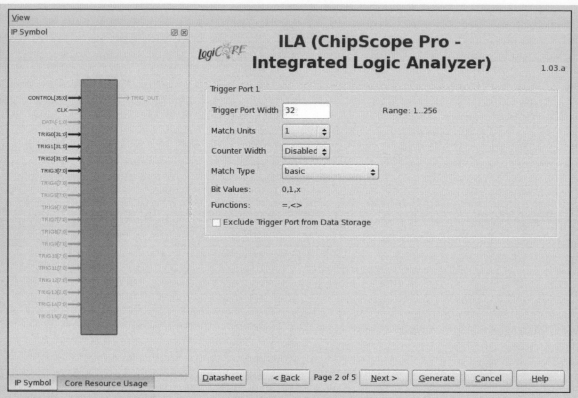

Figure 5.22. Xilinx CoreGen ILA generator wizard trigger port setup page.

```
TRIG1    => slv_reg1,
TRIG2    => slv_reg2,
TRIG3    => slv_reg3(0 to 7)
);
```

ChipScope Pro Analyzer   On the host PC connected to the FPGA via the JTAG programmer, we must launch the Xilinx ChipScope Pro Analyzer. ChipScope is an application that connects to the integrated controller (ICON) of the design running on the FPGA. Once connected, the user can view waveforms from each of the integrated logic analyzers (ILA). The user can also configure how and when each ILA will begin to record data, which is known as setting *trigger conditions*.

To get started, launch the ChipScope Pro Application. In Linux, this is done by

```
% analyzer.sh
```

Once the application finishes loading, the user must connect the FPGA by *Opening the Cable* by clicking the button highlighted in Figure 5.23. This requires the JTAG to be connected and the FPGA to be powered on.

Figure 5.24 shows the message box that is displayed when the JTAG connects to the FPGA.

Now that the JTAG is connected, we have a few options. If the design has not already been loaded (through an ACE file, through iMPACT, XMD, or the SDK), we have the option to download a new bitstream to the FPGA. To do this click the

Figure 5.23. Xilinx ChipScope Analyzer Pro main GUI with "Open Cable" button identified.

JTAG Chain Device Order

Index	Name	Device Name	IR Length	Device IDCODE	USERCODE
0	MyDevice0	System_ACE_CF	8	0a001093	
1	MyDevice1	XC5VFX130T	14	23300093	

Advanced >>

OK   Cancel   Read USERCODEs

Figure 5.24. Xilinx ChipScope Analyzer Pro JTAG device chain display window.

```
Device -> DEV1:MyDevice (XC5FX130T)... -> Configure ... menu option.
```

The device may differ if you are not using the Xilinx Virtex 5 FX130T FPGA. Select the download.bit to be used to program the FPGA. Downloading the bitstream can take a few seconds to a few minutes depending on the JTAG being used. After the download completes, any application initialized into the bitstream's BRAMs will automatically run.

Because the application automatically runs, we might not be able to get the waveform ready to capture data we want right away. We will discuss shortly how we are able to circumvent this problem by using a few lines of C before our actual application runs. The trick we employ is to prompt the user to "Press ENTER" when ChipScope has been set up to capture data when the trigger condition equation is matched. This is done by adding the getchar function to the C application before running the rest of the application.

```
int main() {
  /* Declare and Variables for application */
  char c;
  printf(''Press ENTER when ChipScope has been Triggered\n'');
  c = getchar();
  printf(''Program will now start running\n'');
  /* Rest of the Program will follow */
```

Then, when the program downloads, the user will see the output on the terminal, as seen in Figure 5.25.

```
Press ENTER when ChipScope has been Triggered

Program will now start running
```

Figure 5.25. Program output with inserted code to prompt user to trigger ChipScope.

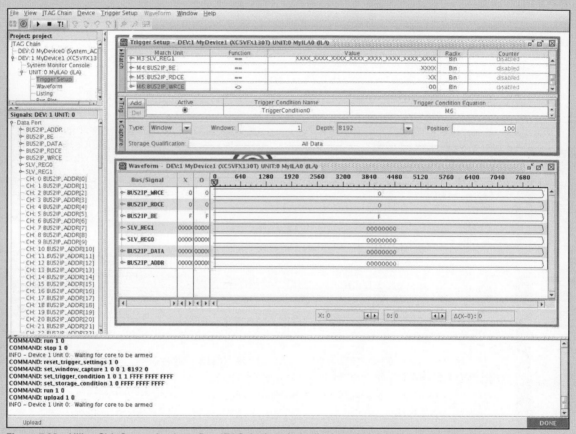

Figure 5.26. Xilinx ChipScope Analyzer Pro JTAG device chain display window.

And, after pressing enter, the application will run as normal. This gives the person testing the design ample time to set up the rest of ChipScope's parameters.

Returning back to the ChipScope Pro Analyzer application, we want to familiarize ourselves with the main layout so that we can quickly become productive with the tool. Figure 5.26 shows the console output window, waveform window, and trigger condition setup window. Also shown are the three key buttons used primarily through the debugging process. The first is the *arm trigger* button, which looks like a VCR/DVD play button. The second is the *disarm trigger* button, which looks like a VCR/DVD stop button. Finally, there is the *T!*, for trigger immediately. This means the ILA will begin to record its samples immediately, not waiting for a trigger condition to be met, which can be used when debugging a system and none of the trigger conditions are met. In this case the trigger immediately button can give the designer a look at signals to identify which state the system is in and maybe help explain why none of the trigger conditions were met.

Initially, the waveform and trigger conditions will not be set up when the project is opened. To start, you will need to import the generated ChipScope Description file (CDC), which is located with the custom hardware core (in the core's `coregen` directory if you are following our instructions). To import this file, simply click the

```
File -> Import menu option
```

followed by navigating to the CDC file and importing it to ILA unit 0.

Once imported, the waveform can be updated to include the new signal names. These will still be default signal names, such as `TRIG0` and `TRIG1`, but they can be renamed by hand to their correct names. These new signals can be dragged and dropped to the waveform, similar to how it can be done using ModelSim.

Finally, we must set up one or more trigger condition equations so that when we trigger our ILA, it actually runs. To do this, expand ILA unit 0. Select *Trigger Setup*, which will open the trigger setup window. We will now set the condition that must be met in order for our ILA to begin recording data and eventually transmitting that data back to the ChipScope Analyzer application. Depending on which signal you want to trigger on, set the *value* and the *function* accordingly. You may want to set the ILA to trigger when data are written into the hardware core. Once this is set, we can arm the trigger and run the application from the terminal.

Before we end our discussion of ChipScope, let us add a few comments regarding common points of confusion. You cannot arm the trigger and then download a new bitstream. This is for the obvious reason that when you arm the trigger, you are telling the on-chip integrated controller when the ILA should begin recording. However, if you redownload the bitstream, it will effectively erase all of that information stored in the ICON and rewrite the entire FPGA configuration. Second, in order to arm the trigger, the trigger setup window must be open; however, the window can be minimized. You may notice if you close the window that the options to arm, disarm, and immediately trigger disappear until you reopen the trigger setup window. Finally, within the trigger setup window is an option to set the *Capture Window*. This sets how much data should be captured before the trigger condition occurs. If set to 1, as it is by default, only 1 sample will be stored prior to the trigger condition being met. If the value is set to 100, then 100 samples will be stored before and 924 samples will be stored after. This is a very useful debugging tool to help explore what the values of signals are shortly before the condition is met.

ChipScope is a tool that requires hands-on practice in order to get more comfortable. There are plenty of tutorials and user guides to read and learn more. Certainly, for experienced readers this is not the only way to implement the ICON and ILAs in the design. It is also possible to include them directly through the XPS GUI. We have opted to cover the direct inclusion method because it provides the programmer with a more direct method to debug signals in the system. We encourage readers to also investigate how to insert the ICON and ILAs directly though the XPS GUI.

Exercises

P5.1. Describe the differences between bit-level parallelism and instruction-level parallelism.

P5.2. Which granularity type is best suited for a Platform FPGA implementation? Why?

P5.3. Given the following dependence relation, generate the exhaustive enumeration approach.

subtask t	$IN(t)$	$OUT(t)$
$s1$ a = (x+y);	$\{x, y\}$	$\{a\}$
$s2$ b = (x-y);	$\{x, y\}$	$\{b\}$
$s3$ r = a/b;	$\{a, b\}$	$\{r\}$

P5.4. Draw the degree of parallelism figure for the following short function:

Time 0   4 threads running concurrently
Time 1   4 threads running concurrently
Time 2   2 threads running concurrently
Time 3   1 threads running concurrently
Time 4   5 threads running concurrently
Time 5   6 threads running concurrently

P5.5. From the degree of parallelism, identify the (a) maximum DOP and (b) average DOP.

P5.6. Given that 50% of the resources are available, meaning only three cores exist, what performance can be expected with this limitation?

P5.7. Replace the following entity description with generic constants.

```
entity test_core is
  port (
    a_input   : in  std_logic_vector(31 downto 0);
    b_input   : in  std_logic_vector(15 downto 0);
    a_be      : in  std_logic_vector(3 downto 0);
    b_be      : in  std_logic_vector(1 downto 0);
    clk, rst  : in  std_logic;
    result    : out std_logic_vector(31 downto 0);
    valid     : out std_logic);
end test_core;
```

P5.8. Using a generate statement, instantiate eight copies of the test_core component.

P5.9. Using CoreGen, generate the ICON and ILA to probe all of the signals in the test_core component.

References

Cytron, R., Ferrante, J., Rosen, B. K., Wegman, M. N., & Zadeck, F. K. (1991). "Efficiently computing static single assignment form and the control dependence graph." *ACM Transactions on Programming Languages and Systems*, 451–490.

Flynn, M. (1972). "Some computer organizations and their effectiveness." *IEEE Transactions on Computers*, 948–960.

Dan Moldovan. (1993). *Parallel processing from applications to systems*. San Mateo, CA, USA: Morgan Kaufmann Publishers, Inc.

Wolfe, M. (1996). *High performance compilers for parallel computing*. Redwood City, CA, USA: Addison Wesley Publishing Company.

Xilinx, Inc. (2009a). *ChipScope Pro ICON data sheet (DS646) v1.04.a.*

Xilinx, Inc. (2009b). *ChipScope Pro integrated logic analyzer data sheet (DS299) v1.03.a.*

Xilinx, Inc. (2009c). *XST user guide (UG627) v11.3.*

6

MANAGING BANDWIDTH

Computation Is Cheap, Bandwidth Is Everything

RCS Lab Mantra

Perhaps the most fundamental issue in building an FPGA computing system is managing the flow of data through the system. The
computational resources in Platform FPGAs are enormous and,
as we have seen in the previous chapter, specific algorithms often
have plentiful parallelism, especially if there is a computation that
is specialized and relatively small. In this case it is very easy to
instantiate a large number of those function units on a single chip.
However, without considering the rates at which various function
units consume and produce data, it is very likely that any potential performance gains may be lost because the function units are
idle, waiting to receive their inputs or transmit their results. Performance issues include both rate of computation and power because
an idle function unit is still using static power. There is also a correctness issue for some real-time systems as well. It is often the
case that data are arriving from an instrument at a fixed rate —
failing to process data in time is frequently considered a fault.

In this chapter the learning objects are related to examining
bandwidth issues in a custom computing system.

- Starting out, Section 6.1 looks at the problem of balancing
  bandwidth, that is, to maximize throughput in a spatial design
  with the minimum amount of resources.
- Then, considering the target is a Platform FPGA, we spend time
  discussing the various methods used to access memory, both
  on-chip and off-chip, along with managing bandwidth when
  dealing with streaming data from instruments off-chip.
- Finally, we close this chapter with a discussion of portability and scaling with respect to a system's performance in
  Section 6.3. As semiconductor technology advances, system
  designers are provided more configurable resources to take
  advantage of. The issue now is not one of "how do I add more
  function units?" but rather "how do I scale the bandwidth of the
  interconnection network?"

6.1. Balancing Bandwidth

Let's begin with a motivating example. Suppose we have a simple computation, $f = xy(x-1)$, that we need to apply repeatedly to a large number of sequential inputs. A simple spatial design of locally communicating function units is shown in Figure 6.1.

In order to maximize the frequency of the design, it is likely that the multipliers chosen for this design will be pipelined, and for this example it is assumed that these multipliers will have four levels of pipelining. We will also assume that the $x-1$ unit completes in a single-cycle operation.

Now let's analyze the timing in this example. At $t = 0$, the inputs x and y are present at the inputs of modules A and B. At $t = 1$, $x-1$ has been computed and the result is present at the input of module C. However, the product $x * y$ is still being computed, so we have to stall the computation. A ***stall*** occurs in two cases: (1) whenever a computation unit has some, but not all, of its inputs or (2) when it does not have the ability to store its output, that is, the unit is prevented from proceeding because it is waiting on one or more inputs or waiting to store a result. Thus, module B is stalled as well because it has to hold its result until module C uses it. These units continue to stall until $t = 4$, at which point module A produces its product, which then allows module C to consume both data and — for exactly one cycle — all three units can proceed in parallel. (Module C is operating on the first two x and y values while modules A and B begin operating on the next two x and y values.)

Clearly, this pattern is going to repeat and modules B and C will stall again until $t = 8$. This is shown graphically in Figure 6.2. So if we have n number of x and y inputs, the network will take $4n$ cycles to compute all of the results. This is undesirable

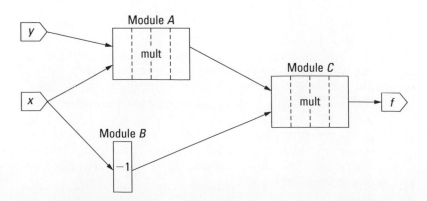

Figure 6.1. A simple network used to compute $xy(x-1)$.

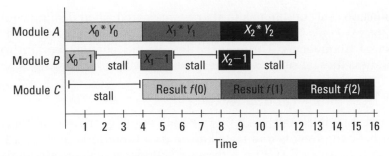

Figure 6.2. Illustration of the stalls example.

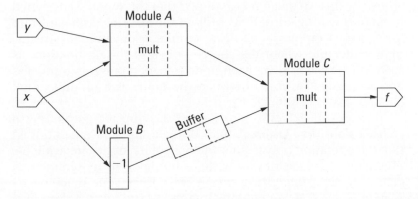

Figure 6.3. High-throughput network to compute $xy(x-1)$.

because one-quarter of our module cycles are spent waiting for data![1]

Even though FPGA resources have been allocated to these units, they are not contributing to the solution when they are idle. Moreover, they are continuously consuming static power — even if the transistors are not changing values. Finally, the throughput (one result every four cycles) is not much better than a time-multiplexed solution and probably slower than a general-purpose processor (which will have a faster clock frequency). What we aim for is a throughput of one result every cycle.

Readers may already recognize the solution to this problem. By adding a four-stage FIFO or buffer into the network, we can make this computation produce a new result every cycle after $t > 4$. There is more than one place to insert these buffers, but Figure 6.3 shows three buffers following the results of module B. The introduction of buffers so that a network of computations does not stall is known as ***pipeline balancing***. Chapter 5 discussed pipelining, both in terms of between compute cores and within individual computations. Implicit is the problem of how to best balance the

---

[1] This would be half of the cycles if the second multiplier was in fact a fast one-cycle unit.

pipeline. This problem is further complicated when dealing with larger systems that span multiple compute cores operating at different frequencies. Through the rest of this chapter we aim to address these problems with respect to Platform FPGA designs.

6.1.1. Kahn Process Network

In the previous stalls example, we built our network of computational components in a style known as a Kahn Process Network (KPN) (Kahn, 1974). In a KPN, one or more processes are communicating through FIFOs, with blocking reads and nonblocking writes. By this we mean it is assumed that the system is designed to support writing into a FIFO without the FIFO becoming full and losing data. Graphically, a KPN resembles a conventional streaming architecture where the source writes to the destination, as seen in Figure 6.4. Here, each node (circle) represents a process and each edge (*A*, *B*, *F*) between the nodes is a unidirectional communication channel.

The network, which is represented as a directed graph, is also called a ***data-flow graph***. As part of the example, we mentioned that the operation only begins when all of its inputs are available *and* there is a place for it to write its output. This requirement is called the ***data-flow firing rule***. A short finite state machine, as seen in Figure 6.5, can be implemented to support the data-flow firing rule.

We already implemented a Kahn process network in Section 3.A when we built the single precision floating-point adder. In that example, before processing data (performing the addition), we had to wait until not only both operands arrived into their FIFOs, but that result FIFO was ready to receive the added result and the adder was ready for new inputs as well. Had we not checked the

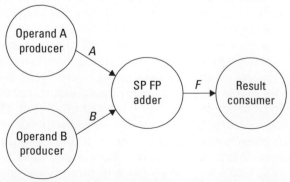

Figure 6.4. Kahn process network diagram of a simple single precision floating-point system.

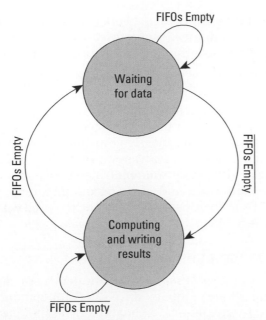

Figure 6.5. Wait/compute states of Kahn process network finite state machine.

status of result FIFO it would have been feasible for the system to have lost data between the inputs and the results by allowing the adder to consume data from the input FIFOs, but not being able to store the results.

An additional feature found in our single precision addition system is the ability to check the status of each FIFO, whether they were full or empty, in parallel to the FIFO read and write operations. This is in contrast to a conventional software implementation where the status of the FIFO would first need to be read prior to any operations occurring. As a result, in a software implementation, these sequential operations can add extra latency to the system. Through the explicit parallelism with FPGAs we can begin to see and understand the effects that something as simple as independent FIFO operations have on the efficiency of the system.

While the Kahn process network supports the unidirectional flow of data from a source to a final destination, it does not offer any mechanism to provide feedback between compute processes. The requirement of "unbound" FIFOs between processes makes for challenging designs in systems where FIFOs are a fixed resource that cannot be reallocated, that is, with FPGAs, when the Block RAM has been allocated as a FIFO and the design is synthesized and mapped to the device, that memory is fixed to that process. Instead, we will see that Platform FPGA designs can quickly incorporate the Kahn process network with fixed on-chip memory resources by including feedback between the compute cores. This approach is discussed within both Section 6.A and Chapter 7.

6.1.2. Synchronous Design

In a systolic data-flow implementation, every operation shares the same clock and data are injected into the network every clock cycle. If we can determine at design time that all data will arrive at the inputs at the right time (as was done in the second implementation of the network in the example), then we can avoid checking the data-flow firing rules and we get maximum throughput. Moreover, the FIFOs used in the example can be replaced with a smaller chain of buffers. This is called a *synchronous design*.

6.1.3. Asynchronous Design

In contrast, the *asynchronous design* strategy uses simple finite state machines to read inputs, drive the computation, and write the outputs. This requires additional resources, but has the benefit that there is no global clock, so the design can be partitioned across different clock domains. If the buffers are added as in the second implementation example, then it too has the benefit of getting maximum throughput.

In both synchronous and asynchronous cases, we have assumed that each operation takes a fixed number of clock cycles. However, the model can be extended to also include variable-length latency operations by using the maximum latency (to always get maximum throughput). The asynchronous design can use the most frequent latency to usually get maximum throughput.

The key to both of these design strategies is that FIFOs have the correct minimum depth. However, we do not want to waste resources — especially memory resources that are relatively scarce on an FPGA device. So the key to this problem is to find the minimum number of buffers needed to maximize throughput. The next section investigates how to best utilize the bandwidth of the FPGA in terms of on-chip communication, off-chip memory, and streaming instrument inputs.

6.2. Platform FPGA Bandwidth Techniques

As with the partitioning problem in Chapter 4, the analytical solution is most useful as a general guide to solving the problem rather than an automatic tool. Unfortunately, just as before, practical issues complicate the clean mathematical analysis. This section considers two places where these techniques can be applied and then considers some practical issues. First, we will look at integrating designs with on-chip and off-chip memory. A variety of memory types and interfaces exist and understanding when each is applicable is important to Platform FPGA designs.

Then, we will consider data streaming into the FPGA from some instrument. For practical embedded systems designs the "instrument" may be a sensor (such as temperature or accelerometer), some digitally converted signal (say from an antenna or radar), or even low-speed devices such as keyboards and mice. In some cases the instrument may need to use off-chip memory as a buffer or as intermediate storage, so while these two sections are presented individually, designs may require that the two functionalities be combined together. Clearly, we cannot discuss every type of instrument; for brevity we focus our attention on high-speed devices, which cause tighter constraints in FPGA designs.

6.2.1. On-Chip and Off-Chip Memory

Most embedded systems will require some amount of on-chip and/or off-chip memory (RAM) because many modern systems are data-intensive. This often means that the designer has to pay some attention to the memory subsystem. As mentioned earlier, the FPGA fabric does not include embedded caches found with modern processors.[2] As a result, the designer must be aware of the system's memory requirements in order to implement a suitable interface to memory so as to not create a memory bottleneck.

Memory requirements differ between software and hardware designs. In software there is less of an emphasis on where data are stored so long as they are quickly available to the processor when they are needed. When writing an application, the programmer does not typically specify in which type of memory (disk, Flash, RAM, registers) data should reside. Instead, more conventional memory hierarchy, depicted in Figure 6.6, is used. This is the typical hierarchy covered in computer organization textbooks. The x axis refers to capacity and the y axis refers to access time. Non-volatile storage such as hard drives provide greater capacity, but at the cost of performance. Access times are typically in the millisecond range. Volatile storage such as off-chip memory provides less storage than disk, but typically at least an order of magnitude faster access times. Cache and registers occupy the peak, providing small but fast access. We mention this because unlike programmers targeting general-purpose processors, who are able to exploit caches and the system's memory hierarchy, Platform

---

[2] Some hard processors may include cache, such as the Xilinx PowerPC 440, although this cache is not accessible by the rest of the FPGA.

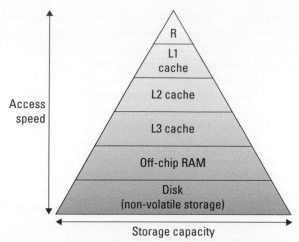

Figure 6.6. Traditional memory hierarchy comparing storage capacity to access times.

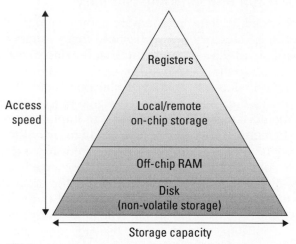

Figure 6.7. FPGA compute core's memory hierarchy with local/remote storage in place of cache.

FPGA designers must either construct custom caches or rely upon a modified memory hierarchy.

From a custom compute core's perspective, we modify the drawing of the memory hierarchy for Platform FPGAs by removing the cache and replacing it with user-controlled local and remote on-chip memory (storage), as shown in Figure 6.7. On-chip memory is considered local storage when the memory resides within a custom compute core, maybe as Block RAM. Remote storage is still on-chip memory, except that it does not reside within the compute core. An example of this is on-chip memory that is connected to a bus, accessible to any core through a bus transaction. There is a fine line between local and remote storage. In fact, the locality is relative to the accessing core. If compute core A needed to read data from compute core B's local memory, we would say that A's remote memory is B's local memory.

Figure 6.8 shows the various memory locations from a compute core's perspective. Here, the register's values are valid to the compute core immediately for read or write access. Local storage access is short, within a few clock cycles, whereas remote storage is longer, within a few tens of clock cycles. In the example the compute core would need to traverse the bus in order to retrieve remote storage data. Finally, we show the off-chip memory controller to provide access to off-chip memory. Access times to off-chip memory range in the tens to hundreds of clock cycles since the request travels off-chip.

While physically there is a difference between cache and on-chip local/remote storage, the concept of moving frequently used data closer to the computation unit remains the same. The difference is the controlling mechanism. For caches, sophisticated controllers support various data replacement policies, such as

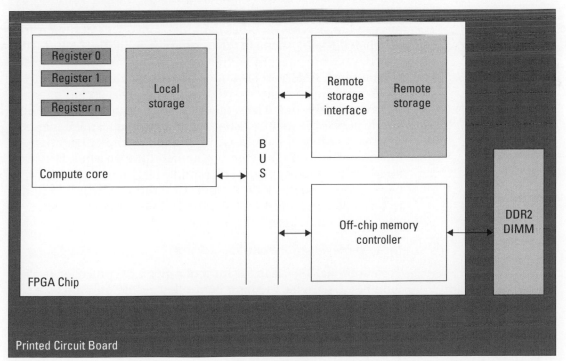

Figure 6.8. Various memory locations with respect to an FPGA's compute core with access times increasing from registers, to local storage, to remote storage, and finally to off-chip memory.

direct-mapped, set associative, and least recently used. In Platform FPGA designs we are left to implement our own controller. This may seem like a lot of additional work, but keep in mind many custom compute cores most likely do not follow conventional memory access patterns or protocols, so creating custom controllers may actually be necessary to achieve higher computation rates.

Unlike most software programmers, hardware designers must be aware of the physical location of data to be processed. We must know whether data are in registers ready to be computed upon, in on-chip memory requiring reads through a standard interface such as a Block RAM, or in off-chip memory requiring additional logic (and time) to retrieve data. In each of these cases, the mechanism to access data may differ. Consider reading data from a register; data are always able to be accessed (and modified), typically within a single clock cycle. The designer references the register by name rather than by address. In contrast, consider how to access data from an on-chip memory, such as Block RAM. When using on-chip memories, we must include a set of handshaking signals. These signals may consist of *address, data in, data out, read enable, and write enable*. In this case, the memory is located within the compute core. This allows for fast access from the compute core, but can limit access by other compute cores. This is similar to a nonshared level 1 cache in terms of accessibility. Read and write access times are typically within one to two clock cycles.

The Block RAM can also be connected to the bus to provide more compute cores access to the memory. This comes at a cost of traversing the bus. The memory is still on-chip, but the interface to access the memory changes to a bus request. We mention this case because it more closely models one implementation for interfacing with off-chip memory, namely through a bus. However, for off-chip memory, instead of a simple bus interface to translate bus requests into Block RAM requests, additional logic is required to correctly signal the physical off-chip memory component.

Off-Chip Memory Controllers

The additional logic is in the form of a memory controller. In Platform FPGA designs, memory controllers reside on-chip as a soft core that interfaces between off-chip memory and the compute cores. The memory controller is responsible for turning memory requests into signals sent off-chip, freeing the designer from adding this signaling complexity into every compute core needing access to off-chip memory and, if multiple cores need off-chip access, having to multiplex between the requests ourselves.

Because the memory controller is a soft core, it is possible to construct different interfaces to both off-chip memory and on-chip compute cores. For example, in Figure 6.9, the memory controller is connected directly to the processor. Alternatively, the memory controller could be connected to a shared bus to allow other cores to access off-chip memory, as seen in Figure 6.10. Similarly, some memory controllers can provide direct access to more than one component; Figure 6.11 shows both the processor and the compute core interfacing to the memory controller directly.

A conventional system may use a bus-based memory controller to provide the greatest flexibility in the design, allowing any master on the bus to read or write to off-chip memory. In Platform FPGA designs this type of access may suffice for many components, but providing a custom compute core with the ability to

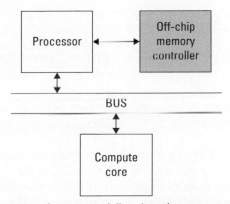

Figure 6.9. The processor is connected directly to the memory controller.

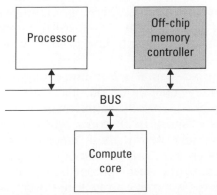

Figure 6.10. The memory controller is connected to a shared bus to support requests by any core on the bus.

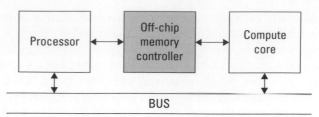

Figure 6.11. The processor and compute core are connected directly to the memory controller.

access memory directly can yield significant performance advantages that are otherwise unrealizable in traditional microprocessor designs. We discuss how memory accesses differ with these physical implementations shortly.

Memory Types

Up until now we have ignored the type of off-chip memory we have chosen to interface with. When purchasing development boards, vendors often populate the boards with one or more types of memory. These may include DDR SDRAM, DDR2 SDRAM, SRAM, Flash memory, EEPROM, etc. For those less familiar with these memories, you may be thinking of how they differ. For example, SDRAM versus SRAM differ in terms of latency, capacity, and maybe most noticeably cost (SRAM costs are significantly greater than SDRAM). SDRAM versus Flash memory differs in terms of long-term volatile versus nonvolatile storage and access time. With such a wide variety of memory, we must familiarize ourselves with the different memory controllers. This task may seem daunting, and for designers who are required to build a memory controller from scratch, it can be. However, in many cases, memory controllers have already been designed that can be instantiated from a repository/library of soft cores. This is one of the strongest benefits when designing with Platform FPGAs, using commodity off-the-shelf components (in our case hardware cores) whenever possible in a design.

When starting a design, it is common to begin with a development board that may contain various necessary and unnecessary peripherals. Vendors typically supplement these development boards with a repository of IP cores, which allows systems to be assembled rapidly. Within these repositories exist different memory controllers that can be instantiated within the design to provide access to the development board's off-chip memory. Ultimately, a design will move away from the development board to a custom design, which may mean a different type (or capacity) of memory.

Fortunately, with some adjustments and modifications to the generics of the memory controller hardware core, different memory can be used [for example, using error correcting code (ECC) logic if the SDRAM DIMM supports it or not]. The goal of this section is not to teach you how to build a memory controller from scratch, but to understand how the different memory controller interfaces can affect a design. Specifically, we are interested in the memory access associated with these different interfaces. This is a shift from traditional processors where all of the computation is done by the processor, so one type of memory access may be all that is necessary. In Platform FPGA designs we can incorporate many different memory access types to support lower resource utilization, lower latency, or higher bandwidth memory requirements when needed. Moreover, we can provide some compute cores with high bandwidth access to memory while limiting memory access to other cores.

Memory Access

How the memory controllers are connected and used can have a dramatic effect on the system, even if a highly efficient memory controller and type of memory is used. It is up to the designer to understand how to connect the memory controller in order to meet the memory bandwidth needs. In Platform FPGA terms, the important trade-off is resource utilization.

Programmable I/O

For starters, the processor can handle transfers between memory and compute cores. With **_programmable I/O_** the processor performs requests on behalf of the compute core. This approach uses a limited amount of resources. The requirement is that each compute core be located on the same bus as the processor. The processor can read data from memory and write it to the compute core, or the processor can read data from the compute core and write it to memory. In this situation, the processor plays the central communication role. As a result, the processor may end up performing less computation while performing these memory transactions on behalf of the compute cores.

The effect is less noticeable as the amount of computation performed in the FPGA fabric increases, reducing the amount of computation being performed by the processor. In practice, there are a variety of ways to pass data among the processor, memory, and compute core. From the processor's perspective, both memory and the compute core are slaves on the system bus with a fixed address range.

```
#include <stdio.h>

#define MEMORY_BASEADDR 0x10000000
#define HWCORE_BASEADDR 0x40000000

int main() {
  int *mem_ptr = (int *)(MEMORY_BASEADDR);
  int *core_ptr = (int *)(HWCORE_BASEADDR);
  int i;

  // Transfer Data from Memory to the HW Core
  for(i=0; i<128; i++) {
    *core_ptr = mem_ptr[i];
  }
}
```

Listing 6.1. Processor performing programmable I/O, reading data from off-chip memory and writing to a custom compute core.

A software approach, such as C/C++, could involve pointers where the processor would access data stored in off-chip memory and then pass it to the compute core. We can also use pointers to provide array-indexed access to off-chip memory. Figure 6.9 depicts the system design used to connect the processor, memory, and compute core to allow data to be transferred from off-chip memory to a custom compute core.

Listing 6.1 provides a simple example of a stand-alone C application to read data from off-chip memory and transfer it to a compute core. We assume that the address space of the off-chip memory controller begins at address 0x10000000 and that the compute core begins at address 0x40000000. The functionality of the compute core is irrelevant for this example; just assume that writing data to its base address is how the processor interfaces with the compute core. Accessing memory as an array, mem_ptr[i], reduces the complexity associated with pointers. In this case, because both pointers were defined as type int (and we assume the processor is a 32-bit processor), incrementing the pointer results in reading from and writing to the next 32-bit data word.

DMA Controller

As a simple solution to the memory transfer problem, the processor doing the work may be acceptable; however, it is certainly not efficient. The time the processor spends reading and writing data is time that could be spent performing other computations. Most modern processors circumvent this problem by introducing ***Direct Memory Access*** (DMA) to perform memory transactions in place of the processor. There are variations on the implementation, but the basic idea is that the processor issues a request to the DMA controller, which then performs the memory transaction. This frees the processor to perform other computations while the DMA controller is operating. This works by allowing the DMA controller to act as both a bus slave and a bus master. As a slave, the DMA controller responds to requests from the processor (or any other bus master) to set up a memory transaction. Then, as a master, the DMA controller arbitrates for the bus and communicates with the memory controller to complete the memory transaction. The order of communication depends on the direction of the transfer (memory read versus memory write), so it is possible to issue a read from a compute core and write data to off-chip memory. To set up the transaction, the DMA controller needs at least the following information: source address, destination address, and transfer length. The source address is where data should be read from, the destination address is where data are to be written to, and the transfer length is the number of bytes in the transfer. Figure 6.12 is a block diagram representation of a simple system, including a processor, DMA controller, memory controller, and compute core.

Bus-Based DMA Interface

While the DMA controller may add an improvement in performance by reducing the overall involvement by the processor, there are still drawbacks. In the programmable I/O implementation the processor's compute capacity is degraded when it performs the memory transactions on behalf of the compute core. Recalling the example (Listing 6.1) where during the `for-loop` the processor is acting as a middleman between memory and the compute core, we can free the processor from much of this responsibility by adding a DMA controller to perform the bulk of the work for each memory transaction. This approach is an improvement, but the processor is still involved and data must still be passed between the DMA controller and the compute core.

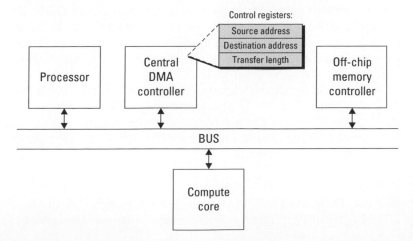

Figure 6.12. Central DMA controller to offload memory transactions from the processor.

We can improve this further by giving each compute core the ability to issue transactions to memory independently. This is still considered DMA, but instead of a centralized DMA controller performing transactions, each compute core can independently issue its own read or write transactions to the memory controller. With a bus connecting the compute cores and memory controller, the arbiter for the bus becomes the communication controller, managing requests for bus access to issue memory transactions. The memory controller still only sees a single transaction at a time (note that some memory controllers can handle more than a single transaction at a time).

To support DMA within a custom compute core, additional logic must be added to the core's interface. Typically, a core can be viewed as a bus slave, responding to requests from other cores on the bus. The example of programmable I/O where the processor writes data to the compute core is an example of this interface. A slave device cannot issue requests on the bus; in order to do so we must make the device a bus master. It is possible for a core to be a slave, a master, or both a slave and a master. Although, practically speaking, a core is usually only a bus slave or both a bus slave and a master. The processor is one of the few bus master-only cores.

That being said, for DMA support we only need a bus master, but we will include a bus slave to allow the processor and other compute cores to still communicate with the core. Because the bus interface depends on the actual bus used in the design, the specifics of the finite state machine needed to incorporate a bus master into a custom compute core will be covered in Section 6.A. From a hardware core's perspective, a master transaction involves asserting the request to the bus arbiter, waiting for the arbiter to grant bus access, and waiting for the request to complete.

We mention bus master information because the memory controller is viewed as a slave on the bus. Once the memory controller responds to the bus transaction, the next core can issue its memory request across the bus. This approach alleviates the processor from performing memory transactions on behalf of compute cores and can improve performance by allowing a transaction to only traverse the bus once (previously, the processor would fetch data and then pass data to the compute core, requiring two trips across the bus). Requests to the memory controller are still performed sequentially by the bus, but any core that can be granted access to the bus can issue requests.

Direct Connect DMA Interface

In some situations it may not be necessary for every core to access off-chip memory. While we could create bus masters and bus slaves accordingly, a larger question should be addressed, namely, why not directly connect compute cores to the memory controller? This approach avoids contention for the bus, especially from other cores that need the bus but do not need access to the memory controller. Overall, this results in lower latency and higher bandwidth transactions to memory. A direct connect to the memory controller does require additional resources, both for the compute core and the memory controller. Some vendors supply memory controllers with multiple access ports (Xilinx offers a multiport memory controller, MPMC). Section 6.A presents one custom direct connect interface for the Xilinx MPMC known as the Native Port Interface.

By circumventing the bus, memory transactions no longer need to arbitrate for the bus, which reduces latency. In designs that require low latency, this may be the only way to meet the timing requirements. In some designs, only a few compute cores may need direct access to memory, and under these conditions is it feasible to directly connect the cores to the memory controller. As the number of cores needing access to memory increases, using a bus-based DMA interface is more appropriate, as each direct connect DMA interface requires resources that would normally be shared by a bus implementation.

One mechanism used to hide latency to off-chip memory is double buffering. **Double buffering** refers to requests that are made while the previous request is still in transit. This allows the memory controller to begin responding to the second request while the first request finishes. During read requests from memory, using double buffering can result in a best-case effective zero latency access for all requests after the first request.

Memory Bandwidth

When implementing designs we must be aware of the memory requirements before building the system. One important consideration mentioned earlier in this chapter is bandwidth. Bandwidth can be calculated based on the operating frequency and data width. For example, a bus operating at 100 MHz over 64-bit data words offers:

$$bandwidth = (100 \times 10^6/\text{sec}) \times 64 \text{ bits} = 6400 \text{ Mbits/sec}$$

Calculating the bandwidth requirements of a compute core can help identify what type of memory controller and interconnect is suitable for the design.

In addition to raw bandwidth, we must consider setup times and latency. When a compute core initiates a transfer, depending on the interconnect, the request may need to wait to be granted

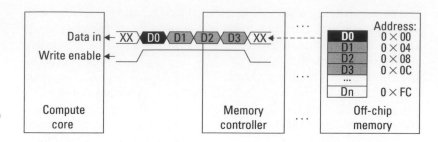

Figure 6.13. Burst transfer of four 32-bit sequential data words from off-chip memory to a compute core.

access to master the bus, wait for the memory controller to complete its current transaction, or wait for data to be returned. To improve this latency would require a more directly connected interface to memory and/or operating the memory controller at a higher frequency, allowing it to perform more operations in the same time period.

Alternatively, a core may issue burst transfers. A ***burst transfer*** allows the compute core to issue a single request to the memory controller to read or write multiple sequential data, as depicted in Figure 6.13. This reduces the number of total memory transactions needed by allowing a single transaction to span more than a single data word and can save a significant amount of time when transferring large amounts of data. In a burst transfer the latency for the first datum is still the same, but each datum after the first arrives in a pipelined fashion, one datum per clock cycle. For example, a burst transfer of four 32-bit sequential data words would take the initial latency time for the first word, denoted as $t_{latency}$, plus three clock cycles for the remaining three words:

$$t_{total} = t_{latency} + 3t$$

There are, unfortunately, two limitations to burst transfers. The first is data must be in a contiguous memory region. A burst transfer consists of a base address and a burst length. Data are transferred starting from the base address up to the burst length. The second is that the length may be limited by an upper bound, requiring multiple burst transactions for such large request lengths. As of this writing, Xilinx imposes a 16-word upper bound on memory requests across a bus and through a direct connect to off-chip memory.

To summarize, what we are trying to emphasize is that unlike traditional microprocessors, Platform FPGAs provide a flexibility to solve complex bandwidth needs. The flexibility sometimes comes at a price, often in terms of resources or programmability. Understanding when and how to chose the right combination is a responsibility the system designer must not take lightly.

6.2.2. Streaming Instrument Data

So far we have focused on interfacing with on/off-chip memory, but many embedded systems designers often face situations where the system must incorporate an application-specific instrument. An exotic example is a system with a number of science instruments flying on a satellite circling the Earth. These instruments are usually "dumb" in that they incorporate very little control and often have very little storage. Once turned on, they begin generating data at a fixed rate. It is up to the computing system to process this data and transmit it to the receiving station within a specific time frame (in this case, when the satellite is over the receiving station on earth). Unlike the off-chip memory problem where the consequence of a poor memory hierarchy is bad performance, the consequence of not being ready to accept data from an instrument is lost data. Data can be lost in two ways: if the device cannot process data arriving from the instruments fast enough or if the device cannot transmit the computed results back, to earth in this example, within the requisite amount of time.

There are plenty of less exotic examples as well. Embedded video cameras and other embedded systems sensors have similar characteristics. Data arrive at some fixed rate and must be processed, otherwise they may be lost. For commercial components, failure to do so results in loss of sales and unsatisfied customers.

These examples essentially break down to bandwidth issues: can the FPGA perform the necessary computation in the specified time period and, if not, how can we buffer data so as to not lose it? This is the focus of this section, how to support high bandwidth instruments. Chapter 7 explores a variety of interconnects that can be used to interface with external devices. In this chapter it is all about bandwidth, how to plan for it, and how to use it. We also are interested in using on-chip and off-chip memory intelligently for short-term storage when the need arises to buffer data.

Many instruments may have a fixed sample rate, say, every 10 milliseconds, and the device must operate on new data. The designer can calculate the exact amount of time allotted to processing data and work to design a compute core accordingly. As the sample rate increases, the amount of time to process data decreases unless the designer prepares the system to be pipelined as described in the previous chapter.

Traditionally, we have considered pipelines as a mechanism to achieve higher overall throughput in general-purpose computers. Here we use them not only to achieve high throughput, but to alleviate the tight timing constraints that may be placed on a designer with a high sampling rate. The requirement on pipelining is that the sampling rate (arrival rate of new data) be less than or equal to the slowest stage in the pipeline. Even still, this requirement can be circumvented if each stage is pipelined as well; however, at this point it becomes a granularity debate as to what is a *stage* in the pipeline. With the use of pipelines we can view the flow of data from the instrument through each compute core and finally to its destination as a *stream*. FPGAs can efficiently handle multiple streams in parallel at very high frequencies, which is what has attracted so many designers away from microprocessors.

When working with more than a single instrument or input, we must now pay attention to the additional complexities. At the beginning of this chapter we spoke about balancing bandwidth with the use of buffers. By knowing the input rate for the instruments, we can calculate the size and depth of the buffers needed to balance the computation. In addition to using on-chip memories as buffers, these memories can be used to cross clock domains.

One challenging aspect of embedded systems design is dealing with different clock domains. This problem may be further complicated when including instruments with different operating frequencies. Using on-chip memory as FIFOs, certain vendors support dual port access to the memory, which enables different read and write clocks (and, in some cases, data widths). Using these buffers with different read and write clocks, a designer can solve two problems with one component, buffering and clock domain crossing. With Platform FPGA designs, this also results in the use of fewer resources than implementing both a buffer and clock logic individually. Here we are able to take advantage of the physical construct of the FPGA and the on-chip memory (i.e., Block RAM), which have both capabilities built into the FPGA fabric.

Where real issues begin to arise is when a compute unit cannot process data fast enough and on-chip memory does not provide sufficient storage space. In this event, it is necessary to use off-chip memory as a larger intermediate buffer. Fortunately, we can incorporate the information presented in the previous section to help solve this problem. By calculating the bandwidth requirement we can determine which memory controller and interconnect are appropriate.

Let's consider three cases where off-chip memory buffer can be useful. These three cases are actually general enough to be applicable for large on-chip memory storage as well, but we present them here for consistency sake. A compute core requiring memory as a buffer may need to:

- store input data arriving faster than can be computed
- retrieve data when it can be computed
- store the computed results

Another unique capability FPGAs provide over commodity processors is the ability to gather data without introducing any instrumentation effect in the computation. Let's say we want to construct a histogram of data arriving from an off-chip source, as seen in Figure 6.14. We can build a second component to perform a histogram calculation based on the input to be processed by the compute core. Unlike a microprocessor design where the histogram computation would require part of the processor's time (slowing down the computation), a parallel histogram hardware core can perform its computation without disturbing the system. In this example, we may also be able to achieve a higher precision in our histogram, as we could sample the input at a higher frequency than a processor, which may not be able to sample and compute quickly enough before the next input arrives.

Inserting data probes into a system can provide a significant amount of insight into the operation and functionality of the design. Often, this information is collected during the research and development stages of the design; however, in some applications these data may prove to be a valuable resource for debugging or improving future designs. These probes can be created to monitor a variety of components, with each running independently. Depending on the functionality, these probes may require additional resources, such as on-chip or access to off-chip memory, and so must be planned for accordingly.

FPGAs have already been shown to be useful in a variety of applications with one or more instruments, but often the FPGA's role is as some glue logic to aggregate all of the instrument signals for the processor. Instead, we emphasize that using the FPGA to do

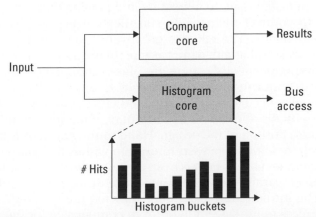

Figure 6.14. Streaming I/O compute core plus parallel data acquisition (histogram) core.

more computation can be of great benefit. To support computation, it may be necessary to introduce buffers, which may be small on-chip memories or large off-chip RAMs. With the application dictating the buffering needs, the designer must carefully consider the system's bandwidth in order to obtain significant gains in performances. These gains are not necessarily over a conventional processor; in fact, the gains may be finer resolution in the computation, more power-efficient designs, or the collection of run-time data that will spawn new and even better designs.

6.2.3. Practical Issues

Perhaps the biggest issue for FPGA-based implementations is the cost to instantiate a FIFO or a buffer. In CMOS transistors, the number of transistors used to instantiate a buffer and FIFO is different. For FPGAs, they are often just different configurations of the same resource. Second, the number of transistors in CMOS is directly proportional to the number of stages in the buffer. The cost model for FPGAs is much more complicated. If the buffer has one stage, the tools will probably instantiate a flip-flop. If there are between two and 16, most likely the tools will use shift registers (SLR16), which essentially cost a function generator and no data storage. For a deeper buffer, the tools might instantiate a few SLR16s but, more likely, would configure an on-chip memory as either distributed RAM or Block RAM. A consequence of this is that the *placement* of the buffers matters a lot in an FPGA system. A design that uses many of the BRAM resources might want to distribute many small buffers throughout the network rather than use one BRAM near the sink. Of course, the analytical formulation just presented does not account for these subtle cost issues.

An associated issue to allocating and placing these memories is how to connect them while still meeting the specified timing requirements. The reliance on advancements in place-and-route technology has aided the designer in meeting more complex design requirements, but there are often timing constraints that the tools cannot resolve. The designer must identify where the longest delay occurs in the design and either introduce a buffer (maybe with the addition of a flip-flop) or, if possible, hand place and route a portion of the design. Hand routing designs is a more advanced topic than we wish to explore within this chapter and because our focus thus far has been on bandwidth, we should consider the constraint from the perspective of solving a bandwidth issue. The introduction of a buffer may require a change to the other buffers in the system in order to compensate for the added delay.

6.3. Scalable Designs

Consider a designer who spends a significant amount of time designing a hardware core to be used in a Platform FPGA-based embedded systems design. The designer has carefully partitioned the design, identified, and implemented spatial parallelism based on a specific FPGA device, such as the Xilinx FX130T FPGA. In many embedded systems, once the design has satisfied its requirements and the product is shipped, the designer may switch to maintenance mode to support any unforeseen bugs in the system. This follows the product development life cycle from Chapter 1.

We now want to shift our focus away from simple maintenance and instead turn our attention to scalability. With each new generation of FPGA devices the amount of available resources increases. When going from one generation to the next, the number of logic blocks, logic elements, and block memories might double. New resources, such as single/double precision floating-point units may appear. And, the number of available general-purpose I/O pins may increase. While these advancements provide the designer with more capacity, capability, and flexibility, it often raises an important question: "How do we modify our design to take advantage of these additional resources?" Simply increasing the number of compute cores may have adverse effects on the system unless careful consideration is made.

6.3.1. Scalability Constraints

At the same time we want to consider how to create a design to be as scalable as possible and to utilize the current FPGA's resources most efficiently. Consider a designer who builds a compute core that utilizes 20% of the FPGA resources. If the design allows for multiple instances of that compute core to be included, we could conceivably instantiate five cores and use 100% of the resources. These calculations are rather straightforward, requiring little math in the computation.

Number of Cores = Available Resources/(Resources/Core)

Practically speaking, this simple formula helps set the upper bound on the actual number of compute cores obtainable on the device. This does not tell the whole story, as we must further analyze the compute core to determine if the system will be saturated past a certain number of cores. In essence we want to find the sweet spot between maximum resource utilization and performance. Arguably, one of the most important scalability constraints is bandwidth. Can we sustain the necessary bandwidth — whether to/from a central processor, between each compute core, or to/from memory — as we increase the number of cores? Simply put, these three components, the processor bus, and memory are at the heart of Platform FPGA design. To see any significant performance gains when scaling the design, we must be considerate of their bandwidth's impact on the system.

Processor's Perspective

Let's start by looking at scalability from the processor's perspective. As we add more compute cores we may be also adding more work for the processor, such as managing each additional core. Of course this analysis is application specific, but the overarching goal of scalable designs is to achieve greater performance, which may be in terms of computation, power consumption, or any number of metrics. To do this with Platform FPGAs we want to offload the application's compute-intensive sections that would normally run sequentially on the processor and instead run them in parallel in hardware. Any resulting responsibilities the processor is left with to control these additional hardware cores is a necessary trade-off for the potential parallelism and performance gains.

We consider the "bandwidth" from the processor's perspective as the amount of "attention" the processor is able to provide a compute core. This vague description is intended to focus the discussion less on data being transferred between the processor and compute core and more specifically on the time the processor is able to interact with the core. We could also consider this to be time spent by the processor to perform some sort of "control" of the hardware core rather than time spent performing its own computations.

To draw on an analogy, imagine a parent feeding a child. The parent feeds the child one spoon at a time until the food is gone and the job is done. The spoon represents the fixed bandwidth between the processor (the parent) and the compute core (the child). The total attention needed is how long it takes to finish the task. Now if we add a second child, the parent is still able to feed each child one spoonful at a time, but the total time to feed both children increases or each child receives only half of the parent's attention. In scalable designs we want to identify the amount of attention a compute core needs and scale the design to not exceed the processor's capacity.

$$Processor\ to\ Core\ Bandwidth = \frac{Available\ Bandwidth\ to\ one\ Core}{Number\ of\ Cores}$$

Compute Core's Perspective

Next, let's consider scalability from the compute core's perspective. The compute core may need to communicate with other cores that may be similar to the processor case we just covered. However, unlike the processor that can only communicate with one core at a time, each compute core can issue requests to any other compute core, presenting a larger constraint on the interconnecting resource, which we referred to earlier as a bus. As we add more compute cores to this bus, we run the risk of saturating that shared resource. The bus has a fixed upper bound on its bandwidth, which is the operating frequency multiplied by the data width:

$$Bus\ Bandwidth = Bus\ Frequency \times Bus\ Data\ Width$$

To calculate the effect of adding additional compute cores onto the bus, we need to take into account what each core's bus bandwidth needs are. To do this we can follow the same formula for the bus bandwidths, but include the core's bus utilization as well.

$$Core\ Bus\ Bandwidth = Core\ Frequency \times Bus\ Data\ Width \times (\%\ of\ Utilization)$$

Here we assume that the core can operate at a different frequency than the bus, but must interface with the bus at a fixed data width. The reason is while a core could issue requests at half of the frequency of the bus, each request requires the entire data width of the bus. (Note: it is possible for more sophisticated buses to circumvent this restriction, and we leave those specific implementation details to our readers.) To calculate the required bus bandwidth we can sum up each core's bus bandwidth. The bus saturates when the required bus bandwidth is greater than the available bus bandwidth.

We can also consider alternative interconnects, such as a crossbar switch, to connect the cores. A crossbar switch allows more than one core to communicate with another core in parallel. This increases the overall bandwidth of the interconnect at the cost of additional resources. The additional resources stem from allowing multiple connections to be made in parallel. You could think of this almost as adding additional buses between all of the cores so that if one is busy, the core could use a different bus. Figure 6.15 shows a four-port crossbar switch with bidirectional channels. Each output port is connected to a four-input multiplexer, totaling in four four-port multiplexers. The multiplexer's select lines are driven by some external controller that decides which input ports are connected to which output ports.

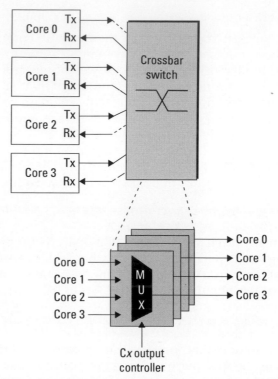

Figure 6.15. A four-port crossbar switch and its internal representation based on multiplexers.

Now when we consider scalability, we do so in terms of the size of the interconnect as well as bandwidth. As we increase the size, we increase the number of cores that can connect and communicate in parallel. If we double the number of connections from four to eight, we are doubling the number of cores that can communicate in parallel, from four to eight. This results in a $2\times$ increase in total bandwidth. Of course, this also doubles the number of multiplexers, which the designer must take into account when scaling the system.

Memory Perspective

Finally, we consider the more classical bandwidth constraint of memory. This includes both on-chip and off-chip memory. We are able to sustain a higher on-chip bandwidth with Platform FPGAs, as we can distribute the memory throughout the system and allow parallel access to the memory. With off-chip memory we have a fixed resource, such as the memory controller and I/O pins connecting the physical memory module. To calculate the available memory bandwidth we consider the operating frequency and data width of the memory controller. We are limited to the memory controller because even if the memory module is capable of faster access or higher bandwidth, the bottleneck is at the memory controller where data are then distributed across the FPGA.

$$Memory\ Bandwidth = Frequency \times Data\ Width$$

Furthermore, we must consider how the memory controller is connected to each component. For scalable designs, a bus allows for the memory controller to be shared more equally between each of the compute cores. While this solution to the scalability problem may appear to be simple on the surface, a bottleneck can be created if the bus bandwidth is less than the memory bandwidth.

The type of memory also plays an important role in the bandwidth problem. A designer needs to consider how the type of memory will affect the system and understand where the potential bottlenecks can occur. As the saying goes, "the chain is as strong as its weakest link," and the slowest component in the system is the weakest link. When purchasing fast SRAM, the performance can be quickly negated by a slow memory controller, slow system bus, or slow compute core. Reducing the number of intermediate steps between memory and the compute core can be beneficial, but at the cost of limiting access to the rest of the system.

6.3.2. Scalability Solutions

Now that we have identified some of the scalability constraints, we want to investigate potential solutions. In some cases these solutions are very specific to only a subset of applications, whereas in other cases these solutions may increase the resource utilization beyond an acceptable level. Our goal is not to solve every problem, but to get readers thinking of novel solutions by being exposed to some of the simpler solutions.

Overcoming Processor Limitations

The processor's computation and control trade-off are where the processor will either perform computation at the expense of idle hardware cores or perform hardware core control at the expense of computation. In the first few chapters of this book we have been less *hardware accelerator* focused and more processor-memory focused. That is, use the FPGA as a platform to reduce the number of chips needed for embedded systems design, but still have a more traditional processor-centric compute model. Moreover, in this model the processor is performing a majority of the computation, sampling inputs and calculating results. In the latter half of this book we are shying away from a processor-centric compute model by considering the processor to act more as a controller of computation performed within the FPGA fabric. We started with partitioning an application and assembling custom compute cores to run in hardware. Then, to increase performance we looked at spatial parallelism of hardware designs.

By moving more of the computation into hardware, we can free up more time for the processor to spend controlling the application flow rather than performing computation. To help alleviate some of the processor's burdens, we use direct memory access to allow hardware cores to retrieve memory. For applications with instruments streaming data into the FPGA, we can perform as much computation within hardware before interrupting the processor. We use these interrupts instead of polling to reduce bus traffic, letting the hardware core tell the processor when its computation is complete, rather than having the processor check its status continuously.

When considering how to improve the performance of a Platform FPGA-based system from the processor's perspective, we want to:

- move more of the application's computation into hardware
- allow hardware cores to independently access memory through DMA
- use interrupts instead of polling to communicate with the processor

Overcoming Bus Bandwidth Limitations

To improve the bus bandwidth the common approaches are to improve the two contributing factors to the bandwidth calculation, that is, the operating frequency and data width of the bus. By doubling the frequency or the data width, the bus bandwidth is doubled. Running the bus at a higher frequency forces each core connecting to the bus to interface at that same frequency (it is possible using clock domain crossing logic to have the bus interface operate at one frequency and the core operate at a different frequency). Likewise, increasing the data width increases the required resources to implement the wider bus.

Alternatively, to satisfy the bus bandwidth requirement, it may be possible to partition the design into multiple buses. This depends on the communication patterns of the cores on the buses; however, it can be a viable solution to the bus bandwidth limitation. It is also possible to bridge each of the buses together to provide access to cores on other buses at the cost of additional latency through the bridges.

By partitioning the buses, what we are actually trying to accomplish is segregating cores that need to communicate with each other on a single bus. These cores may not need to communicate with every other core in the system, so by moving them onto a separate bus, we can improve the total bus bandwidth across each bus.

In a scalable design where the same hardware core is replicated numerous times, the separate bus approach may not be suitable. This is especially true if the core being replicated only communicates with the processor and off-chip memory. In these cases, it may be possible to replicate the logic within a hardware core and share a single interface to the bus. The result is a larger signal hardware core but potentially more efficient bus utilization, as the hardware core itself will need to prioritize and arbitrate requests rather than the arbiter. Requests within the hardware core can also be grouped together to form a more efficient burst transaction.

Using burst transfers whenever possible can result in an improvement in the utilization of the bus by transmitting more data over the course of fewer requests. Because the bus arbiter will see fewer requests being issued by each of the cores, the arbiter is able to spend less time arbitrating over the requests and spend more time granting access to the bus. The hardware cores will also benefit by receiving (or sending) data at a higher bandwidth.

Therefore, a quick recap on the proposed solutions to improve bus bandwidth limitations includes:

- increasing operating frequency of the bus
- increasing data width of the bus
- segmenting the bus into smaller, more localized buses
- using burst transfers to communicate more efficiently

Overcoming Memory Bandwidth Limitation

This chapter has already covered many ways to overcome the memory bandwidth limitation. Initially, this chapter focused on bandwidth in general, which is a core concern for any design. The shift to off-chip memory bandwidth is due to the processor-memory model and contention for memory bandwidth with the increasing number of custom hardware cores. A bus provides a simple mechanism to access the shared resource, off-chip memory. Unfortunately, the bus may be inadvertently degrading performance by sharing on-chip communication — say between two hardware cores — with off-chip memory requests. When the bus is granted for on-chip communication, the memory controller may sit idle, waiting for the next request.

Giving priority to memory requests can alleviate some of these idle times by allowing the request to be issued to the memory controller. Then while the request is being fulfilled, the bus can be used for on-chip communication, until data are ready to be transmitted back to the hardware core. Of course, balancing the requests can be challenging and the result could have a negative impact on on-chip communication.

A more ideal solution may be to separate on-chip communication from off-chip memory requests. At the expense of additional logic, a hardware core can be responding to an on-chip request through one interface (such as a bus) while issuing an off-chip memory request through a separate interface. If a core has a direct connection to the memory controller, then the memory bandwidth is not restricted to the bus bandwidth, and requests do not need to wait for another core's transaction to complete before being issued.

Finally, by using double buffering we can hide the latency of subsequent requests by overlapping requests with the transmission of data. We already mentioned this approach earlier, but it is important to emphasize the role double buffering can play in improving memory utilization.

In summary, how memory is accessed can have a significant effect on the performance of the system. When designing with bandwidth in mind to achieve high performance efficiently we must:

- provide support for direct memory access
- separate interfaces for on-chip and off-chip communication
- use double buffering of requests whenever possible

Chapter in Review

This chapter's focus has been on bandwidth, a critical consideration when designing systems for Platform FPGAs. Beginning with the problem of balancing bandwidth, we investigated methods and techniques such as the Kahn Process Network to maximize performance with a minimum amount of resources. This was followed by a thorough discussion of on-chip and off-chip memory, including interfaces and controllers, the differences in memory access, and performance. With embedded systems, streaming instruments play a large role in the final product; therefore, we also spent time discussing various methods to efficiently support complex on-chip communication bandwidth requirements. Finally, we discussed designs from a portability and scalability perspective. Our interests are motivated by the ever-advancing semiconductor technology in hopes of reusing existing hardware components and scaling systems to yield greater functionality on future FPGA devices.

In the gray pages that follow, we further investigate these bandwidth questions with respect to the Xilinx Virtex 5 FPGA. We consider bandwidth in terms of on-chip memory access from FIFOs and BRAMs along with off-chip memory access. Included in these gray pages are examples to help readers implement the various memory interfaces in their own design. The final demonstration covers the Xilinx Native Port Interface (NPI) integration into a custom compute core to provide efficient, high bandwidth transfers to and from off-chip memory.

Practical Expansion: Managing Bandwidth

Practically speaking, there are a number of ways to integrate memory into a design. What we aim to do in these gray pages is highlight some of the more common methods to incorporate memory in a design. By memory we mean both on-chip and off-chip memory. We also look at memory as being a buffer, perhaps only a few elements deep to balance a computation, and as a storage space, such as off-chip memory.

How memory is accessed in Platform FPGA designs is an important design concept. We will look at on-chip memory access, such as FIFOs and RAM, as well as off-chip memory. With off-chip memory we have a variety of choices to consider when accessing memory. These choices can have a dramatic impact on the performance and resource utilization of the system.

6.A. On-Chip Memory Access

Including on-chip memory is almost essential for bandwidth-sensitive designs. More important than simply using on-chip memory is using it efficiently. In designing systems for scalability, a poorly allocated resource such as memory can severely limit the scalability of the system. We would not want to waste an entire BRAM as a buffer if we only needed a single register. Of course there are designers who may only be familiar with a single solution, so every problem that looks similar will be solved the same way. We want to present a few more options to try to emphasize the importance of using resources efficiently to solve the problem. We cover two ways to use on-chip memory, in the form of FIFOs and random access memory.

6.A.1. FIFOs

A FIFO, or queue, is useful as a buffer to capture data as it arrives from the producer and to allow the consumer to retrieve data, in order, at its convenience. We have already shown one way to use on-chip memory in the form of FIFOs in Section 3.A. Using CoreGen we can quickly configure and generate a FIFO to be used in our system; however, it is necessary to understand some of the configuration options to best utilize the available resources in the design.

The CoreGen FIFO Generator wizard presents a number of choices as to how a FIFO will be implemented in the FPGA fabric. The choices include Block RAM, distributed RAM, a shift register(s), and built-in FIFO(s) if supported by the device. These different memory types offer different features and, as a result, different reasons for choosing one over another.

FIFO Memory Types   Block RAM is the most efficient use of FPGA resources when needing to store large amounts of data in on-chip memory and can support different read and write clock rates and data widths. BRAMs can be useful when needing to aggregate two 32-bit inputs into a single 64-bit output or supporting crossing a clock domain by writing data into the FIFO at one frequency and reading data out at a different frequency. Depending on the FPGA device, the storage capacity of one BRAM can vary. The Virtex 5 BRAM can store up to 36 K bits of data per BRAM, which can be configured as a 1-bit wide and 32,768 deep RAM to a 36-bit wide and 1024 deep RAM. The summary page of the FIFO Generator wizard estimates the number of BRAMs required to support the specified data width and depth.

Distributed RAM uses LUTs from the memory slices (SLICEM) to create a synchronous RAM. Not all slices in an FPGA can be used for distributed RAM because not all slices are of type SLICEM. Due to the limitation on resources, distributed RAMs are recommended for use when the width and depth of the FIFO are smaller than a single BRAM, although the exact trade-off between when to use a BRAM and when to use distributed RAM depends on the FPGA. Distributed RAM can also be used when a design requires more storage space than is available in BRAM. This requires that memory slice resources are available. Distributed RAM can be used with different read and write widths, but unlike with BRAM, the same clock must be used for both read and write ports.

Shift registers are also implemented in memory slices. A shift register FIFO does not support independent read and write clocks or data widths. The shift register is more resource efficient than distributed RAM for small FIFOs with depths less than 32 elements. Shift register FIFOs are well suited for these small buffers for helping balance on-chip computation bandwidth.

Virtex 4, 5, and 6 devices also support built-in FIFOs in place of using Block RAM. The built-in FIFOs are embedded within the FPGA fabric like BRAM, however, with a fixed purpose. These FIFOs can support independent read and write clocks, but have fixed data widths of 4, 9, 18, and 36 bits.

FIFO Configuration Options   Once the memory type has been selected, there are configuration options that can be set to provide additional support to the FIFO. The first choice is the read mode; it can be either standard FIFO or first-word fall-through FIFO. In *standard* read mode when the FIFO is not empty, issuing a read from the FIFO will produce valid data the next clock cycle, that is, a one clock cycle read latency. In *first-word fall-through* the head of the FIFO is driven to the data_out port of the FIFO so as to provide the ability to peek at the first element without being required to dequeue the element. The added advantage with this read mode is that the read latency is reduced to zero clock cycles, as data are valid the same clock cycle the read enable is asserted. However, not all memory types support first-word fall-through.

Data read and write width and depths can also be set depending on the memory type. The FIFO Generator provides guidance as to what width and depths are supported based on the memory type selected. Along with data there are optional flags and handshaking signals that can be included in the generated FIFO. These include FIFO status indicators, such as *almost full* or *almost empty*. Both almost full and almost empty signals are asserted when they are one data element from the FIFO being full or empty, respectively. A common use for these signals is for flow control, to pause writing data into the FIFO or to pause reading data out of the FIFO.

Handshaking signals between read and write ports of the FIFO can make integration of the FIFO into a design easier. In addition to the default read and write enable signals are optional signals, *data valid* and *write acknowledge*. *Data valid* is asserted when read data are valid on the data output port, and *write acknowledge* is asserted after data have been written into the FIFO. There are also options to count data in the FIFO and output the count value as a port. The specific use of the counter is application specific, but it can save the designer time by directly including it within the FIFO instead of writing HDL or instantiating a separate counter component.

To clarify, it is possible to instantiate your own FIFO based on any of the memory types mentioned previously without the use of the Xilinx CoreGen tool. FIFOs can be inferred by the synthesis tool based on the written HDL, or precisely specified by instantiating the exact memory primitive. These more advanced methods can result in more efficient designs, in terms of both resources and performance. We encourage readers to become comfortable and familiar with the supplied tools while at the same time trying to understand which primitives are being instantiated by the wizard.

6.A.2. Block RAM

As mentioned earlier, BRAM is a useful resource when needing to store data on-chip and implementing FIFOs with varying read and write data widths and clock rates. In addition to being used as FIFOs, BRAMs can be used as either an on-chip RAM or ROM. Implementing an on-chip RAM or ROM does not require use of a BRAM.

In fact, just as with FIFOs there are trade-offs to implementing a RAM or ROM in different types of memory. The type of memory can be specified in three ways. First, using HDL the specific memory primitive can be instantiated along with supporting HDL to include handshaking signals not directly included within the primitive. Second, writing HDL to let the synthesis tool infer the type of memory, which can make the decision process for the designer easier, alleviates the sometimes difficult task of selecting one type of memory over another. The disadvantage of this approach is that it requires the designer to write the HDL in a way the synthesis tool can correctly infer the correct memory type. In this case, ugly HDL can often lead to inefficient resource utilization. The last common way to select the type of memory to be used is through the Xilinx CoreGen tool. We have already used CoreGen to configure the FIFOs, so using CoreGen to generate RAMs and ROMs should be relatively straightforward.

CoreGen allows us to generate RAMs and ROMs from either Block RAM or distributed RAM. With the Virtex 5 the distributed RAM is limited to a maximum depth of 65,536 data words and a maximum data width of 1024 bits, depending on the architecture. The depth must also be a multiple of 16 words. However, BRAMs can be combined to create upwards of megabytes of on-chip storage. The limitation is based on the number of available BRAMs on-chip. BRAMs are more resource and power efficient than distributed RAMs as well. On Virtex 5 designs (and Virtex 6) the BRAMs can be configured as a single 36K BRAM or as two independent 18 K BRAMs.

In addition to physical memory, there are choices between the different interface types. These include whether the RAM/ROM is a single port or dual port memory. A single port consists of one address, data, and read/write enable ports. This is useful for lookup tables or when a single controller is reading or writing to memory. Dual port memory can be either a *simple* or a *true* dual port memory. In a simple dual port memory, port A is write only and port B is read only. This is most useful when sharing data from a source to a destination. In the event that both ports need to have read and write access, a true dual port memory is required. CoreGen also supports initializing memory through a memory coefficient (COE) file. When the bitstream is programmed to the FPGA, these data are loaded into the on-chip RAM. For lookup tables, this feature is very useful to avoid the need to add additional circuitry to initialize the memory.

6.A.3. LocalLink Interface

As seen with both FIFO and BRAM interfaces, accessing memory can vary depending on the application's needs. One important interface to briefly cover is not necessarily associated with on-chip memory, but more so with data transmission and communication. LocalLink is a unidirectional point-to-point interface standard created by Xilinx and is being used in an increasing number of Xilinx IP Cores. Data are transferred synchronously in packets (called frames) from a source to a destination. Both the source and the destination have flow control, allowing each to pause transmission of data in the event there are no new data to send or the destination is unable to receive and process new data.

The standard (shown in Table 6.1) consists of a set of required signals; any signal ending with _n is negative logic where '0' is asserted and '1' is deasserted:

LocalLink Signals	Signal Description
CLK	Synchronous clock signal
RST_N	Reset signal aborts frame
DATA	Data transmitted from source to destination
SRC_RDY_N	Source is ready to transfer data
DST_RDY_N	Destination is ready to receive data
SOF_N	Start of frame indicator for first byte of frame
EOF_N	End of frame indicator for last byte of frame

Table 6.1 Xilinx LocalLink Signals.

A transfer is initiated by the source by asserting the signals src_rdy_n and sof_n along with the first data word in the transfer. When the destination is ready to receive data it will assert dst_rdy_n. Only when both src_rdy_n and dst_rdy_n are asserted are data valid to the destination and should the source start to assert the next data word. Data are transferred in frames, with the start-of-frame and end-of-frame signals used as indicators to the beginning and end of the frame of data. The start-of-frame and end-of-frame signals should only be asserted during the transmission of the first and last byte of the frame, respectively.

In order to add a LocalLink interface to an existing hardware core we must add these signals to the entity description. Because LocalLink is unidirectional, if a hardware core is only sending data to a destination or only receiving data from a source, a single instance of the LocalLink ports is needed. In some cases there is a need for a hardware core to communicate either bidirectionally or receive data from a source and produce data to another destination. Under this circumstance, both transmit and receive LocalLink ports are needed. Keep in mind the directionality of the ports, that is, a source will transmit (output) data, start-of-frame, end-of-frame, and source ready signals, whereas the receiver will input these signals.

To help illustrate this point we will show how to add the LocalLink interface ports to an existing hardware core via the Microprocessor Description file. We will assume that the hardware core has two LocalLink ports, a transmit port and a receive port. Using the Xilinx bus standard XIL_LL_DMA to create bus interfaces, we must specify the transmit port as an *initiator* and the receive port as a *target*. In the Xilinx Platform Studio these two types result in a point-to-point bus that is similar to how the PLB Block RAM controller connects to the Block RAM component.

```
BUS_INTERFACE BUS = TX_LLINK0, BUS_STD = XIL_LL_DMA, BUS_TYPE = INITIATOR
BUS_INTERFACE BUS = RX_LLINK0, BUS_STD = XIL_LL_DMA, BUS_TYPE = TARGET
```

Connecting the LocalLink ports to the bus becomes fairly straightforward. The LocalLink bus standard connects transmit (source) ports to receive (destination) ports. As a result, the transmit ports for the hardware core are connected by default to the transmit ports of the LocalLink transmit bus. However, the receive ports are also connected to the transmit ports, but for the receive bus. This effectively crosses the transmit and receive signals to allow the receive port to input data that is output from the transmit port.

```
## LocalLink Transmit Port
PORT ll_tx_data    = LL_Tx_Data,   DIR = O, BUS = TX_LLINK0, VEC = [31:0]
PORT ll_tx_sof_n   = LL_Tx_SOF_n,  DIR = O, BUS = TX_LLINK0
PORT ll_tx_eof_n   = LL_Tx_EOF_n,  DIR = O, BUS = TX_LLINK0
PORT ll_tx_src_rdy_n = LL_Tx_SrcRdy_n, DIR = O, BUS = TX_LLINK0
PORT ll_tx_dst_rdy_n = LL_Tx_DstRdy_n, DIR = I, BUS = TX_LLINK0, INITIALVAL = VCC
# LocalLink Receive Port
```

```
PORT ll_rx_data     = LL_Tx_Data,    DIR = I, BUS = RX_LLINK0, INITIALVAL = GND, VEC = [31:0]
PORT ll_rx_sof_n    = LL_Tx_SOF_n,   DIR = I, BUS = RX_LLINK0, INITIALVAL = VCC
PORT ll_rx_eof_n    = LL_Tx_EOF_n,   DIR = I, BUS = RX_LLINK0, INITIALVAL = VCC
PORT ll_rx_src_rdy_n = LL_Tx_SrcRdy_n, DIR = I, BUS = RX_LLINK0, INITIALVAL = VCC
PORT ll_rx_dst_rdy_n = LL_Tx_DstRdy_n, DIR = O, BUS = RX_LLINK0
```

The LocalLink interface will be addressed shortly when we discuss a specific bus master implementation, although the LocalLink port is not connected through the MPD file because the bus master uses the LocalLink between components within the custom hardware core. The MPD file is used more directly in Chapter 7 when we implement interface components off of the FPGA.

6.B. Off-Chip Memory Access

Accessing memory can be critical in a design and knowing how the memory can be accessed is an important design consideration. Throughout this section we aim to cover more specific details regarding various ways to access memory as well as different ways to interface with memory. Because memory bandwidth can be a precious commodity in systems design, being aware of the different memory access choices as a designer can affect the performance of the system.

6.B.1. Programmable I/O

The white pages presented a short C-code snippet that reads data from main memory and writes it to the compute core. Programmable I/O arguably provides the simplest interface between a hardware core and memory while requiring the least physical resources. This of course is at the expense of performance. The application may dictate how memory should be accessed. We only address the topic so as to alert readers to alternative solutions to accessing memory.

A specific Xilinx implementation for programmable I/O does not differ from the C-code already presented in Chapter 6's white pages. The requirement that both the memory controller and the hardware core are addressable by the processor means that both are slaves on a bus that the processor is a master on. It is possible that the memory controller and hardware core reside on two different buses, as long as the processor has access to both buses.

Because the objective of this chapter is to discuss different memory access methods, we focus our attention on these alternative approaches.

6.B.2. Central DMA Controller

An improvement on the performance of programmable I/O, where the processor performs the entire request, is to use a DMA controller. Xilinx provides a *Central DMA Controller* [Xilinx, Inc., 2009f] as a core in its EDK IP Core Repository. The processor issues a DMA request to the controller, which in turn handles the DMA transaction on behalf of the processor. The DMA controller can support:

- processor initiated read from memory and write to hardware core
- processor initiated read from hardware core and write to memory
- hardware core initiated read from memory and write to hardware core
- hardware core initiated read from hardware core and write to memory
- hardware core to hardware core transfers

It is important to understand that *memory* in the term "direct memory access" does not necessarily mean off-chip memory. In fact, transfers between the hardware core's on-chip memory are also possible. We will discuss in more detail how a hardware core can initiate a request to the DMA controller shortly.

For now, we will focus on how the central DMA controller works and, more specifically, will learn how to initiate a DMA request. Start by adding the central DMA controller to a previously created base system. The central DMA controller includes both a master and a slave bus interface. The slave interface is for receiving DMA requests, for example, from the processor. The master interface is used to issue an actual memory request to the memory controller or compute core.

The central DMA controller consists of a set of control registers used to set up the DMA request. In C, we can create a structure to access these registers more easily.

```
typedef struct{
    unsigned int rst_reg;     // Reset Register
    unsigned int dmac_reg;    // DMA Control Register
    unsigned int sa_reg;      // Source Address Register
    unsigned int da_reg;      // Destination Address Register
    unsigned int len_reg;     // Length Register
    unsigned int dmas_reg;    // DMA Status Register
    unsigned int isr;         // Interrupt Service Register
    unsigned int ier;         // Interrupt Enable Register
}cdma_regs;
```

To set up a DMA request, the processor will need to write into the source address register, destination address register, and, finally, the length register. The source address register is where the central DMA controller will read data from. The destination address is where read data will be written to. The length register contains the number of bytes to read from the source. The length register should always be written last because writing to it triggers the central DMA controller to begin the DMA request. Once the processor writes the length register it can either poll the DMA status register to determine when the request has completed or, if interrupts are used it can wait until the interrupt occurs to indicate the transaction has finished. In either case, the processor is free to perform any computation in parallel with the DMA transaction as long as the computation does not alter the contents of memory that is part of the DMA request until the transaction completes. We include a short C-code example here:

```
int main() {
    // Central DMA Registers
    volatile cdma_regs *cdma_core = (cdma_regs*)(CDMA_BASEADDR);
    // Hardware Core Register(s) - Can be any interface
    volatile hw_regs *hw_core = (hw_regs*)(HWCORE_BASEADDR);
    // Data located in memory to be read by DMA Controller
    volatile unsigned int data[TEST_SIZE];
    int i;

    // Write Data into Memory
    for(i=0;i<TEST_SIZE;i++) {
        data[i] = i;
    }
    // Set DMA Source Addresses
    cdma_core->sa_reg = (unsigned int) &data[0];
    printf("Source Reg = 0x%08x\n", cdma_core->sa_reg);
```

```
    // Set DMA Destination Addresses
    cdma_core->da_reg = (unsigned int) &hw_core->slv_reg0;
    printf("Dest Reg   = 0x%08x\n", cdma_core->da_reg);

    // Set DMA Length Register & Start DMA Transfer
    cdma_core->len_reg = TEST_SIZE * sizeof(int);
    printf("Length Reg = 0x%08x\n", cdma_core->len_reg);

    // Perform Parallel Computation - if any

    // Verify Results - Read from Slave Register
    printf("slv_reg0 = 0x%08x\n", hw_core->slv_reg0);

    printf("Test Complete\n");
}
```

In order to receive burst transactions from the DMA controller, the hardware core must be set to support burst and cache-line transfers. This can be done when creating the hardware core through the Create and Import Peripheral Wizard that is part of the Xilinx EDK. Without this support, only single transfers can be made to the core.

The central DMA controller is a useful tool in offloading memory requests from the processor. The example just given illustrates this by transferring data from memory to a slave register on a hardware core. The final `printf` statement should reveal the contents of `slv_reg0` to be the last element transmitted from memory. Because each application's need differs, we assume that writing to this single port is sufficient.

6.B.3. Bus Master

Adding bus master capability to a hardware core can further improve performance by allowing the core to issue its own memory requests rather than relying on the processor and DMA controller. It is still possible to integrate a bus master hardware core with the central DMA controller; however, it would be more efficient for the hardware core to directly issue memory requests to the memory controller rather than have the DMA controller act as a middleman. This is due to the fact that the DMA controller requires at least three separate bus transfers to set the source, destination, and length registers, whereas the bus master can issue a single request to the memory controller.

Because each bus implementation differs, we will continue to use the Processor Local Bus (PLB) for our demonstrations. Starting with the Xilinx Create and Import Peripheral Wizard, refer back to Section 2.A; we can add bus master support with an additional check box, "User Logic Master." This will generate a template for a master interface on the PLB. To better understand how the master works, we will look at the new interface signals and logic added to the `user_logic.vhd` file.

Bus Master Signals   The Xilinx IP Interconnect (IPIC) (listing in Table 6.2) is used to help reduce the number of signals the hardware core needs to interface with when issuing requests across the PLB. The hardware core can be a bus slave, bus master, or both a slave and a master. Because we have covered a bus slave already, we will focus our attention on the master signals, which are denoted by _Mst in the signal name. Signals beginning with `IP2Bus_` indicate an output signal from the entity to the bus. Signals beginning with `Bus2IP_` indicate an input signal coming from the bus into the entity. Signals ending with _n indicate negative logic signals.

From these lists we can see there are more than just input and output signals. In fact, there are two other groups of signals. The first group contains the LocalLink signals, which follow the LocalLink specification for transmitting data to and

IPIC Signals	Signal Description
Bus2IP_Mst_CmdAck	Bus Command Acknowledgment
Bus2IP_Mst_Cmplt	Bus Transfer Complete
Bus2IP_Mst_Error	Bus Error
Bus2IP_yMst_Rearbitrate	Bus Rearbitrate
Bus2IP_yMst_Cmd_Timeout	Bus Command Timeout
IP2Bus_MstRd_Req	Issue a Master Read Request
IP2Bus_yMstWr_Req	Issue a Master Write Request
IP2Bus_yMst_Addr	Address to Read or Write Data
IP2Bus_Mst_BE	Write Data Byte Enable
IP2Bus_Mst_yLength	Number of Bytes for Request
IP2Bus_Mst_yType	Transfer Type
IP2Bus_Mst_Lock	Lock Bus during Transaction
IP2Bus_Mst_Reset	Issue a Bus Reset
Bus2IP_yMstRd_d	LocalLink Read Data
Bus2IP_yMstRd_rem	LocalLink Read Data Remainder
Bus2IP_MstRd_ysof_n	LocalLink Read Start of Frame
Bus2IP_MstRd_eof_n	LocalLink Read End of Frame
Bus2IP_MstRd_src_yrdy_n	LocalLink Read Source Ready
Bus2IP_yMstRd_src_dsc_n	LocalLink Read Source Discontinue
IP2Bus_MstRd_ydst_rdy_yn	LocalLink Read Destination Ready
IP2Bus_MstRd_ydst_dsc_n	LocalLink Read Destination Discontinue
IP2Bus_MstWr_d	LocalLink Write Data
IP2Bus_yMstWr_rem	LocalLink Write Data Remainder
IP2Bus_yMstWr_sof_n	LocalLink Write Start of Frame
IP2Bus_MstWr_eof_yn	LocalLink Write End of Frame
IP2Bus_MstWr_ysrc_rdy_n	LocalLink Write Source Ready
IP2Bus_MstWr_src_ydsc_n	LocalLink Write Source Discontinue
xBus2IP_yMstWr_dst_rdy_n	LocalLink Write Destination Ready
Bus2IP_MstWr_dst_dsc_n	LocalLink Write Destination Discontinue

Table 6.2 Xilinx IPIC signals for PLB master.

from the hardware core. Because LocalLink is a unidirectional specification and we want to support bidirectional transfers, read and write transfers, there are two sets of LocalLink signals in the master interface.

The second group contains the bus control signals. As with the LocalLink signals there are two directions: input from the bus and output from the hardware core. On the bus side are command acknowledgment, transfer complete, error, arbitrate, and command timeout signals. These signals are used to relay bus information to the hardware core. If the bus needs the hardware core to reissue a request, it will assert the bus rearbitrate signal. On the hardware core side are read request, write request, address, data byte enable, length, type, lock, and reset signals. The hardware core asserts these signals to indicate a read request from a specific address and, if the request is a burst request, the length of the transfer. These are considered the control signals for the bus transaction.

Bus Master Logic Now that we have a better idea of what the bus master interface signals are, we will look at the master logic within the `user_logic.vhd` file. The template provided by Xilinx includes three finite state machines for the bus master. One state machine is for the LocalLink read requests. The second state machine is for the LocalLink write requests. The third state machine acts as a controller over the bus requests and read and write state machines.

Xilinx also provides a set of master registers to allow simple testing with a software application. Because each application may differ in how the bus master is driven (by a software application, by a custom finite state machine, etc.), we will instead focus on the three finite state machines and allow readers to experiment with alternative interfaces to the master control registers. Comments generated by the Xilinx wizard provide a helpful starting point for software-based master requests. In our experience, once the three state machines are understood, augmenting them to meet the needs of the design is trivial.

For those familiar with the earlier version of the Xilinx EDK software, the tools prior to version 10.1 did not include a LocalLink interface. The older master interface relied on the instantiation of a bus slave with the bus master to support transfers to and from the hardware core. Since version 10.1, this has been replaced with the LocalLink interface. Now the hardware core does not need to instantiate a bus slave if only a bus master is required, resulting in fewer resources being used whenever possible.

Control Finite State Machine We have briefly discussed a finite state machine for the bus master, which interacts with the bus control signals and the read and write finite state machines. For more details regarding the control finite state machine, see Figure 6.16.

The `CMD_IDLE` state waits for either a read or a write request (this can be from the master control register provided by the template or a custom finite state machine). Once a read or write request has been issued, the corresponding LocalLink finite state machine is started and then transitions into the `CMD_RUN` state.

To indicate that the request has been received by the bus interface, the `Bus2IP_Mst_CmdAck` (command acknowledgment) signal is asserted. This also notifies the control finite state machine to deassert the bus request signals because the request is in transit and keeping the signals asserted risks issuing a second identical request. In the event a timeout or error occurs, the control finite state machine should respond based on the requirements of the system. The template asserts status signals to be read by the software application. This may suffice for your application or more logic may be necessary to support reissuing the request.

When the command acknowledgment signal is asserted it may be asserted with the `Bus2IP_Mst_Cmplt` (command complete) signal. The command complete signal indicates that all requested data have been transferred to or from the hardware core. If both signals are asserted at the same time, the control finite state machine should transition to the `CMD_DONE` state. Under some circumstances the command acknowledgment signal is asserted without the command complete, in which case the control finite state machine can deassert the request signals, but must wait for the command complete signal. This is done in the `CMD_WAIT_FOR_DATA` state. Finally, the `CMD_DONE` state returns to the `CMD_IDLE` state to wait for the next request.

Read Request Finite State Machine The read request finite state machine controls read requests where data are read from a core on the bus and returned to the hardware core via the read LocalLink interface signals. Because data are being delivered to the hardware core there are only a few signals to interact with. The read request is triggered by the control finite state machine, so until a read request is asserted the default state is the `LLRD_IDLE` state. Once a request is asserted the read request finite state machine transitions to the `LLRD_GO` state. In this state, when both the `Bus2IP_MstRd_src_rdy_n` (source ready) and the `IP2Bus_MstRd_dst_rdy_n` (destination ready) signals

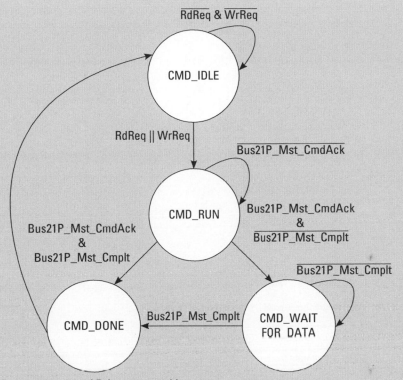

$\overline{\text{RdReq}}$ & $\overline{\text{WrReq}}$

CMD_IDLE

RdReq || WrReq

$\overline{\text{Bus2IP\_Mst\_CmdAck}}$

CMD_RUN

Bus2IP_Mst_CmdAck
&
$\overline{\text{Bus2IP\_Mst\_Cmplt}}$

Bus2IP_Mst_CmdAck
&
Bus2IP_Mst_Cmplt

$\overline{\text{Bus2IP\_Mst\_Cmplt}}$

CMD_DONE

Bus2IP_Mst_Cmplt

CMD_WAIT
FOR DATA

Figure 6.16. Bus master control finite state machine.

are asserted, data are flowing from the source to the destination. The hardware core can pause the transfer by deasserting the `IP2Bus_MstRd_dst_rdy_n` signal. This can be useful in streaming applications where data arrive after they can be processed or data are written into a FIFO for internal buffering and the buffer becomes full. The read request completes when the `Bus2IP_MstRd_eof_n` (end-of-frame) signal is asserted, meaning that the last word in the transfer has arrived. The read request finite state machine returns to the `LLRD_IDLE` state waiting for the next read request. Figure 6.17 illustrates the LocalLink read request finite state machine.

Write Request Finite State Machine The write request finite state machine controls write requests where data are written from the hardware core to another core on the bus via the write LocalLink interface signals. Similar to the read request finite state machine, the write request finite state machine waits in the `LLWR_IDLE` state until a write request is issued. If burst transfers are supported, allowing more than a single word of data to be transmitted per each transaction, then the next state can be either `LLWR_BURST_INIT` or `LLWR_SNGL_INIT`, indicating a burst or single transaction.

During a single word write transaction the write request finite state machine must wait until both the destination and the source are ready, as indicated by the `Bus2IP_MstWr_dst_rdy_n` and `IP2Bus_MstWr_src_rdy_n` signals. At this time both the `IP2Bus_MstWr_sof_n` and the `IP2Bus_MstWr_eof_n` signals must be asserted as well to signify that the LocalLink transfer is a single word. The write request finite state machine can then return to the `LLWR_IDLE` to wait for the next write request.

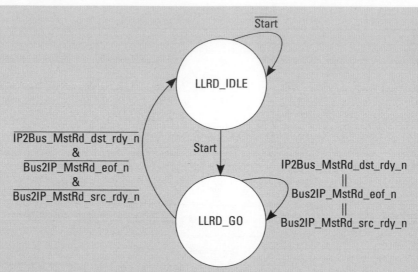

Figure 6.17. Bus master read request finite state machine.

Conversely, a burst write transaction requires the write finite state machine to issue multiple words of data to the destination core. The transfer length is identified by IP2Bus_MstWr_Length. A burst transfer must still adhere to the requirement that a valid transfer requires both source and destination ready signals to be asserted. The start-of-frame signal must be asserted during the transfer of the first word in the burst, while the end-of-frame signal must be asserted during the transfer of the last word. A counter can be used to help keep track of the amount of data left to be transferred in the burst transaction. A last state LLWR_BURST_LAST_BEAT is used to assert the end of frame accordingly. When the transfer completes, the write finite state machine returns to the LLWR_IDLE to wait for the next write request. Figure 6.18 illustrates the LocalLink write request finite state machine.

Adding bus master support to a hardware core can enhance the functionality of the hardware core greatly, allowing it to act much more independently than when just a bus slave. Not all applications may require this functionality; however, understanding how it is implemented is a useful design tool.

6.B.4. Native Port Interface

When hardware cores need high bandwidth and low latency transfers to off-chip memory, Xilinx provides a custom interface to their Multi-Ported Memory Controller (MPMC) known as the Native Port Interface (NPI). This is different than a bus master in that it is a direct connect interface to the memory controller instead of a connection to a shared bus. Because the interface is not shared it does not require arbitration, allowing for lower latency transfers. Also, due to the direct connection, the hardware core can receive data at the memory controller's operating frequency, which can be higher than the bus operating frequency, providing a higher bandwidth.

Unlike the bus master, Xilinx does not directly support NPI integration into a hardware core through generated templates. This may change in future releases of the tools, but as of now the designer is required to make some modifications to a hardware core needing the NPI. Also, the NPI does not follow the LocalLink standard. If there is a need to use the LocalLink interconnect, the MPMC already supports the LocalLink directory through the SDMA port. However, the LocalLink interface is limited to a maximum data width of 32 bits. With NPI we can operate at 64-bit widths for 2× the bandwidth.

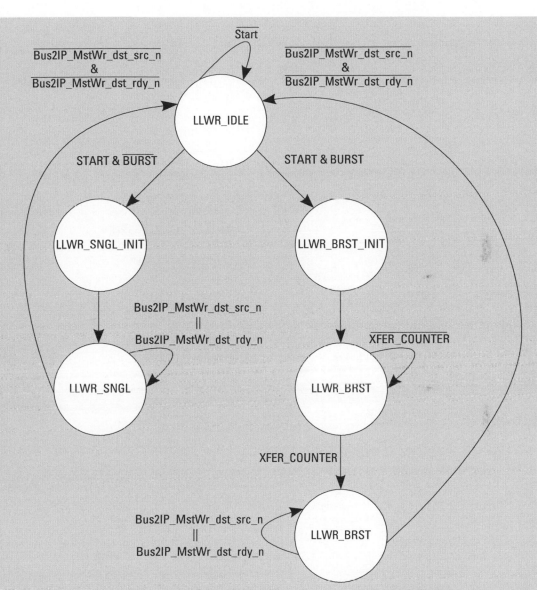

Figure 6.18. Bus master write request finite state machine.

NPI Signals First we must discuss the signals used by the native port interface, shown in Table 6.3. These signals will need to be added to the hardware core's entity port listing. The interface consists of three groups of signals. The first group contains the request and control signals: clock, reset, address, address request, address acknowledgment, read not-write, size, read-modify-write, and initialize done. The second group contains the write FIFO signals, which transfer data out to the MPMC. The last group contains the read FIFO signals, which bring in data from the MPMC.

NPI Signals	Direction	Signal Description
XIL_NPI_CLK	IN	Clock
XIL_NPI_RST	IN	Reset
XIL_NPI_Addr	OUT	32-bit Address
XIL_NPI_AddrReq	OUT	Address Request
XIL_NPI_AddrAck	IN	Address Request Acknowledgment
XIL_NPI_RNW	OUT	Request Type Request
XIL_NPI_Size	OUT	Request Size (Burst or Single)
XIL_NPI_RdModWr	OUT	Read Modify Write
XIL_NPI_InitDone	IN	Memory Initialization Done
XIL_NPI_WrFIFO_Data	OUT	64-bit Data Output
XIL_NPI_WrFIFO_BE	OUT	8-bit Data Output Byte Enable
XIL_NPI_WrFIFO_Push	OUT	Data Output Write Enable
XIL_NPI_WrFIFO_Empty	IN	Data Output FIFO Empty
XIL_NPI_WrFIFO_AlmostFull	IN	Data Output FIFO Almost Full
XIL_NPI_WrFIFO_Flush	OUT	Data Output FIFO Flush (Reset)
XIL_NPI_RdFIFO_Data	IN	64-bit Data Input
XIL_NPI_RdFIFO_Pop	OUT	Data Input Read Enable
XIL_NPI_RdFIFO_RdWdAddr	IN	4-bit Read Write Address
XIL_NPI_RdFIFO_Empty	IN	Data Input FIFO Empty
XIL_NPI_RdFIFO_Flush	OUT	Data Input FIFO Flush (Reset)
XIL_NPI_RdFIFO_Latency	IN	2-bit Read FIFO Latency Indicator

Table 6.3 Xilinx Native Port Interface signals.

The clock signal must be the same clock connected to the MPMC. We will discuss how to connect this clock along with connecting the hardware core to the MPMC shortly. The read and write FIFOs exist physically within the MPMC. The NPI signals interact with these FIFOs. It is not necessary to instantiate separate read and write FIFOs within your hardware core, except when required by the design.

NPI Little Endian   The native port interface and multi-port memory controller data are little endian, which may be in conflict with the application running on the Xilinx PowerPC (which can interface with memory as big endian). As a result, it may be necessary to do an endian swap when reading data from and writing data to memory. This can be accomplished through byte aligning, which simply flips the bytes from one endian to the other.

```
INPUT_ENDIAN_SWAP_PROC : process ( NPI_RdFIFO_Data ) is
begin
  data_from_mem(0 to 7)   <= NPI_RdFIFO_Data(7 downto 0);
  data_from_mem(8 to 15)  <= NPI_RdFIFO_Data(15 downto 8);
  data_from_mem(16 to 23) <= NPI_RdFIFO_Data(23 downto 16);
  data_from_mem(24 to 31) <= NPI_RdFIFO_Data(31 downto 24);
  data_from_mem(32 to 39) <= NPI_RdFIFO_Data(39 downto 32);
  data_from_mem(40 to 47) <= NPI_RdFIFO_Data(47 downto 40);
```

```
    data_from_mem(48 to 55) <= NPI_RdFIFO_Data(55 downto 48);
    data_from_mem(56 to 63) <= NPI_RdFIFO_Data(63 downto 56);
end process INPUT_ENDIAN_SWAP_PROC;
```

NPI Data Transfers The NPI request transfer size (length) is set with `XIL_NPI_Size`, which encodes the number of bytes to read or write in a 4-bit signal. The smallest transfer consists of a data word, while the largest is fixed to 16 words. This is the same limitation set on the bus master for a single transaction, except with the bus master the length signal specifies the number of bytes to transfer. Because we are using the NPI to transfer 64-bit data words (8 byte), the smallest transfer is 8 Bytes and the largest is 128 Bytes. Table 6.4 lists the encoding for `XIL_NPI_Size` necessary to transfer the specific number of bytes. It is possible to transfer less than 8 Bytes through the use of the `XIL_NPI_WrFIFO_BE` (byte enable) signal along with the `XIL_NPI_RdModWr` (read-modify-write) signal. We will discuss how to accomplish this shortly.

With burst transfers, the latency of the first word is the longest, then every consecutive word thereafter arrives one clock cycle after the previous word. As mentioned in Section 6.2, we can improve the performance through the use of *double buffering*. This is more useful for read requests, which are covered in the next section.

One additional constraint NPI imposes on burst transfers is on the address (`XIL_NPI_Addr`), which must be aligned to the transfer size. That is, each address must be on the boundary of the transfer length. For a single word transfer the address must end with 0x0 or 0x8, meaning aligned to the 8-Byte boundary. For a double word transfer the address must end with 0x0 or a multiple of 0x10 and so forth up to the 128-Byte boundary, requiring the address to end in 0x00 or 0x80.

NPI Logic Because each hardware core may need to interact with the NPI differently, for brevity, we will just cover the logic necessary to instantiate the NPI. Whether the design calls for integrating the NPI into a Xilinx-generated hardware core on the PLB or another custom core, this NPI instantiation information is targeted toward how to use the NPI. Adding higher levels of control (as is done with the bus master template via the software addressable master control register) is left to the individual implementation of the design.

To begin, add the NPI signals to the entity list of the hardware core being designed. Be sure to match the direction and bit vector widths accordingly (if there is no indication of the width, assume 1 bit). After we discuss the necessary logic required to implement the NPI, we will explain how to modify the MPD file and connect the NPI to the MPMC within Xilinx Platform Studio.

As with the bus master, we will use three finite state machines to perform memory transactions with NPI. The control finite state machine will be used to issue requests to the memory controller. The read request finite state machine will

XIL_NPI_Size	Bytes Transferred
0000	8
0001	16
0010	32
0011	64
0100	128

Table 6.4 Native Port Interface data transfer size encoding.

handle receiving data from the memory controller. The write request finite state machine will handle writing data to the memory controller. We divide the NPI logic into these three functionalities in order to support parallel transactions. While it is not possible to issue a read and a write request at the exact same time, it is possible to issue one request and, once acknowledged, issue a second request. Depending on the application, this may or may not be necessary; however, with the desired goal of designing a high bandwidth interface to memory, supporting as many parallel operations as possible will help us achieve our goal.

Control Finite State Machine   Unlike the bus master control finite state machine where only one read or write request can be in process at a time, we have implemented the NPI control finite state machine to support nearly parallel transfers of read and write requests. We say *nearly* because we are not able to support the starting two requests concurrently. We can issue one request directly followed by a second request (while the first request is still being processed). To support this functionality we have moved much of the read and write control logic to the read and write request finite state machines, respectively. The result is a simpler two state control finite state machine, as seen in Figure 6.19.

The control finite state machine waits in the CMD_IDLE state until a request is issued by either the read or the write request finite state machine. When a request is received, the control finite state machine passes the request on to the MPMC. The request signals include the address, type (read or write request), and transfer size.

The control finite state machine transitions to the CMD_ACK state, waiting for the XIL_NPI_AddrAck signal. The address acknowledgment indicates that the MPMC has received the request and is beginning to process the request. During this same clock cycle the NPI control finite state machine must deassert the request signals and return to the CMD_IDLE state so as to not unintentionally issue a second identical request.

In the event that both a read and a write request signal are received by the control finite state machine at the same time, a priority is set to issue read requests before write requests to support designs needing low-latency reads from

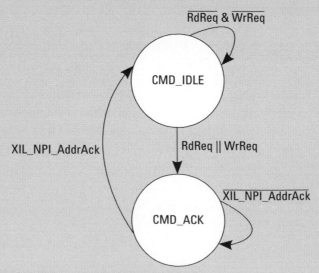

Figure 6.19. Native Port Interface control finite state machine.

memory and to avoid stalls in the computation. Of course the two requests are still able to be issued *nearly* in parallel since once request will still be in process while the other request can be made.

Read Request Finite State Machine   The read request finite state machine waits for a request from some external component (or controlling mechanism). We have specifically chosen not to build the issuing read request logic into these examples because it is very application specific. As with the bus master, it is possible to include a control register to trigger read and write requests or design any number of custom components. Instead we focus our efforts on the finite state machines needed to interact with the native port interface and multiport memory controller. Because we have moved the request logic out of the control finite state machine and into both read and write request finite state machines, we need to begin by explaining how these state machines are integrated.

The read request finite state machine (Figure 6.20) waits in the RD_IDLE state until it receives a start signal to issue a read request. Once the request is received the read request finite state machine transitions to the RD_ISSUE_REQ state, where it passes the request to the control finite state machine. The control finite state machine is actually where the requests are issued to the MPMC. The read request finite state machine waits in this state until the request has been acknowledged by the MPMC, indicating that data will be returned shortly to the NPI. Once the acknowledgment is received, the state machine transitions to the RD_WAIT_FOR_DATA state.

The data's arrival to the NPI is indicated through deassertion of the XIL_NPI_RdFIFO_Empty signal. Data are buffered in a read FIFO (RdFIFO) within the MPMC to allow the NPI to read data at its convenience through the use of XIL_NPI_RdFIFO_Pop. There is a read latency associated with the RdFIFO. It is a fixed latency (although the requests can be pipelined), which is denoted by XIL_NPI_RdFIFO_Latency. The latency depends on the parameters of the MPMC and because it is not possible to pass parameters between independent cores, it must be issued as a signal. This 2-bit signal is also necessary because there is no *data valid* signal to indicate when data from the FIFO are valid. If the latency is two clock cycles, then when the XIL_NPI_RdFIFO_Pop is first asserted to retrieve data (assuming the RdFIFO is not empty) the first data will be valid after two clock cycles.

The transfer is finished when the XIL_NPI_RdFIFO_Empty signal is asserted. Therefore, the read request finite state machine transitions to RD_DONE to signify that the transfer is done, followed by transitioning back to the RD_IDLE state to wait for the next request.

To achieve a high bandwidth, consecutive read requests can be issued to the MPMC, as long as the RdFIFO does not become full, which poses a risk of losing data. With each transaction being limited to a maximum of 128 Bytes, any larger transfers must be split into multiple smaller transfers no larger than 128 Bytes each. These transfers can be chained together and issued one after the other.

Write Request Finite State Machine   The write finite state machine (Figure 6.21) operates similarly to the read finite state machine in that it waits in the WR_IDLE state for a start signal to trigger a write request. Unlike a read request, which waits for data to arrive in the RdFIFO, a write request must first store data into the Write FIFO (WrFIFO) and then issue the request to the MPMC. The reason is that the MPMC assumes when it receives a write request that all data to be transferred are already stored in the WrFIFO.

In the event of a single transfer the write request finite state machine must issue a single word transfer, done in the WR_SNGL state. This requires asserting the XIL_NPI_RdModWr (read-modify-write) signal. Working with the XIL_NPI_WrFIFO_BE (byte enable) signal, if writing less than 8 Bytes, data must be read from memory and then modified and written back. The read-modify-write signal must be asserted with the address and address request signals after data have been written to the WrFIFO.

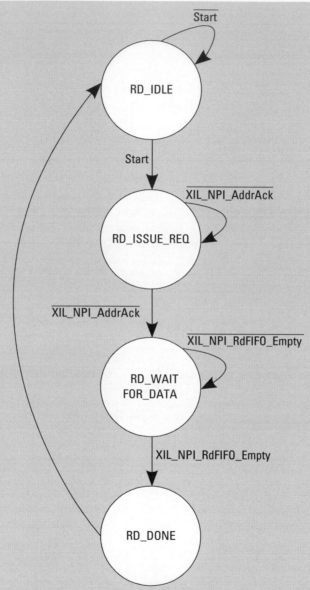

Figure 6.20. Native Port Interface read request finite state machine.

If the transfer is a burst request, data need to be written first to the WrFIFO followed by issuing the write request. This is done in the WR_BURST state. The address must also be aligned to the transfer size to avoid unpredictable performance (such as writing to the wrong address range). The size of the burst is specified by the XIL_NPI_Size signal. If a request does not fit within these burst sizes, it must be split into smaller requests. For example, writing a total of 88 Bytes would require a burst of 64 Bytes plus 16 Bytes plus a final transfer of 8 Bytes. This problem is further complicated by the need

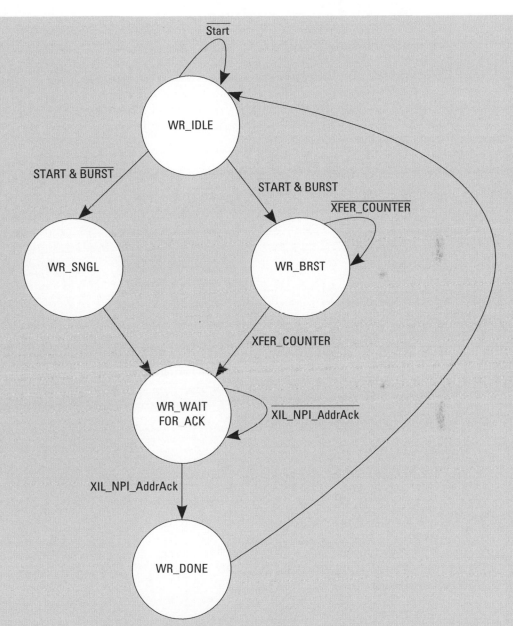

Figure 6.21. Native Port Interface write request finite state machine.

to align the address on the data boundary. A simple write request finite state machine can ignore these requests and pass the requirement on to the component issuing the requests (which could be the processor or some other compute core). A more complicated solution would involve the write request finite state machine managing these requirements. For now we rely on the request to be aligned properly.

Regardless of the transfer type, the next state, WR_WAIT_FOR_ACK, waits until the MPMC acknowledges the request. Once the acknowledgment is received the write request finite state machine deasserts the request and transitions to the WR_DONE state, signaling that the transfer has finished, and returns to the WR_IDLE state.

NPI Microprocessor Peripheral Description   After the NPI logic has been written and tested in simulation, it is time to modify the microprocessor peripheral description (MPD) file. We do this to support connectivity between the hardware core and MPMC in Xilinx Platform Studio. We start with the bus interface, which indicates that the NPI signals are part of the XIL_NPI bus standard.

```
## Bus Interfaces
BUS_INTERFACE BUS = XIL_NPI, BUS_TYPE = INITIATOR, BUS_STD = XIL_NPI
```

Now all of the NPI ports can be added and associated with XIL_NPI bus. These are all of the necessary NPI ports that are listed in the hardware core entity description. The default assignment is used to connect each signal to the corresponding bus signal. If left unassigned, the signal will be unconnected during PlatGen, potentially resulting in a synthesis error.

```
## Ports
PORT XIL_NPI_Addr = "Addr", DIR = O, VEC = [31:0], ENDIAN = LITTLE, BUS = XIL_NPI
PORT XIL_NPI_AddrReq = "AddrReq", DIR = O, BUS = XIL_NPI
PORT XIL_NPI_AddrAck = "AddrAck", DIR = I, BUS = XIL_NPI
PORT XIL_NPI_RNW = "RNW", DIR = O, BUS = XIL_NPI
PORT XIL_NPI_Size = "Size", DIR = O, VEC = [3:0], BUS = XIL_NPI
PORT XIL_NPI_WrFIFO_Data = "WrFIFO_Data", DIR = O, VEC = [63:0], ENDIAN = LITTLE, BUS = XIL_NPI
PORT XIL_NPI_WrFIFO_BE = "WrFIFO_BE", DIR = O, VEC = [7:0], ENDIAN = LITTLE, BUS = XIL_NPI
PORT XIL_NPI_WrFIFO_Push = "WrFIFO_Push", DIR = O, BUS = XIL_NPI
PORT XIL_NPI_RdFIFO_Data = "RdFIFO_Data", DIR = I, VEC = [63:0], ENDIAN = LITTLE, BUS = XIL_NPI
PORT XIL_NPI_RdFIFO_Pop = "RdFIFO_Pop", DIR = O, BUS = XIL_NPI
PORT XIL_NPI_RdFIFO_RdWdAddr = "RdFIFO_RdWdAddr", DIR = I, VEC = [3:0], ENDIAN = LITTLE, BUS = XIL_NPI
PORT XIL_NPI_WrFIFO_Empty = "WrFIFO_Empty", DIR = I, BUS = XIL_NPI
PORT XIL_NPI_WrFIFO_AlmostFull = "WrFIFO_AlmostFull", DIR = I, BUS = XIL_NPI
PORT XIL_NPI_WrFIFO_Flush = "WrFIFO_Flush", DIR = O, BUS = XIL_NPI
PORT XIL_NPI_RdFIFO_Empty = "RdFIFO_Empty", DIR = I, BUS = XIL_NPI
PORT XIL_NPI_RdFIFO_Flush = "RdFIFO_Flush", DIR = O, BUS = XIL_NPI
PORT XIL_NPI_RdFIFO_Latency = "RdFIFO_Latency", DIR = I, VEC = [1:0], BUS = XIL_NPI
PORT XIL_NPI_RdModWr = "RdModWr", DIR = O, BUS = XIL_NPI
PORT XIL_NPI_InitDone = "InitDone", DIR = I, BUS = XIL_NPI
PORT XIL_NPI_Clk = "Clk", DIR = I, BUS = XIL_NPI
PORT XIL_NPI_Rst = "Rst", DIR = I, BUS = XIL_NPI
```

Xilinx Project Modifications   With all of the modifications to the HDL and MPD file it is now possible to connect the hardware core to the MPMC in the Xilinx Platform Studio. To do this the MPMC must first be modified to include an additional port of type *NPI*.

From within XPS, under the *System Assembly View*, select the MPMC (by default the instance name is DDR2_SDRAM_DIMM0 for ML-510 base systems) and choose *Configure IP*.

The MPMC supports up to eight ports. The default configuration is for Port 0 to be configured as PLBV46. What this means is that Port 0 can be connected as a slave to a single instance of the Processor Local Bus. We need to add a second port of type NPI. Once set, our MPMC will resemble Figure 6.22.

Close the MPMC configuration window and expand MPMC interfaces to see SPLB0 connected to plb_v46_0 and MPMC_PIM1 listed as *No Connection*. Selecting the drop-down menu, we can change the connection to the NPI bus of the hardware core.

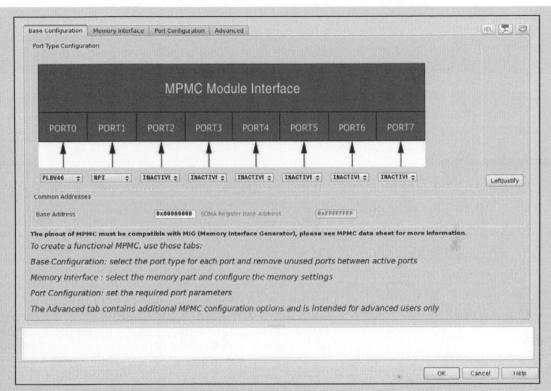

Figure 6.22. Xilinx Multi-Port Memory Controller configured with two ports: PLB and NPI.

An important note regarding ML-510 designs: by default the Base System Builder wizard sets the memory controller for DIMM0 to the PPC440 Memory Controller. This is not the same as the Multi-Port Memory Controller. With this memory controller you will not be able to add additional ports, such as the Native Port Interface.

As is evident by the amount of material covered within this section, there are a wide variety of ways to not only implement memory, but access memory as well. It is hoped that these examples act as a catalyst to explore and develop efficient memory access interfaces within your designs. Whether implementing the simpler programmable I/O interface, using a central DMA controller, or supporting memory access from within the hardware core, it is necessary to recognize the trade-offs between resources and performance.

Exercises

P6.1. Considering Figure 6.1, if the amount of time to perform a multplication was reduced by a factor of 2, draw the corresponding system, including any necessary buffers to maintain a pipeline balance.

P6.2. Draw the interface between two components where one is producing data and the second is consuming data. Be sure to follow the Kahn Process Network model.

P6.3. Describe the similarities and differences between synchronous and asynchronous design.

P6.4. Draw the interface for a FIFO. Compare this interface to that of a BRAM.

P6.5. Write the VHDL that will instantiate a BRAM and function like a FIFO. Include in this design a mechanism to report to the user the number of elements currently stored in the FIFO.

P6.6. Write an application that will use the Central DMA controller to read 512 Bytes of data from DDR2 and write it to a hardware core. Assume data are stored at 0×0123000 and must be written to $0 \times B001000$.

P6.7. Compare the performance of a system that is operating in CMDA mode with that of a system running as a bus master that is able to access memory on its own.

P6.8. Implement a Native Port Interface design that will support direct memory access reads and writes to the MPMC without the aid of the processor.

References

Kahn, G. (1974). "The semantics of simple language for parallel programming." In: Rosenfeld, J. L. (Ed.), *Information Processing '74* (pp. 471–475). *Proceeding of the IFIP Congress*. North Holland, Amsterdam.

Xilinx, Inc. (2005). *LocalLink interface specification (SP006) v2.0.* San Jose, CA: Xilinx, Inc.

Xilinx, Inc. (2009a). *Answer record 24912 — 11.1 EDK, MPMC v5.00.a — How do I create an NPI Core and connect it to MPMC in EDK?.* http://www.xilinx.com, last accessed June 2010.

Xilinx, Inc. (2009b). *Block memory generator data sheet (DS512) v3.3.* San Jose, CA: Xilinx, Inc.

Xilinx, Inc. (2009c). *Distributed memory generator (DS322) v4.3.* San Jose, CA: Xilinx, Inc.

Xilinx, Inc. (2009d). *FIFO generator data sheet (DS317) v5.3.* San Jose, CA: Xilinx, Inc.

Xilinx, Inc. (2009e). *Multi-port memory controller (MPMC) v5.04.a.* San Jose, CA: Xilinx, Inc.

Xilinx, Inc. (2009f). *XPS central DMA controller (DS579) v2.01b.* San Jose, CA: Xilinx, Inc.

## 7

OUTSIDE WORLD

The Network Is the Computer^{TM}

(company slogan)
Sun Microsystems, Inc.

Some have predicted that computers will "disappear" in the future. This is not some neo-Luddite forecast but rather the suggestion that computers will continue to shrink in size and increase in number such that all of our computing functions will be ubiquitous and pervasive. That is, they will disappear into our everyday articles. The idea is that there will be many embedded computing systems distributed throughout things such as our clothes, entertainment devices, home appliances, packaging, public infrastructure, and so on — making the need for conventional, general-purpose computers moot. The strength of these small devices, however, will not be in their individual ability but rather in their collective functionality. Hence, these small systems will need to communicate with many kinds of devices and via a variety of networking protocols.

Until recently, embedded systems designers were often forced to use very light weight, custom codes and protocols because hardware capabilities of the target machine were limited. Today, more capable hardware allows the embedded systems programmer to use much more sophisticated protocols, such as the Universal Serial Bus (USB) and even the Internet Protocol (IP) that underlies the World Wide Web. Indeed, standard protocols such as HTTP and free software such as Linux and a Web server such as Apache are becoming increasingly important for even the simplest electronic devices.

There are several compelling reasons to use these standards. First, using existing protocols means that a custom protocol does not need to be created. Furthermore, using existing protocols often means that reference software already exists — which, if it can be reused, can save a considerable amount of development cost. Indeed, the cost savings include the initial investment in developing the software for the embedded system, future improvements, *and* the client software used to communicate with the embedded system. For example, consider a power distribution unit (PDU) in a server rack containing an embedded Web server. If this is the case, then a system administrator only needs a Web browser to control power to her servers. In general, building an embedded system that uses standard protocols whenever possible increases interoperability in the new era of ubiquitous, persuasive embedded systems.

Of course, computer networking is itself a huge subject and multivolume books have been written about it — there is no way one chapter can provide a comprehensive treatment. However, we can survey the basic technologies starting with the low-speed serial technology and protocols that emerged in the 1970s and have been common in embedded systems up to the latest high-speed technologies and sophisticated protocols used in high-end systems today. Rather than be comprehensive, the goals of this chapter are to enumerate the different technologies and protocols, discuss the common use cases, and provide references to more detailed information.

7.1. Point-to-Point Communication

A decade ago, a subset of the RS-232 serial communication standard was the de facto mechanism to connect embedded systems to other computing devices [Electronics Industries Association, 1969]. It was often used for debugging during development and user configuration in the field. This asynchronous serial communication protocol required three wires — transmit, receive, and a common ground. It was simple and the IC device that implements the protocol (a UART — a Universal Asynchronous Receiver/Transmitter) was inexpensive.

RS-232 interfaces have been almost completely replaced with Universal Serial Bus (USB) [USB Implementers Forum (USB-IF), 2010a,b] interfaces in modern consumer electronics because USB offers higher speeds, more capabilities, and a smaller physical form factor than RS-232. Unfortunately, with these advantages comes considerable complexity in terms of hardware and software. Using USB on a client device usually requires either buying a controller chip or integrating a USB controller into the FPGA fabric. However, the biggest disadvantage for using USB on educational embedded systems projects is the drivers needed on the host PC. Given that all modern operating systems include the necessary drivers and terminal emulation software for RS-232-like terminal devices and the examples presented in this book only require terminal access to the embedded system, there is no need for the extra overhead of USB.

7.1.1. RS-232

Among existing embedded systems, the most common serial communication protocol in use is RS-232. This standard has evolved significantly over time and is the "COMx" port that used to be standard on desktop PCs in the 1980s and 1990s. It began as a protocol to connect dumb terminals (Data Terminal Equipment or DTE) to acoustic modems (Data Communication Equipment or DCE). The standard was not developed for peers to communicate but rather there is a specific orientation. To attach two DTEs together one has to use a "null" modem that effectively cancels two DCEs by interchanging the transmit and receive lines (i.e., the transmit of one end is connected to the receive of the other and vice versa). Along with these two lines, a common ground is also required. The standard goes on to define many more pins but most embedded systems only use this subset. (Sometimes hardware flow control is used as well and that includes two more wires.)

The basic hardware component to support the RS-232 is called a UART — a Universal Asynchronous Receiver/Transmitter. This device is essentially a parallel-to-serial shift register. The protocol involves the transmission of a series of spaces and marks, which represent 0 and 1, respectively. The protocol defines voltages that range from +25 to −25 V with a voltage between +3 and +25 V representing a space and −3 to −25 V representing a mark. Note that this is drastically different from the 0 to 5 V that is common in Transistor-Transistor Logic (TTL) so often line drivers (a separate IC) have to be used to convert TTL signals to RS-232 signals.

Because there is no common clock between the two devices, a receiver does not know when a message may begin. The protocol specifies that the transmitter holds a space (say, +15 V) until it is ready to transmit a word. The transmitter then sends one mark (to signal that a message is coming) followed by data followed by an optional parity bit. The protocol specifies that the message must be terminated by one or two stop bits, which are spaces. At this point the transmitter holds a space on the line until the next message is ready to be transmitted.

The biggest issue one has to deal with when using RS-232 is that all of the options have to be agreed upon from the beginning. Both sides have to be configured in advance to know the baud rate; how many data bits are in a message; if odd, even, or no parity encoding is used; and how many stop bits to expect. This set of parameters is often abbreviated in a format such as `9600 8N1`, which represents one of the most common RS-232 parameter sets: 9600 baud with 8 data bits, no parity bit, and 1 stop bit.

Newer protocols such as USB include negotiation on things such as signal rates so that the fastest transfer rate that both the host and the client can support is used.

7.1.2. Other Low-Speed Communication

Between RS-232 and USB in terms of complexity, a number of simple protocols are used to connect components within an embedded system. Two common examples of these are Interintegrated Circuit (abbreviated IIC or I^2C) and Serial Peripheral Interface (SPI). Both of these protocols share the concept that one device is the master (or host) and that there are one or more slaves (or peripherals).

The I^2C technology was developed by Philips Semiconductor, and the complete bus specification is available on the Web [Philips Semiconductor, Inc., 2009]. I^2C uses two wires: serial data and serial clock. SPI uses four wires: chip select, serial clock, data in, and data out. Both are commonly used to interface temperature or voltage sensors in embedded systems. While there are some trade-offs of which the low-speed communication protocol is better in terms of power usage and physical space, the decision to use one protocol over the other is typically driven only by what slave devices you need to connect in this embedded system. If the best analog-to-digital converter for your project uses an SPI protocol, then you should include an SPI master in your system.

This also leads to one significant advantage of using an FPGA for an embedded system over off-the-shelf microprocessors. If the project requires an I^2C, an SPI, two UARTs, Ethernet, and other interfaces, it is very easy to add these components to an FPGA-based design. However, if you were to use an off-the-shelf microcontroller, it may be difficult to find a chip with exactly the right number and kind of interfaces required.

7.2. Internetworking Communication

7.2.1. Concepts

As just mentioned, the range of technologies and protocols that encompass what is called "computer networking" is enormous. Within the field, individuals often specialize into very different subfields, ranging from Web application programming to Internet router specialists. Also contributing to the large knowledge base is the fact that different technologies can be mixed and matched. For example, when people think Ethernet, many assume TCP/IP. However, other protocols run over (and co-exist with TCP/IP) on an Ethernet network. Furthermore, TCP/IP can also be used on local area networks other than just Ethernet. This section focuses specifically on the Internet Protocol (IP) and the most common higher level protocols that use Transmission Control Protocol (TCP). The chief distinction is that IP provides a technology-independent packet-based mechanism to send information, whereas TCP is used to deliver an in-order stream of information from one host to another. The other sometimes used component of this family is UDP, which provides a TCP-like application interface but is oriented around packet-based communication.

The Internet Protocol was designed to span multiple computer networks, all of which may have different characteristics (in terms of reliability, speed, and technical specifications). To manage this complexity and to maximize the reuse software, the networking functionality is often divided into layers (and the collection of layers is called a network protocol stack). Figure 7.1 shows the network protocol stack for a typical TCP/IP system.

In networking terms, a single machine is called a ***host*** and we assume that there are two applications running on two hosts that wish to communicate. If the two hosts are directly connected to one another, then all the layers of the protocol stack are probably overkill. However, very often it is the case that the two hosts are not directly connected and so intermediate devices are needed to relay information. Figure 7.2 shows two hosts, *A* and *B*, that are communicating through an intermediate device. If the two hosts are on

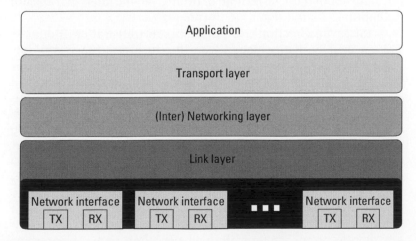

Figure 7.1. A network protocol stack.

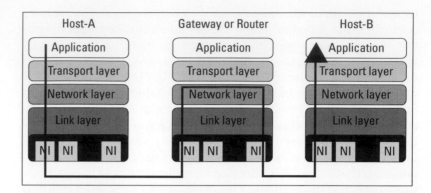

Figure 7.2. Two hosts communicating with TCP/IP.

the same network, the intermediate device is called a *switch* (or, in some cases, a *bridge*). If the communication is across two different technologies, or two different networks, the intermediate device is called a *router*. Finally, a gateway is a router that generally connects a network to the rest of the Internet. (If the address of a host is known but its location is not, the usual behavior is to send the packet to a gateway device that forwards it on.)

The network protocol stack is organized by functionality. The lowest level — the *physical layer* — is responsible for generating a real-world, physical signal. It might use a line driver to create physical voltages (such as +15 and −15 V, as is the case of RS-232) or a transceiver to pulse light into a fiber-optic cable. The components that create these signals are often designated PHY (pronounced *fy* and rhymes with pie) and these components may be part of a Platform FPGA (as is the case with the multi-gigabit transceivers on a Virtex 5) or a separate integrated circuit. The PHY does not know anything about data, its format, or timing. Its sole job is to translate a computer's digital signals to the analog world.

At this layer, direction is important so the transmitters (denoted TX) are usually distinguished from the receivers (RX). Also, because it is so closely associated with the physical world, these components are almost exclusively physical, analog components. The receiver and transmitter together are collectively known as the network interface.

As Figure 7.1 suggests, it is not uncommon to have multiple network interfaces. Indeed, many devices today might support multiple network interfaces. For example, unique devices such as Earth-orbiting satellites typically have at least one network interface to communicate with ground stations and may have another one to talk with the satellite's payload, typically an instrument of some sort. A more down-to-earth example would be a modern mobile phone. It typically will have one network interface to talk with a cell tower. But increasingly, phones have a second Bluetooth interface that is used for short-range wireless connections between the phone and a headset.

Before we leave the lowest layer, it is worthwhile to make mention of network topology. The ***network topology*** is nothing more than the underlying point-to-point relationship between each host on the network, which — as any reader who has made it this far into the book recognizes — can be modeled as a graph. Previously, most topologies are ***regular***, which means that from the point of view of any vertex in the graph, the relative positions of every other vertex in the network are the same. However, this regularity is an artifact of the way we have built networks for the last 50 years. That is, engineers have imposed a structure to make the system manageable. As we lurch toward ubiquitous, embedded computing, this regularity breaks down. For example, with mobile phones the network is now extended by many little "one-hop" links over a Bluetooth connect. Likewise, the space-based example of the previous paragraph can be extended to an emerging commercial technology of launching nanosatellites. These satellites are typical low mass — which makes them easy to launch — but limit what kind of instruments can fly as payload. So one current proposal suggests that we launch constellations of these satellites that work together. Because their exact, relative position is difficult to maneuver, they naturally form *ad hoc* networks where the edges cannot be determined before the satellites are in orbit. Dealing with the network that arises from circumstances is handled at a high layer but it occurs because we have multiple network interfaces.

Onward! We have what is known as the (data) ***link layer***. In Figure 7.1, the link layer is depicted as being closely integrated with the physical layer. This is because some technologies (such as the early versions of Ethernet) in essence required that it be implemented in hardware. This layer had the responsibility of not only communicating with the physical layer, but also observing the physical layer for anomalies or errors. For example, the earliest versions of Ethernet relied upon a common media — a coax cable — as a shared resource for communication. The purpose of the link layer is to allow two directly connected hosts to communicate. It manages a link between two hosts in terms of checking for transmission errors and flow control.

Often the link layer is divided into upper and lower halves. The lower half — called the Media Access Controller (or MAC) — is responsible for talking to the PHY. In modern versions of the Ethernet standard, several well-defined interfaces exist between the MAC and the PHY. The original MII is TTL/CMOS (0 to +5.5 V) signals and was common in 10/100 (FastEthernet) networks. GMII and RGMII were introduced for Gigabit Ethernet because additional bandwidth was required (RGMII is a variation that uses half the number of data lines but transfers data on both rising and falling edges of the clock). Most recently is Serial Gigabit Media Independent Interface or SGMII, which uses high-speed serial transceivers style of communication between the lower half of the link layer and the PHY. Details of these protocols are not critical to the Platform FPGA designer, but one has to be aware of them because often the MAC is implemented in the FPGA and PHY is off-chip so one has to match the right protocol to the off-chip hardware. Logical Link Control (LLC) is the upper half of the link layer. It is responsible for flow control between two nodes in the network.

Above the link layer is the ***network layer***. The purpose of this layer is to provide connectivity beyond two directly connected hosts. For example, in Figure 7.1, there are three network devices. In order for A to communicate with C, A transmits packets to B via the link layer. Those packets are passed to the network layer, which makes decisions about how to advance the packet toward its destination. Often this means retransmission through the link layer to another device. The most common network layer is the Internet Protocol (the IP in TCP/IP). This standard includes the usual data packet plus control packets used to manage the network. At this layer, communication can bridge different physical communication channels. For example, a packet can arrive over a link layer that interfaces to a serial communication channel and forward the packet to a link layer that interfaces to Ethernet. Novel's IPX is another example of a network layer protocol.

However, applications usually do not interface with the network layer directly. In the Open Source Interface (OSI) model, there are several layers in between the application and the network layerISO/IEC JTC1 (1984). However, the best known approach is the Internet Protocol and it has one layer — the ***transportation layer***. There are multiple interfaces to the transport layer associated with IP. The most common is the Transportation Control Protocol or TCP. This protocol provides an error-free end-to-end stream service. That is, data are written to the stream on one end and arrive (in order) at the other end. These data may have been broken up into packets, fragmented, reassembled, duplicated, and traversed different routes along the way but the network layer hides all of that and provides a reliable data path for the application. Another network layer interface is the User Datagram Protocol (UDP). This provides unreliable, packet-based communication between applications. It uses the Internet Protocol, and while a user datagram will always be delivered intact, it may have been fragmented and reassembled en route. As mentioned, this is an unreliable service and some datagrams might not be delivered without notice to the application.

7.2.2. Application Interface

Application access to the network layer is most often through the Berkeley socket interface. Our goal is not to provide a comprehensive guide here, rather it is to provide enough information about a basic server and client to jump start any project that needs a custom protocol. Also, a basic knowledge of how an Internet application works will help demystify existing applications and their configuration files.

The simplest Internet application is a TCP/IP client/server. The basic idea is that the client will initiate contact with a server by referring to the service's IP address and a port number. Because a port is associated with no more than one server on a specific machine, this provides a unique address for a server. Not every machine will associate the same port number with same server, but most common servers have a conventional port number. For example, an HTTP server is usually found on port 80, a secure shell server uses 22 by default, and so on. Sometimes the port will be explicitly changed to avoid unwanted, automated probes of a system. A file /etc/services documents (potential) services on a specific host and a library call getservbyname allows a program search for a port associated with a service on the local machine. By using the socket interface described next, the operating system on the client's machine and the operating system on the server's machine set up a two-way communication. The network protocol stack described previously provides an error-free stream- (versus packet-) based communication channel.

The programming interface uses file descriptors, just like many of the system calls we are already familiar with (open/read/write). However, one can use the C library call freopen to get an fprintf or fscanf style interface.

A typical application-specific server application will ordinarily start during the boot process. It generally does not require root privileges unless it uses a port below 1024.[1] The server's first task is to request a specific port from the OS and configure it. This is accomplished with the socket, bind, and listen system calls. The way they are typically used is illustrated in Listing 7.1. The

---

[1] This was a weak form of security employed when there were just a handful of Internet hosts and the administrators of hosts could be trusted but not necessarily individual users. Now it is simply a legacy requirement.

```
int sockfd ;
struct sockaddr_in serv_addr ;
int n ;

/* create file descriptor */
sockfd = socket(AF_INET, SOCK_STREAM, 0) ;
if( sockfd < 0 )
    error("ERROR opening socket") ;

/* clear address, then set family (Internet),
   and port, specify client IP address that can
   connect (any), and inform OS */
bzero((char *) &serv_addr, sizeof(serv_addr)) ;
serv_addr.sin_family = AF_INET ;
serv_addr.sin_port = htons(portno) ;
serv_addr.sin_addr.s_addr = INADDR_ANY ;
if( bind(sockfd,(struct sockaddr *)&serv_addr,
        sizeof(serv_addr)) < 0 )
    error("ERROR on binding") ;

/* set queue length */
listen(sockfd,5) ;
```

Listing 7.1. Setting up a server.

first system call (socket) creates a file descriptor to manage client connections to the server. The next two, bind and listen, tell the operating system what port the server would like to be associated with (bind) and sets a queue length (listen). In this example, five clients can line up to access the server before the OS immediately rejects client requests. (We will explain the htons call shortly.)

The next step is for the server to wait for a client connection. This is typically performed inside of an endless loop. Listing 7.2 illustrates the basic algorithm. The accept system call will block the server process until some client has initiated a connection. The first argument is the file descriptor that was just configured and a structure to hold the client's address. The return value is a *new* file descriptor that is used for communication with this client. (The sockfd can then be used again in another accept system call.) When communication with the client has finished, the file descriptor is closed and the communication channel is formally torn down.

Listing 7.3 shows the client application also uses the socket system call to create a socket file descriptor. Unlike the server, this is the only file descriptor needed to communicate with the server. To get initiate communication with the server, the client (1) creates the socket and gets the server IP address with a gethostbyname library call, which is copied to the socket address structure. The

```
int newsockfd, clilen;
struct sockaddr_in cli_addr;

while( 1 ) {
    clilen = sizeof(cli_addr) ;
    newsockfd = accept(sockfd,(struct sockaddr *)&cli_addr,&clilen) ;
    if (newsockfd < 0)
        error("ERROR on accept") ;
    do_work(newsockfd) ;
    close(newsockfd) ;
}
```

Listing 7.2. Main server loop.

```
struct sockaddr_in serv_addr;
struct hostent *server;

sockfd = socket(AF_INET,SOCK_STREAM,0) ;
server = gethostbyname(hostname) ;
bzero((char *)&serv_addr,sizeof(serv_addr)) ;
serv_addr.sin_family = AF_INET ;
bcopy((char *)server->h_addr,
            (char *)&serv_addr.sin_addr.s_addr,
            server->h_length);
        serv_addr.sin_port = htons(portno) ;
connect(sockfd,(struct sockaddr *)&serv_addr,
            sizeof(serv_addr)) ;
```

Listing 7.3. Client application.

last step is to call the connect system call, which will set up the communication channel. On return, the sockfd file descriptor can be used to read and write to the server.

In the host and client, we use a library call, htons, which handles the byte order of an integer. The name is short for host to network (16-bit integer). Likewise, there is an htonl, which is short for host to network (32-bit integer). When the host machine uses the same byte ordering as the network standard, these functions do nothing. If they differ, the subroutines reverse the byte order. There is also an ntohs and ntohl used for converting network standard to host byte ordering of integers, if needed.

An important variant of the client/server application described here is a UNIX dæmon. In this system, the client operates exactly the same. What is different is how the server is started. As described previously, we assume the server was started when the machine was booted. This is perfectly appropriate for an application-specific embedded system. However, for many UNIX-based machines, we do not want to start a server for every single service that the host might have a server for. This led to "super server," which is a single application whose job is to set up a socket for a set of servers (essentially doing the steps illustrated in Listing 7.1). When a client appears on any of the ports that the super server is watching, the super server accepts the connect and then starts the real server with the communication socket as the standard input/output. In doing so, the server process only uses system resources when there is a client requesting a connection. It is easy to build a server that, based on a command line argument, will start either as a dæmon or as an ordinary server. There are multiple implementations of the "super server." The original was called `inetd` (Internet dæmon super server). The package `xinetd` is a popular variation. Busybox also has a simple implementation as well. We mention this because both styles of server are common. Also, it is worth pointing out the difference between the `/etc/inetd.conf` and the `/etc/inittab`. The latter is for starting processes at boot; the former launches Internet servers on demand.

Once a communication channel is established, any data can be streamed across this channel. Even though binary data can be transmitted, older protocols were often 7-bit ASCII based and organized around "lines" that were terminated with a newline. Newer protocols tend to use a syntax similar to extended Markup Language (XML). For very simple protocols, the former can be easy to parse (read line by line from the file descriptor). For more complex, structured data, XML works very well but usually requires additional software to parse the format. Regardless, any data sent over a communication channel using Berkeley sockets as described here are *plaintext*. That is, data are not encrypted and it is very easy for an anonymous, third party to view these data. In some cases, such as public Web sites, there is no need for privacy but if there is any sensitive information that needs to be transmitted, encryption is essential.

On a final note, very often a custom protocol is not needed. There are a number of well-defined protocols with open source software available that will support many of the common tasks. Moreover, they often have the benefit of using encrypted transmission or at least offer an encrypted variant. We review the most common ones for embedded systems next.

7.2.3. Higher Level Protocols

By itself, the IP protocol stack is convenient but does not provide much more than connectivity for an embedded system. However, on top of these protocols (most often over TCP but sometimes

UDP), several higher level protocols have evolved. Many of these protocols were not formally standardized but interoperability was achieved through an informal "request for comment" (RFC) memo and the sharing of source code. Often the first software implementation became the reference code. The low-level details of these protocols are beyond the scope of this chapter. What is important for the embedded systems designer is to know what protocols exist, their high-level characteristics, and what is required to use them. Hence, we want to spend a little bit of time surveying the protocols found most often in embedded systems.

Telnet

One of the earliest protocols developed using TCP/IP is called `telnet`. It was proposed in RFC15 in 1969 and then extended over time; it was one of the first standards adopted by the Internet Engineering Task Force (IETF), which formalized the RFC process. Telnet uses TCP/IP to provide an (insecure) bidirectional stream between a client and a telnet server (or dæmon) running on the embedded device. This server can either ask for authentication (username/password) or simply start a command line based configuration or diagnostic application. This protocol uses port 23 on a standard UNIX implementation and has been largely replaced by Secure Shell (ssh and sshd) for workstations and servers because telnet passes all data (including usernames and passwords) unencrypted over the network. However, for embedded systems — where a password might not be needed — this is a great, lightweight protocol. For this reason, it is often the first protocol implemented in a "network-ready" embedded system. For example, adding an Ethernet interface to a printer makes it a "network printer." Now that the printer is on the network, one could use telnet to check toner levels, manually set the network configuration, etc. Again, it must be stressed that telnet should not be used for sensitive data.

The telnet server is implemented as Internet dæmon. On an embedded Linux project, the first step is to make sure the network interface has been configured. This is usually done when the system is booted (see next section). Next, we need to tell the network super server to watch for connections to this service. Most sample `inetd` configuration files have this required line commented out. Simply edit `/etc/inetd.conf`, find the line below, and remove the leading pound sign.

```
#telnet stream  tcp     nowait  root    /usr/sbin/telnetd telnetd -i
```

If a client tries to connect to port 23, the `inetd` process will accept the connection and then start the "real" telnetd.

World Wide Web

The World Wide Web (WWW) evolved out a sequence of efforts in the late 1980s and early 1990s to make scientific (and other) data publicly accessible. Much earlier Internet efforts resulted in the File Transfer Protocol (FTP), which, like telnet, was a simple, insecure mechanism used to transfer files from one host to another over TCP/IP. With the addition of no-password accounts, this allowed the transmission of raw data but there was no imposed structure to these data. The consumer of these data had to know the file name and directory organization conventions used by the producer. What the World Wide Web did was introduce a server protocol, Hyper-Text Transfer Protocol or HTTP, that transmitted files to a file in the Hyper-Text Mark-up Language or HTML. The HTML files included display as well as references to non-HTML data files. This allowed a client to not just copy or display the raw data but present it to the user. This quickly led to innovations that allowed the server to interact with a user client. The "client" for World Wide Web servers is commonly known as a Web browser.

This technology is a great boon for embedded systems developers. Custom protocols — whether over a direct communication link such as RS-232 or the Internet — mean that the embedded systems designer has to develop both the server (running on the embedded system) *and the client*. This puts the designer in the untenable position of either supporting the client on multiple potential hosts or limiting the embedded systems potential market. With HTTP and HTML, the designer simply has to follow the prevailing standards and their embedded system works with any Web browser.

There are a number of Web server implementations. Perhaps the most common is Apache. It is a vast project that has spawned a number of supporting projects. It is feasible to cross-compile Apache and run it on a Platform FPGA design. However, there have been numerous implementations that run the gamut from purely academic exercises to implementations carefully designed to run on limited resource-embedded systems. Often, embedded systems do not have the resources to support a full Apache server. There are many lightweight servers such as lighttpd, `http://www.lighttpd.net/`, and thttpd, `http://www.acme.com/software/thttpd/`, that may be more appropriate for an embedded system. Also check with your FPGA vendor for servers that are already included with their tools.

Simple Network Management Protocol

Perhaps the next most important network protocol for embedded systems is the Simple Network Management Protocol (SNMP). It is also unique because it is one of the few common protocols based on the UDP (packet-based versus TCP's stream based) transport layer.

The protocol was originally developed to interrogate and manage autonomous network devices such as routers and switches. For example, one might want to know the number of packets sent and received on a particular switch port or the device as a whole. Alternatively, one might want to know what speed a port autonegotiated to when a host was connected. SNMP allows an administrator to do this through a client over IP. It includes a "call back" mechanism whereby the network device can inform the administrator that some event has occurred. It turns out that the design was general enough to support a wide range of equipment, and a number of different network-ready devices use the protocol. For example, one can monitor the current flowing through a Power Distribution Unit using SNMP.

Network File System

The Network File System (NFS) protocol has been used for more than two decades to make a remote filesystem (running on a server) appear like a local filesystem. To use this protocol requires a certain degree of trust between the server and the client. As such, it is generally not useful for embedded systems in the field.

However, it can be extremely useful during development. The end product might be limited in the amount or speed of secondary storage. Having the system under test on a network and configured to mount a home directory or the root filesystem from a server in the lab is a huge benefit. Once the operating system is configured and stable, the embedded system can be left running continuously. On a workstation, the application work can be repeatedly cross-compiled and tested without (manually) copying the executable between workstation and embedded system. Moreover, it allows the filesystem to grow arbitrarily large while under development. Then, when everything is working, it can be cleaned up to fit the space constraints of the embedded system's native filesystem.

Network Time Protocol

One of the servers that the DHCP protocol can provide is a Network Time Protocol (NTP) server. (The DHCP protocol is used to set up an embedded system on an unknown network; it is described next.) This server (along with an array of public servers on the

Internet) gives the embedded system an opportunity to determine the time automatically. In the 1970s and 1980s, Video Cassette Recorders (VCRs) all across the world endlessly flashed 00:00 for the time. No matter how simple the interface — it was difficult to get users to set the time, let alone reset it after power was lost! But there are many situations when knowing the time is a very useful thing for an embedded system. Compiling an NTP client that gets the time (and periodically checks it for consistency) is a valuable software feature.

Secure Socket Layer

To make plaintext network traffic secure, there is the Secure Socket Layer (SSL). With SSL, a client and server connect via a hand-shaking protocol to configure the secure connection. This is done through the exchange of shared public keys and certificates, which are used to generate a session key. Once connected, the client and server can communicate securely through the encoding and decoding of the encrypted data based on the session key. Open source library implementations, such as OpenSSL, can provide an encryption layer between the application and the transport layers. OpenSSL is maintained by a large community of developer volunteers and is based on the SSLeay library. Because OpenSSL is open source, its use requires acceptance of the dual license, which covers both conditions of the OpenSSL License and the original SSLeay license.

7.2.4. Operating System Configuration

Most of the time, an embedded system will want to bring up a network interface (i.e., configure and activate it) when the system boots and bring it down (deconfigure) during an orderly shutdown of the system. Traditionally, this has been accomplished with two commands: `ifconfig` and `route`. The first configures a network interface and requires one argument followed by one or more options. The second command updates the kernel's network routing table. However, to do so means that the embedded systems developer has to know in advance what network the device will be on, the assigned IP address, a gateway to the rest of the Internet, etc. So, in practice, these commands are handy during development but in the field, high-level protocols (such as DHCP described later) are needed.

Generally, these commands are being phased out. As networking has become more complex, two programs (called `ip` and `tc`) have evolved to handle that complexity, advanced routing, and traffic shaping. The `ip` command, which comes as part of the iproute2 package, can do everything that `ipconfig` can and more.

The interface also directly supports a number of network configurations that didn't exist when ifconfig was written. If you don't want to install the whole iproute2 package, Busybox provides an `ip` command that makes porting start up scripts from desktop machines a little easier.

The Dynamic Host Configuration Protocol (DHCP) and BOOTP protocols allow a deconfigured network interface to broadcast a request for information about its network parameters. These protocols rely on a server existing somewhere on the network that watches for such broadcast messages. With Ethernet networks, every network interface has a MAC address, also known as its hardware address. Ethernet equipment providers acquire blocks of these addresses and each product receives a unique address. Based on the server configuration and the hardware address found in the broadcast Ethernet packet, the server can (1) assign an IP address, (2) choose to assign an address from a pool of available addresses, or (3) ignore the request. The information that the server can provide can be as simple as the IP address (and length of time, after which it has to broadcast another request). But it can also provide other useful network information, such as a netmask and a gateway host to the Internet. It also usually provides the IP address of other useful servers: a domain name server, a network time server, and so on.

Nowadays, DHCP is a requirement for embedded systems. Without it, the end user would have to know and be able to enter the network information. In the case of embedded systems destined for a home network, end users might not have any idea how their home router works. (They just followed the three-step setup and it "just worked.") In the case of mobile devices, the end user might not know anything about the coffee shop's network. In both cases, the DHCP server exists — it's just up to the embedded system to use it.

Finally, it is worth mentioning the ifupdown package. It provides two commands — `ifup` and `ifdown` — that read a `/etc/network/interfaces` configuration file. It still uses the primitive (ifconfig and route) commands but has three advantages: (1) if there are multiple interfaces, all of the information about how to set up the network is collected into one configuration file, (2) it integrates nicely with dynamic changes — like someone plugging in a USB wireless network interface, and (3) the package "knows" about many of these low-level networking systems.

Chapter in Review

Two decades ago, relatively few embedded systems had networking or communication requirements. Simple serial protocols were sufficient for configuration and sharing of data. Nowadays, embedded systems often have multiple network interfaces and, out of market necessity, have to support a rich set of complex high-level protocols. With Open Source implementations available, embedded systems designers do not have to understand the details of these protocols. However, designers do have to understand how the pieces fit together and when it is appropriate to use existing software. The goal of this chapter was to provide an overview of this enormous topic. First, we discussed a few of the common point-to-point serial communication protocols such as RS-232 and I^2C. Next, we described the Internet Protocol and the role of various protocol layers. If an application needs a custom protocol, an IP server needs to be developed. A bare bones simple server was presented. Finally, several existing protocols and commands used to configure the network were presented.

Practical Expansion: Outside World

Communication off-chip can be a big part of embedded systems design. FPGAs have shown their capabilities for years as being a significant player in integrating off-chip communications. However, until more recently, the role FPGA played in the communication has been somewhat limited. As designers shift their focus away from the glue-logic support to platform support of FPGAs, there comes increasing support for off-chip communication. These practical pages discuss two such types of off-chip communication. First is high-speed communication in the form of Xilinx RocketIO through the use of high-speed serial transceivers. The second topic focuses on low-speed communication with the use of an integrated I2C bus within a system. These two examples are meant to highlight the similarities and differences in high-speed and low-speed designs.

7.A. High-Speed Serial Communication

More recently, FPGA vendors have started to include high-speed serial transceivers on the FPGA as diffused IP. These transceivers allow FPGAs to interface directly with emerging communication standards such as PCI-Express, Serial ATA, InfiniBand, and 10 Gigabit Ethernet. Serial interfaces require fewer traces on a PCB, which has allowed smaller and more dense PCB layouts. However, the serial traces have to be designed with tightly controlled impedences for the transceivers to operate correctly. There is extra complexity on the FPGA logic side of the transceiver, as the components have many more configuration settings than a typical I/O pad.

7.A.1. RocketIO

The Xilinx Virtex series of FPGAs from the Virtex-II Pro through the Virtex-6 have included some form of RocketIO high-speed serial transceiver. These transceivers have maximum data rates of 3.125 Gbps on the Virtex-II Pro to 11 Gbps on some Virtex-6 FPGAs. Typical devices have between 8 and 24 of these full-duplex transceivers on a chip, which can be used to connect several high-speed devices or to create custom networks of FPGAs. While it is possible to directly instantiate the transceiver primitive in a design, Xilinx recommends using the Coregen utility to abstract the complexities of the transceiver and provide the user with a simpler data link layer interface.

7.A.2. Aurora Example

Aurora is a data link layer protocol provided by Xilinx. It provides the user with a simple stream or frame-based interface to send and receive data. Aurora can also be used to bond several transceivers together to make a single higher bandwidth link. Aurora is able to handle transceiver-specific operations such as inserting clock-correction sequences into the data stream.

As an example, consider connecting two ML-510 development boards via their Personality Module (PM) interface present on each board. Eight of the 24 GTX RocketIO transceivers on the FPGA are connected to the PM interface. We can use these eight transceivers — each transmitting and receiving data at 5.0 Gbps — to achieve 40 Gbps of bandwidth in each direction, a total of 80 Gbps of bandwidth between the two FPGAs.

To achieve this, launch the CoreGen wizard. Start a new project and select the xc5vfx130t-2ff1738 FPGA. Generate an Aurora 8B10B core (found under Communication & Networking — Serial Interfaces) to have eight lanes, with each lane being 4 bytes wide. We will use the 250-MHz SGMII reference clock. Select the streaming interface to reduce the complexity of the logic needed to send data between the two boards. Note that framing is a convenient way to represent

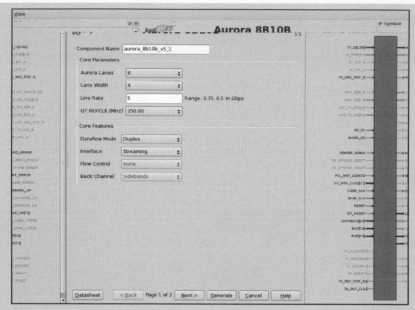

Figure 7.3. Aurora configuration.

packets from a network layer but streaming is better for simple data-flow style applications. The first page of the coregen wizard is shown in Figure 7.3.

The second page of the wizard lets the user select which GTX transceivers to use for the eight lanes of the aurora channel. From the ML-510 schematic you can determine that the transceivers connected to the personality modules are `GTX_X0Y3` through `GTX_X0Y6` (each GTX tile contains two transceivers). Map the lanes to transceivers as shown in Figure 7.4. Coregen will generate an entire system design that uses this aurora core. The top level of this example project can be found in:

```
component_name_example_design.vhd
```

The following sections describe the important components of this example project.

7.A.3. LocalLink Interface

The Aurora component uses the LocalLink interface as described in Section 6.A.3 to send and receive data from the user logic — in this case a simple data generator and error detector. As we selected the streaming interface the start-of-frame and end-of-frame LocalLink signals are not used. Also notice that in this example the data signal is 256 bits wide instead of the standard 32.

7.A.4. Clock Correction

It is common in serial communication to only send data between the transmitter and the receiver, not a copy of the clock. In low-speed communication such as RS232, both sides agree on a data rate and generate their own internal reference clocks. However, in high-speed serial communication it is more common for the receiver to infer the reference clock from data by

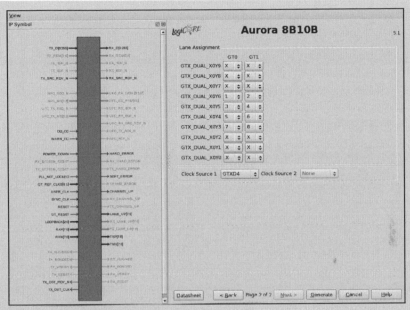

Figure 7.4. Lane to transceiver mapping.

examining transitions on the data line itself. This system works well, except if long periods go by without a transition; the receiver's inferred clock can drift from the transmitters.

We use two methods to handle this problem. First we use data encoding like 8B10B, which takes 8 bits of data from the user and encodes it as 10 bits with more transitions to send across the wire. Second, we periodically stop sending user data and insert a clock correction sequence with many transitions to make it easy for the receiver to infer the correct clock. Xilinx recommends sending these clock correction sequences at every 10,000 clock cycles. The cc_manager component in the example design handles the insertion of these sequences.

7.A.5. Error Testing

When debugging a communication system like this it is very useful to have a component that generates a stream of data along with a second component that receives that data and checks for errors. Xilinx has provided these components in the example design: frame_gen and frame_check. These components use a 16-bit linear feedback shift register with an XNOR feedback to generate a pseudo-random data sequence to be transmitted and received.

7.A.6. Loopback

This example may be infeasible to build and test on a student's budget. It would require two ML-510 development boards and a custom PCB that would connect the two personality module connectors with controlled impedance transmission lines between the transceivers on each board. However, you can test this on just one ML-510 without a custom PCB by using the loopback feature of the transceivers. The transceivers have two internal loopback paths, either before or after the final output driver, to allow testing on a single FPGA. You can set the loopback_i signal in the example design to any nonzero value to use these loopback paths instead of requiring a second ML-510.

7.B. Low-Speed Communication

While the first section focused on high-speed serial communication, the focus in this section is to support a popular low-speed communication protocol, namely I2C. We use the I2C bus to read and write to a connected EEPROM that is physically located on the Xilinx ML510 development board. Once we have the ability to interact with this device, it will become academic to interface with other devices on the I2C bus.

7.B.1. Generating the Hardware Base System

Using the Base System Builder wizard from within XPS we can quickly generate a hardware base system for the Xilinx ML510 with the Virtex 5 FX130T FPGA. Create the following base system:

Board Selection	Xilinx Virtex 5 ML510 Evaluation Platform Revision C
System Configuration	Single-Processor System
Processor Configuration	400-MHz PowerPC processor with 100-MHz Bus
Peripherals	refer to Figure 7.5

With this short example we do not need to include every peripheral in the system. This will save us greatly on our build times, allowing us to try different implementations without spending a lot of time sitting idle by the computer.

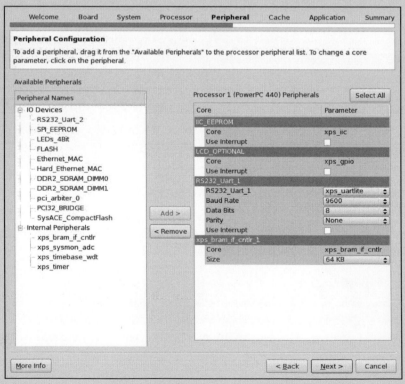

Figure 7.5. Peripherals needed for the I2C EEPROM example on the Xilinx ML510 development board.

Selecting the I2C EEPROM peripheral in the base system builder wizard will instantiate and instance the Xilinx component IIC_EEPROM. Because the wizard has already instantiated the component, we will walk through the Xilinx XPS project files to highlight the changes that would have been made had we needed to add the I2C component to the project without the aid of the wizard.

User Constraints File   We begin with the user constraints file located within the project directory:

```
% <XPS_PROJECT>/data/system.ucf
```

Looking at the schematic for the ML510 development board we can see that the I2C bus consists of two pins, SDA and SCL, for the data and clock lines, respectively. On the ML510 development board this EEPROM is connected to the I2C bus while the I2C bus connects to the FPGA via pins K8 and L7. We can specify through the user constraints file that SDA will connect to K8 and SCL will connect to L7.

```
Net fpga_0_IIC_EEPROM_Sda_pin LOC=K8   | IOSTANDARD=LVCMOS33;
Net fpga_0_IIC_EEPROM_Scl_pin LOC=L7   | IOSTANDARD=LVCMOS33;
```

Microprocessor Hardware Specification File   In order for the system to interface with the SDA and SCL pins of the I2C bus that have been set in the user constraints file, we must first specify at the top of the MHS file the pin connections and directions. This will allow us to use these pins to connect to the I2C component shortly.

```
PORT fpga_0_IIC_EEPROM_Sda_pin = fpga_0_IIC_EEPROM_Sda_pin, DIR = IO
PORT fpga_0_IIC_EEPROM_Scl_pin = fpga_0_IIC_EEPROM_Scl_pin, DIR = IO
```

Now we are able instantiate our component. In the MHS file this is done with the following syntax:

```
BEGIN xps_iic
 PARAMETER INSTANCE = IIC_EEPROM
 PARAMETER C_IIC_FREQ = 100000
 PARAMETER C_TEN_BIT_ADR = 0
 PARAMETER HW_VER = 2.02.a
 PARAMETER C_BASEADDR = 0x81600000
 PARAMETER C_HIGHADDR = 0x8160ffff
 BUS_INTERFACE SPLB = plb_v46_0
 PORT Sda = fpga_0_IIC_EEPROM_Sda_pin
 PORT Scl = fpga_0_IIC_EEPROM_Scl_pin
END
```

A few points can be made from this declaration. First off is the address map of the component. We are connecting this device to the same PLB that the PowerPC is connected to. Reading and writing from this address range (0x81600000–0x810ffff) provide us with the ability to read and write effectively to the EEPROM. That is, the I2C EEPROM component acts as a bridge between the system on-chip and the external world. The second point to make is that the two pins, SDA and SCL, are connected from this component through the top-level MHS pin declarations defined previously and finally connected to the user constraints file pin declarations. This same strategy can be employed for connecting external components to the FPGA.

Microprocessor Software Specification File   Finally, we need to specify within the MSS the software libraries that support our interfacing with the I2C component. This is necessary to allow the SDK to generate the necessary address space and supporting interface that we will discuss shortly.

```
BEGIN DRIVER
 PARAMETER DRIVER_NAME = iic
 PARAMETER DRIVER_VER = 1.16.a
 PARAMETER HW_INSTANCE = IIC_EEPROM
END
```

Next we must *Export Hardware Design to SDK*, which will synthesize and generate the system's bitstream along with creating the SDK project and launching the SDK.

I2C Software Application   Once the SDK launches we must generate a new software platform. Because we are not including Linux in this example, we can generate a simple `standalone` software platform. We will also need to generate a new C application using this stand-alone software platform.

Within our C application we must include a set of libraries to support interfacing with the I2C bus.

```
#include <stdio.h>
#include <string.h>
#include "sleep.h"
#include "xbasic_types.h"
#include "xparameters.h"
#include "xiic.h"
#include "xiic_l.h"
#include "xgpio_l.h"
#include "xtime_l.h"
```

Next, we can set some C #define values for more human readable code and try to eliminate any magic numbers in our code.

```
#define START_ADDR      1024
#define END_ADDR        2047
#define IIC_BASEADDR    XPAR_IIC_EEPROM_BASEADDR
#define IIC_SIZE        (END_ADDR-START_ADDR+1)
#define IIC_ADDR        0xA0
#define IIC_DELAY       5000 /* useconds */
```

At this point we can use a `struct` to set up what will be the contents of the EEPROM. Initially, the EEPROM will be empty; however, we can write new contents based on the storage needs of the device. In this example we will set some board information that could be used later to identify the board, say, during a boot procedure.

```
typedef struct iic_eeprom_struct {
  /* Generally used parameters */
  /* 0x000 to 0x010 Plain text ID of which board */
  char which_board[17];
  /* 0x011 to 0x015 Plain text Board Rev (A, B, C, etc) */
```

```
    char board_rev[5];
    /* 0x016 to 0x01A Plain text minor board rev (001, 002, etc) */
    char minor_board_rev[5];
    /* 0x01B to 0x02E Plain text which FPGA is on the board (main FPGA if multiple) */
    char which_FPGA[19];
    /* 0x02F to 0x037 Plain text Serial Number of board */
    char board_sn[9];
    /* 0x038 to 0x044 Plain text MAC Address for this board */
    char board_mac_id[13];
    /* 0x045 to 0x050 Plain text last date that tests were run (DD-MMM-YYYY) */
    char last_test_date[12];
    /* 0x051 to 0x05C Plain text Manufacture Date (DD-MMM-YYYY) */
    char manufacture_date[12];
    /* 0x05D to 0x06D Plain text Manufacture ID (Name) */
    char manufacture_id[17];
    /* 0x06E to 0x080 Plain text set to 'Xilinx Virtex-X Based MLxxx' (?19?) */
    char tested_before[19];
}iic_eeprom_struct;
```

Next, we will want to set some function prototypes to support easier read and write access to the I2C bus.

```
/* Function Prototypes */
/* For IIC Read / Write */
static void send(Xuint32 addr, Xint8 *data, Xuint32 len);
static void receive(Xuint32 addr, Xint8 *data, Xuint32 len);
```

Finally, we can write the actual main function that will read and set the values of the EEPROM on the I2C bus.

```
int main() {
  char *eeprom_ptr;
  iic_eeprom_struct *eeprom;

  /* Pointer to the base of the EEPROM Struct */
  eeprom_ptr = (char*)eeprom;

  /* Read EEPROM for all current contents */
  receive(START_ADDR, eeprom_ptr, IIC_SIZE);

  /* Print out contents of EEPROM */
  printf("Board ID: \n", eeprom->which_board);
  printf("Board Rev: \n", eeprom->board_rev);
  printf("Minor Board Rev: \n", eeprom->minor_board_rev);
  printf("Which FPGA: \n", eeprom->which_FPGA);
  printf("Board SN: \n", eeprom->board_sn);
  printf("Board MAC ID: \n", eeprom->board_mac_id);
  printf("Last Test Date: \n", eeprom->last_test_data);
  printf("Manufacture Date: \n", eeprom->manufacture_date);
  printf("Manufacture ID: \n", eeprom->manufacture_id);
  printf("Test Before: \n", eeprom->test_before);

  /* Write new contents to EEPROM Structure */
  eeprom->last_test_data = "01-JAN-2010";
```

```
  /* Write out all contents to EEPROM */
  send(START_ADDR, eeprom_ptr, IIC_SIZE);

  printf("Program Complete\n");
  return 0;
}
```

In order to support both reading and writing across the I2C, we must include the read and write subroutines.

```
/*******************************************************************
** function: send
** purpose:  Write data across I2C
** input:    addr - Address to Write on I2C buss
**           data - Data to be written to address (addr)
**           len - length of data to be written
*******************************************************************/
static void send(Xuint32 addr, Xint8 *data, Xuint32 len) {
  Xint8 sendBuf[34];
  Xuint32 pos, wlen;
  Xuint32 ret;

  wlen = 32;
  for (pos=0; pos < len; pos+=32) {
    if ((len - pos) < 32)
      wlen = len-pos;

    sendBuf[0] = (Xint8) ((addr+pos) >> 8);
    sendBuf[1] = (Xint8) (addr+pos);
    memcpy(&sendBuf[2], &data[pos], wlen);
    ret = XIic_Send(IIC_BASEADDR, IIC_ADDR>>1, sendBuf, wlen+2, XIIC_STOP);
    usleep(IIC_DELAY);
  }
}


/*******************************************************************
** function: receive
** purpose: Read from EEPROM
** input: address (addr), data pointer (data) and length (len)
** output: Data read from EEPROM stored at data pointer address
*******************************************************************/
static void receive(Xuint32 addr, Xint8 *data, Xuint32 len) {
  Xint8 address[2];
  Xuint32 ret;

  address[0] = (Xint8) (addr >> 8);
  address[1] = (Xint8) addr;
  ret = XIic_Send(IIC_BASEADDR, IIC_ADDR>>1, address, 2, XIIC_STOP);
  ret = XIic_Recv(IIC_BASEADDR, IIC_ADDR>>1, data, len, XIIC_STOP);
}
```

7.B.2. Testing the Design

Now all that is necessary is to save and program the application to the FPGA. This can be done by either creating the ACE file as we did in Section 1.A or through the SDK and a JTAG as was done in Section 3.A. If this is the first time the program runs, the EEPROM may not be initialized to any values. If that is the case you will need to write more than simply the `last_test_data` value. Of course, you are not limited to this information. In fact, you can program the contents you would like to the EEPROM. Because the EEPROM is nonvolatile, when the FPGA loses power and reboots the contents will remain in the memory. As stated earlier, this can be very helpful when designing systems.

Exercises

P7.1. Contrast RS-232 and USB in terms of end-user experience and in terms of design complexity.

P7.2. How does a TCP/IP server differ from a UNIX dæmon?

P7.3. How is the physical layer different from the rest of the layers in the protocol stack?

P7.4. What does the link layer do?

P7.5. What does the network layer do?

P7.6. Why does a TCP/IP server use two socket file descriptors instead of one?

P7.7. What helps determine whether to use I2C versus SPI?

References

Electronics Industries Association (1969 August). *EIA standard RS-232-C interface between data terminal equipment and data communication equipment employing serial data interchange.* International Organization for Standardization, Geneva, Switzerland.

ISO/IEC JTC1 (1984). *Open system interconnection — OSI reference model (ISO 7498).* International Organization for Standardization, Geneva, Switzerland.

Philips Semiconductor, Inc. *The I2C-Bus specification, Version 2.1* (2009). `http://www.semiconductors.philips.com/acrobat/various/I2C_BUS_SPECIFICATION_3.PDF`.

USB Implementers Forum (USB-IF). *USB 2.0 specification* (2010a January). `http://www.usb.org/developers/docs/`.

USB Implementers Forum (USB-IF). *USB 3.0 specification* (2010b January). `http://www.usb.org/developers/docs/`.

GLOSSARY

abstraction is the act or an instance of abstracting or taking away; an abstract or visionary idea

application layer the network layer closest to the user's application

basic block is a maximal sequence of sequential instructions with a Single Entry and Single Exit (SESE) point

bitgen the process after routing which takes a placed and routed netlist and generates a configuration file known as a bitstream that will be used to configure the FPGA device

bitstream is the file that includes a header and frames of configuration information needed to set the SRAM cells of the FPGA in order to set the functionality of the device

block a hard block is a core that has been implemented in CMOS transistors, and a soft block is a core that has been implemented in the function generators and memories of an FPGA device

Block RAM (BRAM) a Xilinx-defined term for a fix amount of on-chip memory within the FPGA fabric

bottom-up approach is a form of system design where a design is constructed starting with simple components that are used to build small systems, and those small systems are used to build bigger systems, and so on

burst transfer consists of issuing a single request to read or write multiple sequential data elements; typically used by compute cores during DMA transactions to access on-chip/off-chip memory

CLB see configurable logic block

coarse-grain parallelism refers to relatively large subtasks, such as an entire subroutine

cohesion is a means of measuring the abstraction; if the details inside of a module come together to implement an easily understood function, the module is said to have cohesion

computing machine a device consisting of a control or processing mechanism that responds to inputs by signaling its outputs; implicit in the machine is an encoding that gives meaning to the inputs and outputs

configurable logic block (CLB) is the Xilinx-defined term for a logic block, which is a group of slices and interconnect circuitry

control flow graph (CFG) is graph $G = (V, E)$ where the vertices (or nodes) V of the graph are basic blocks and the directed edges indicate all possible paths that program could take at run time

core see IP Core

correctness refers to a system that has been (mathematically) shown to meet a formal specification

coupling is a measure of how modules of a system are related to one another

cross-compiler is a high-level language translator that runs on one platform (a specific processor, a C library, and an operating system) but produces executables for another platform

data-flow firing rule the operation only begins when all of its inputs are available and there is a place for it to write its output

data-flow graph the network represented as a directed graph

degree of parallelism (DOP) can refer to: the number of concurrent operations at a single moment of time; a time-varying function over the entire application; or the average of a time-varying function over some task T which may not include the entire application

design is the digital circuit that is programmed (or configured) into an FPGA

diffused core see hard core

direct memory access (DMA) alleviating the processor from the task of transferring memory to or from a compute core through the use of a intermediate DMA controller or by allowing the compute core to directly access memory

double buffering refers to requests that are made while the previous request is still in transit

embedded computing system a computing system that is integral to a larger, enclosing product; in contrast to general-purpose computing, the embedded computing system is not intended to be an end-product in and of itself

evolvability refers to changes that are due to new features

external criteria are characteristics of a design that an end user can observe

feature a portion of software that is a candidate for hardware implementation (by candidate we mean the hardware implementation is much faster than the software implementation)

fine-grain parallelism means that the subtasks t_i are relatively small, maybe the equivalent of a few instructions

formal functional description of a module meaning the behavior is either described mathematically (in terms of sets and functions) or otherwise codified, such as a completed C subroutine or a behavioral description in VHDL

formal interface is the module's name and an enumeration of its operations including, for each operation, its inputs (if any), outputs (one or more), and name

FPGA Field-Programmable Gate Array

general interface includes the formal interface and any additional protocol or implied communication (through shared, global memory, for example)

general-purpose computing system is a product itself and, as such, the end-user directly interacts with it

hard core is an FPGA-specific term for an IP core where the logical operations and interconnection have been specified and mapped to the components of a particular device; thus, the core has a fixed shape and location on the chip

hardware description language (HDL) is a programming language used to describe the behavior of hardware and is the most common form of design entry today

hardware refers to the physical implementation of a computing machine

HDL see hardware description language

high speed serial transceivers are devices that serialize and deserialize parallel data over a serial channel

host see host machine

host machine (cross-compiling) when cross-compiling a program, the system that will run the compiler tools for a particular target

host machine (networking) the name for a single machine, in networking terms, that is connected to the Internet through a local network

implementation of a module is some realization of the module's intended functionality

implicit functional description of a module refers to the a module's description that is universally understood, such as the module "FullAdder"

informal functional description of a module meaning its intended behavior is described in comments, exposition, or narrative

instance of a module is a use of an implementation

instantiate means to make a copy of an instance

instrumentation when a compiler or synthesis tool adds extra functionality to a task to help determine its characteristics within an application; for example, tasks are often instrumented to measure the amount of time spent completing the task or how much time is spent idle

interface is the set of rules that govern this interaction between two entities (for example, the interaction between a processor and an I/O device)

internal criteria are characteristics that are inherent in the structure or organization of the design, but not necessarily directly observable by the user

Internet Protocol a technology-independent, packet-based protocol used for transferring data across multiple networks

interoperability refers to a system design that works well with other devices

IP Internet Protocol

IP see IP core

IP core intellectual property that refers to a hardware specification which, depending on how it is expressed, can be used to manufacture an integrated circuit (hard core) or configure the resources of an FPGA (soft core); diffused IP core is a hard core embedded in an FPGA integrated circuit

legal order of subtasks to be an ordering that produces the same results as total order T for all inputs.

link layer (data link layer) is depicted as being closely integrated with the physical layer; it has the responsibility of not only communicating with the physical layer but also monitoring the physical layer for anomalies or errors

logic block is a group of several logic cells along with special-purpose circuitry, such as an adder/subtractor carry chain

logic cell is a low-level building block of an FPGA which consists of a look-up table and a D flip-flop storage element

maintainability refers to the ability to fix problems that arose from unspecified behavior

mapping (also known as map) the process of grouping gates and assigning functionality to FPGA primitives

marshaling hardware and software implementations have different views of the state of the machine; marshaling is the process of explicitly organizing the transfer of state between the two

module any self-contained operation that has an interface and some functional description

module is any self-contained operation that has: a name, an interface, and some functional description; it can refer to hardware, software, or even something less concrete

monitor is a primitive type of debugger that is interrupt-driven; either the processor is interrupted or the application being debugged traps to the debugger; also, a monitor usually only supports the most basic functionality, such as reading/writing absolute addresses, setting breakpoints, and manipulating registers

netlist is a computer representation of a collection of logic units and how they are to be connected

network layer provides connectivity beyond two directly connected hosts, the most common being the Internet Protocl in TCP/IP

network topology is the underlying point-to-point relationship between each host on the network which can be modeled as a graph

parallel multiple instances of components are used and some operations may happen concurrently

partial order when no specific order is imposed for subtasks

partitioning the process of decomposing an application into multiple implementations (such as hardware and software implementations)

physical layer is the lowest layer of the network protocl stack and is responsible for transmitting and receiving real-world, physical signals from outside the device

pipeline balancing refers to the introduction of buffers so that a network of computations does not stall

placement the next step in the back-end tool flow after map, during which the FPGA primitives in the netlist are assigned to specific blocks on a particular FPGA

platform FPGA an FPGA device that includes sufficient resources and functionality to host an entire system on a single device

portability refers to a system design that can move to new hardware or software platforms

processor is hardware that implements the sequential execution model

program software written in a specific programming language that is the representation of desired machine behavior

programmable I/O the processor performs all requests on behalf of the compute core; specifically related to transfers between memory and the compute core

refactoring is the task of looking at an existing design and rearranging the groupings and hierarchy without changing its functionality

regression testing is a sometimes automated testing procedure that might be simulation driven (i.e., test benches), or it may be a set of systems that wraps around the component and exercises its functionality

regular refers to the point of view of any vertex in the graph; the relative positions of every other vertex in the network are the same

reliability in terms of hardware means that the system behaves correctly in the presence of physical failures

repairability refers to fixing situations where the behavior was specified but the implementation was incorrect

resilience accepts the fact that there will be errors and the design "works around" problems even if it means working in a degraded fashion

router is a device to support communication across two similar technologies but different networks

routing (also known as route) is the process after placement, during which wire segments are selected and passing transistors are set

sequential the same components are reused over time and the the results are the accumulation of a sequence of operations

simulation the process of translating an HDL source into a form suitable for a general-purpose processor to mimic the hardware

slice is the Xilinx-defined term for a group of logic cells

soft core is an FPGA-specific term for an IP core where the logical operations and interconnection is specified but these operations have not been mapped to a particular device

software is a specification that describes the behavior of the machine, generally written in a programming language (such as C, MATLAB, Java, etc.)

software reference design is a piece of software where the functionally mimics the behavior of the whole system, even hardware modules that have not been implemented yet

spatial in hardware designs, multiple threads of control are able to operate simultaneously

stall occurs in two cases: whenever a computation unit has some, but not all, of its inputs, or when it does not have the ability to store its output

stand-alone C is a program that runs without the support of any additional system software

state (of an application) is all of the data associated with an application's data and part of the system's storage; this includes all of the data stored in registers, program counter, condition codes, RAM, and configuration registers in I/O devices

switch is an intermediate device used to conenct two or more hosts together to support communication

switch box is used to route between the inputs/outputs of a logic block to the general on-chip routing network along with passing signals from wire segment to wire segment

synthesis the process of generating the logic configuration for the specified device from an HDL source

system software refers to any software that assists the application, usually by adding a software interface to access the hardware

target see target machine

target machine in cross-compiler terms, the processor and operating environment for the embedded system

top-down approach is a form of system design where a design is described at the top level as consisting of smaller subcomponents, which are described by still smaller subcomponents, down to the smallest component

totally ordered a composed sequence of n subtasks $\langle t_0, t_1, ..., t_{n-1} \rangle$ in which t_i is started before t_{i+1} for all i

transport layer provides support between the application and the network layer; it provides a specific type of service such as a reliable, continuous connection or a packet-based interface

verifiability is the degree to which parts of the system can be formally verified, that is, proven correct

INDEX